LIQUID SCINTILLATION COUNTING

LIQUID SCINTILLATION COUNTING

Volume 2

Proceedings of a Symposium on
Liquid Scintillation Counting
organized by
The Society for Analytical Chemistry
Brighton, England
September 13–16 1971

Editors: M. A. Crook (Polytechnic of the South Bank)
P. Johnson (Wellcome Research Laboratories)
B. Scales (Pharmaceuticals Division, ICI Ltd.)

London · New York · Rheine

Heyden & Son Ltd., Spectrum House, Alderton Crescent, London NW4 3XX
Heyden & Son Inc., 225 Park Avenue, New York, N.Y. 10017, U.S.A.
Heyden & Son GmbH, 4440 Rheine/Westf., Münsterstrasse 22, Germany

Library of Congress Catalog Card No. 70-156826

ISBN 0 85501 067 3

Printed in Great Britain by J. W. Arrowsmith Ltd., Bristol BS3 2NT.

Contents

III. METHODS OF SAMPLE PREPARATION OF INORGANIC MATERIALS INCLUDING CERENKOV COUNTING

IV. METHODS OF SAMPLE PREPARATION OF ORGANIC MATERIALS INCLUDING BIOLOGICAL SYSTEMS

Preface

This volume is an account of the Proceedings of an International Symposium on Liquid Scintillation Counting, held at Brighton, England, September 13th–16th, 1971, and organized by the Radiochemical Methods Group of the Society for Analytical Chemistry (now Analytical Division, The Chemical Society). The undoubted success of the symposium (17 countries represented by over 200 delegates), was due in no small part to the committee of the Radiochemical Methods Group, the Plenary Lecturers (Drs. Birks, Rapkin, Fox, Dyer and Spratt) and the Chairmen of the five sessions (Drs. Lyle, Birks, Dobbs, Scales and Johnson). Success was also due to the excellence of the cooperation and support of the various manufacturers:

Beckman
Camlab
E.M.I.
Intertechnique
LKB Instruments
Koch-Light
Nuclear Chicago
Nuclear Enterprises
Panax
Packard
Philips (Pye Unicam)
Tracerlab
Twentieth Century Electronics

and particularly Messrs. Ginger, Moore and Raven who gave invaluable help as co-opted members of the organizing committee.

These proceedings form Volume 2 of a series of which Volume 1 records a Liquid Scintillation Counting Symposium at the University of Salford in 1970. It is significant that although only one year elapsed between the two symposia, it was possible to assemble a programme at the second meeting of entirely different emphasis to that at the first. Thus, bio-medical applications are emphasized in this volume, and the section on data handling involving quench correction techniques and computer processing received more submitted papers than could be accommodated. The lack of meetings in Europe to discuss, from the user's view point, the diverse application of this still growing technique means that there is room for further meetings of scientists from various disciplines who have as a common aim and need the exploitation of liquid scintillation counting.

September 1972

M. A. Crook
P. Johnson
B. Scales

SECTION I

BASIC THEORY OF THE CHEMISTRY AND PHYSICS OF SCINTILLATORS AND SCINTILLATION COUNTING

Chapter 1

Liquid Scintillators

J. B. Birks and G. C. Poullis

The Schuster Laboratory, University of Manchester, Manchester, England

INTRODUCTION

Liquid scintillation counting is widely used for the radioassay of biological and other materials labelled with carbon-14, tritium or other radioisotopes. Since the discovery[1] of liquid organic solution scintillators in 1949, many combinations of solvents and solutes have been tested, and a great variety of scintillator 'cocktail' recipes have been prescribed and used. Many of these liquid scintillator solutions do not have optimum performance, although they may be adequate and economical for some applications.

The main criterion (apart from price) used in the comparison of different scintillator solutions is the *relative pulse height* (RPH) which depends on the following factors:

1. *the nature and purity of the solvent.* (These determine the solvent fluorescence lifetime and spectrum and its scintillation yield (G-value) of excited solvent molecules/100 eV of energy expended by the ionizing particle);
2. *the nature of the primary solute.* (This determines its fluorescence quantum efficiency, lifetime and spectrum);
3. *the absorption spectrum and concentration of the primary solute.* (These determine the quantum efficiency of solvent–solute energy transfer);
4. *the nature and concentration of any secondary solute.* (These determine the secondary solute absorption and fluorescence spectra and its fluorescence quantum efficiency and lifetime);
5. *the nature and energy of the ionizing radiation.* (For β-particles or Compton electrons of energy $E > 100$ keV, the RPH is proportional to E, but for lower energy electrons and for protons, α-particles and other heavy ionizing particles, the scintillation efficiency is reduced);
6. *the concentration of dissolved oxygen in the solution, which depends on the temperature and the external atmospheric pressure.* (This determines the degree of oxygen quenching of the solvent and solute molecules at a given temperature);
7. *the form and concentration in which the radioactive specimen is introduced into the solution.* (These determine the magnitude of the impurity (chemical) quenching and of any colour quenching);
8. *the size and shape of the scintillator vial.* (This determines the optical path length for

radiative transfer from the primary to the secondary solute and for any colour quenching attenuation of the RPH);

9. *the material of the vial.* (This determines its transmission spectrum for the scintillation emission);
10. *the nature of the reflector and light collection system.* (This determines the fraction of the scintillation emission incident on the photomultiplier cathode, which is a function of the scintillator emission spectrum); and
11. *the spectral response of the photomultiplier cathode.* (This converts the scintillation photons into electrons, which are accelerated and multiplied in the dynode system, yielding anode pulses which are fed to amplifiers, discriminators, channel analysers, scalers, etc. and are ultimately recorded as counts).

PREVIOUS STUDIES

The original studies of liquid scintillators, which led to the adoption of *p*-terphenyl (TP), the oxazoles (e.g. PPO), the oxadiazoles (e.g. PBD) and the substituted *p*-quaterphenyls (e.g. BIBUQ) as primary solutes, and the choice of toluene, *p*-xylene and *p*-dioxan containing 100 g/l naphthalene as standard solvents, have been described by Birks.[2]

The Los Alamos group, who were responsible for many of these developments, adopted a standard technique for comparing the RPH of different scintillator solutions.[3] The 1 ml solution specimen was contained in a quartz cell, the base of which was optically coupled to the photomultiplier window, and it was excited externally by 624 keV internal conversion electrons from caesium-137. The light was collected by a hemispherical aluminium reflector (aluminium has a higher reflectivity than titanium dioxide paint in the near ultraviolet), and it was observed by a DuMont 6292 photomultiplier, chosen to have an 'average' S11 spectral response, with peak sensitivity at about 440 nm wavelength. Air-equilibrated toluene was used as the standard solvent. Due to the high altitude (6000 ft = 1830 m) and consequent reduced atmospheric pressure, the average Los Alamos oxygen quenching factor of 18% is less than the corresponding factor at sea level and normal atmospheric pressure. The RPH of an 8 g/l TP solution in toluene (= the RPH of a 3 g/l PPO solution in toluene = 100) was adopted as a reference standard. The RPH of each solution was measured as a function of concentration, and the maximum RPH recorded at the optimum solute concentration c_0, or at the maximum solubility c_s, if $c_s < c_0$.

In 1955 the Los Alamos group[4] reported the RPH, c_0 (or c_s) and mean emission wavelength $\bar{\lambda}$ for 102 compounds in toluene solution. They found 60 primary solutes, all except six of which are oxazole or oxadiazole derivatives, with RPH $\geqslant 75$ and 27 with RPH $\geqslant 100$. The best primary solutes included PBD, PBO, PPO and TP.

The Los Alamos group[3] also compared the RPH of 49 carefully purified, but air-equilibrated, solvents each containing 3 g/l PPO. The best solvents were *p*-xylene, isopropyl biphenyl, *m*-xylene, phenylcyclohexane and toluene. *p*-Dioxan is a further useful solvent for internal liquid scintillation counting, since it has a high solubility for many materials and it is completely miscible with water. Its efficiency as a scintillator is considerably enhanced by adding up to 100 g/l naphthalene, which functions as a secondary solvent, and increases the solvent–solute energy transfer efficiency.

The Los Alamos group[5] also compared a number of secondary solutes, using the standard conditions described above, except that a titanium dioxide reflector was used as an alternative to the aluminium reflector, to provide a better indication of large volume scintillator performance. The best secondary solutes included POPOP, BBO, PBO and α-NPO.

The energy transfer from the primary to the secondary solute is primarily radiative,[6] unlike the solvent–solute transfer which is due to solvent excitation migration and diffusion and which is non-radiative. The increase (or decrease) in RPH on adding a secondary solute to a binary solution depends critically on the following spectral factors:

1. the primary solute emission spectrum;
2. the solution volume (the radiative transfer efficiency increases with optical path length);
3. the secondary solute absorption spectrum and concentration (at higher concentrations non-radiative transfer may become significant);
4. the transmission spectrum of the vial (which limits the fraction of primary solute emission detected);
5. the reflection spectrum of the reflector and light collection system; and
6. the spectral response of the photomultiplier.

The function of a secondary solute is to optimize the match between the scintillator emission and the photomultiplier spectral response. Due to the several factors involved, the RPH values for solutions containing secondary solutes are sensitive to the experimental conditions.

THE NEED FOR A NEW LIQUID SCINTILLATOR SURVEY

The pioneer studies described above have provided a reasonable basis for the choice of liquid scintillator solutions, although later recipes have sometimes been introduced on the basis of RPH measurements made under less precise experimental conditions. Nevertheless, the measurements suffer from several defects when they are applied to modern liquid scintillation counting practice:

1. the current quartz-windowed bialkali photocathode tubes have a spectral response (peak at about 380 nm) which differs markedly from that of the earlier tubes (peak at about 440 nm);
2. the early measurements were made with external radiation sources, while in the internal counting technique the source is incorporated into the scintillator;
3. the light collection geometries used in the early studies differ markedly from those used in modern instruments;
4. the extensive Los Alamos data were obtained in high altitude conditions of reduced oxygen concentration. The solute oxygen quenching factor depends on the solvent viscosity, the solute fluorescence lifetime, and the oxygen concentration, and it is not independent of the solvent and solute as assumed by the Los Alamos group. Hence, the order of merit of air-equilibrated specimens at sea level is not necessarily the same as at higher altitudes;
5. several new compounds have come into use, usually backed by commercial claims of superiority over existing materials;
6. the early measurements were made with single photomultipliers, while current liquid counting instruments usually employ two photomultipliers operating in coincidence to reduce background noise.
7. many of the early measurements involved integration of the scintillation intensity or pulse, so that both the prompt and delayed scintillation components contributed to the RPH. In the Philips instrument the coincidence resolving time is 10 ns, so that only the prompt component contributes to the RPH. In other instruments, which employ slower

photomultipliers, the coincidence resolving time is considerably increased, so that the delayed scintillation component may then contribute to the RPH.

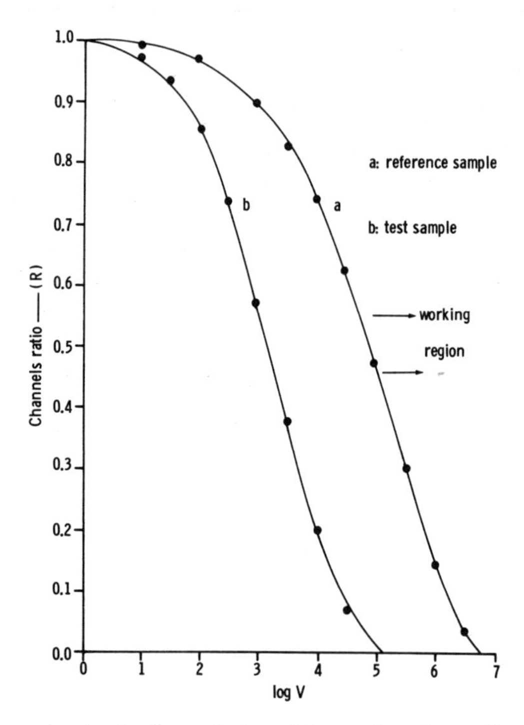

Fig. 1. Ratio R of count rates in channels 1 and 2 as a function of log V.

EXPERIMENTAL METHOD

For all these reasons it was considered desirable to measure the RPH of a wide range of scintillator solutions under conditions which correspond closely to those used for current internal liquid scintillation counting. The measurements were made with a Philips automatic liquid scintillation analyser, which utilizes two 56 DUVP photomultipliers (quartz window, bialkali cathode) operating in coincidence (resolving time 10 ns). The measurements were made on air-equilibrated solutions at ambient temperature (about 20°C) and normal atmospheric pressure (about 760 torr). The specimens were contained in standard 20 ml vials, made of low potassium content glass, which give a low background of < 20 c.p.m. A small quantity of ^{14}C-labelled hexadecane (which is a non-quencher) was added to each solvent to provide an internal source of carbon-14 β-rays of about 20000 d.p.m.

The RPH was determined from the integral pulse height spectrum, using a channels ratio method to determine the latter. This method has the advantage of being independent of the carbon-14 activity. Channels 1 and 2 were initially set to accept all counts above noise and background, and gave equal count rates. A variable discriminator was used to set the minimum height V of pulses recorded in channel 2, and the ratio R of the count rates in channels 2 and 1 was measured as a function of log V (Fig. 1). The RPH was taken to be proportional to the value of V (= $V_{0.5}$) at which $R = 0.5$. The curve of R against log V is linear in the region of $R = 0.5$, so that $V_{0.5}$ can be evaluated by interpolation between observations of log V at R above and below 0.5. The logarithmic discriminator error was $< 2\%$, and the probable statistical error was about 1%. The RPH is expressed relative to that of a 3 g/l solution of PPO in toluene (= 100), the same reference standard as the Los Alamos group.

MATERIALS

The solvents were:

(i) benzene
(ii) toluene
(iii) xylene
(iv) *p*-xylene
(v) mesitylene, and
(vi) *p*-dioxan containing 100 g/l naphthalene,

and were supplied by Koch-Light Laboratories Ltd. The mesitylene was passed through an alumina chromatographic column and fractionally distilled and the middle fraction was collected. The other solvents were used without further purification.

The primary solutes were:

(i) TP (*p*-terphenyl)
(ii) PPO (2,5-diphenyloxazole)
(iii) BBOT (2,5-bis(5′-*tert*-butyl-2′-benzoxazolyl)-thiophene)
(iv) PBO (5-phenyl, 2-(4-biphenylyl)-oxazole)
(v) butyl-PBD (2(4′-*tert*-butylphenyl),5-(4″-biphenylyl)-1,3,4-oxadiazole)
(vi) PBD (2-phenyl, 5-(4-biphenylyl)-1,3,4-oxadiazole)
(vii) BIBUQ (4,4-bis(2-butyloctyloxy-*p*-quaterphenyl)),

and were supplied by Koch-Light Laboratories Ltd. and used without further purification.

The secondary solutes were:

(i) PBBO (2-(4-biphenylyl)-6-phenylbenzoxazole)
(ii) POPOP (1,4-bis-[2-(5-phenyloxazolyl)]-benzene)
(iii) dimethyl POPOP (1,4-bis-[2-(4-methyl-5-phenyloxazolyl)]-benzene)
(iv) BBO (2,5-di-(4-biphenylyl)-oxazole)
(v) bis-MSB (*p*-bis-(*o*-methylstyryl)-benzene),

supplied by Koch-Light Laboratories Ltd., and

(vi) α-NPO (2-(1-naphthyl), 5-phenyl-oxazole)
(vii) DPH (1,6-diphenylhexa-1,3,5-triene)

supplied by Nuclear Enterprises Ltd., and they were used without further purification.

For the quenching studies carbon tetrachloride supplied by Eastman-Kodak Co. was used without further purification.

BINARY SOLUTIONS

Measurements of RPH against primary solute concentration (g/l), were made for binary solutions in benzene, toluene, xylene, *p*-xylene, mesitylene and *p*-dioxan containing 100 g/l naphthalene, of the following primary solutes: TP (this solute gives negligible RPH in the *p*-dioxan/naphthalene solvent because of the absence of solvent–solute energy transfer) (Fig. 2), PPO (Fig. 3), BBOT (Fig. 4), PBO (Fig. 5), butyl-PBD (Fig. 6), PBD (Fig. 7), and BIBUQ (Fig. 8).

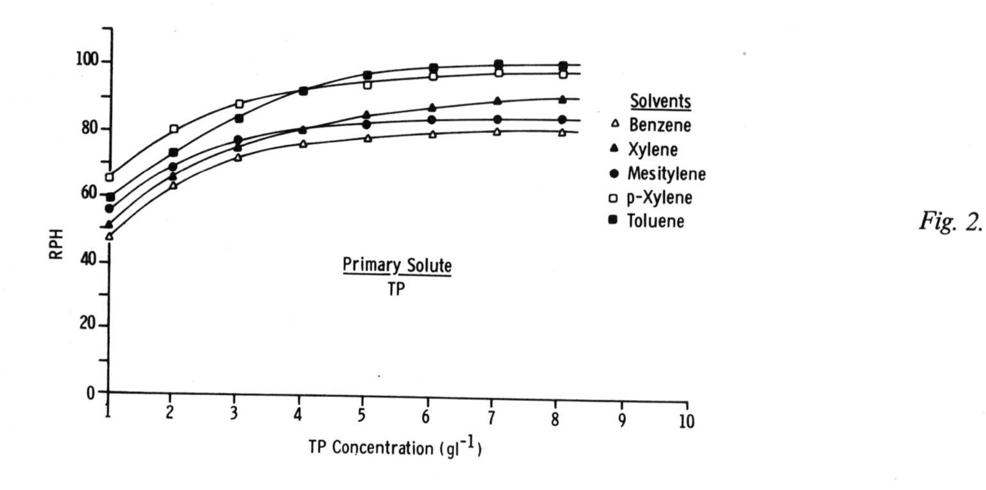

Fig. 2.

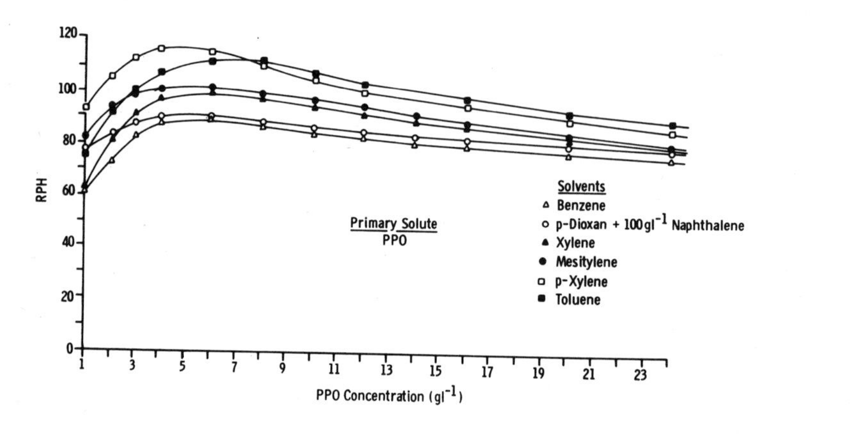

Fig. 3.

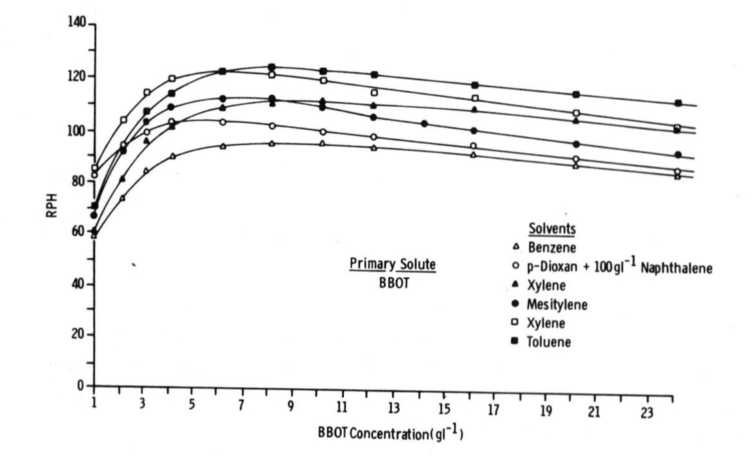

Fig. 4.

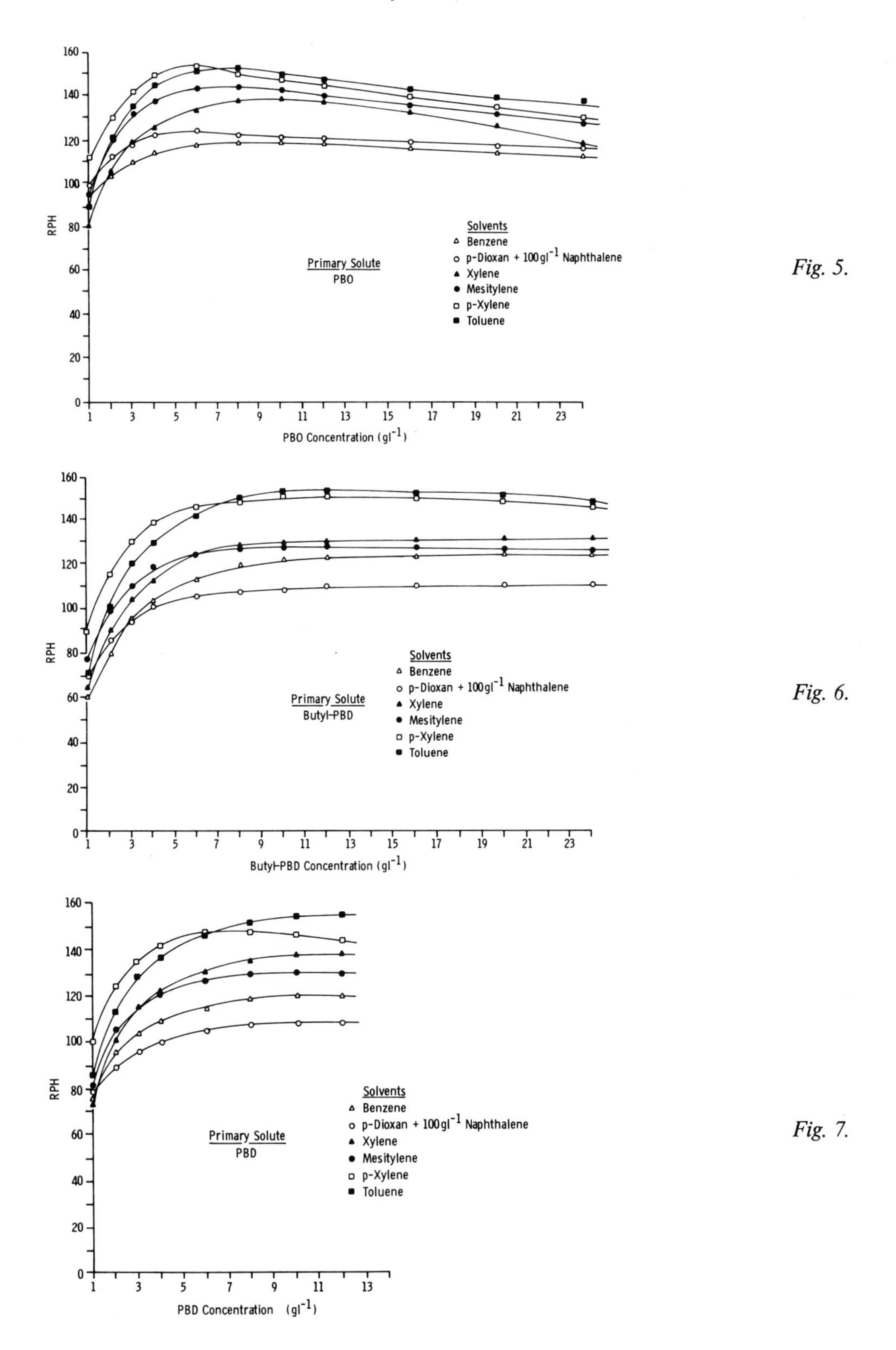

Fig. 5.

Fig. 6.

Fig. 7.

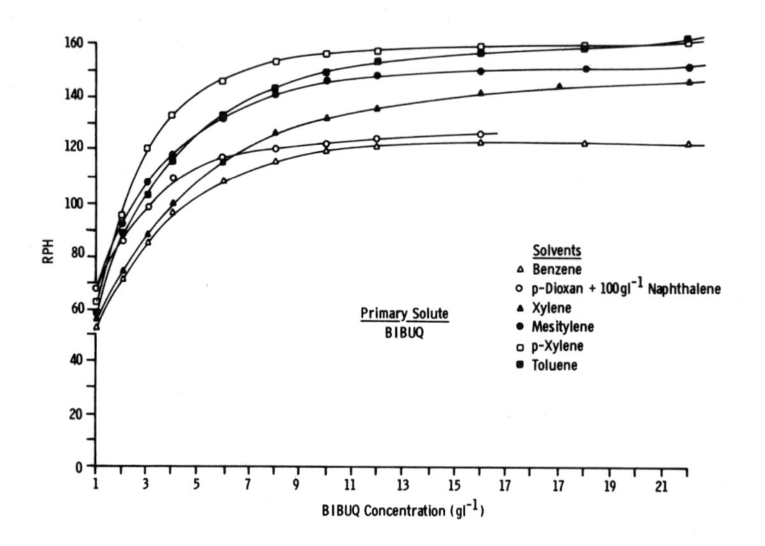

Fig. 8.

Table 1. Binary solutions – maximum RPH and optimum concentration c_0 (g/l).

Solute	Benzene		Toluene		Xylene		p-Xylene		Mesitylene		p-Dioxan + 100 g/l naphthalene	
	RPH	c_0	RPH	c_0	RPH	c_0	RPH	c_0	RPH	c_0	RPH	c_0
TP	81	8	101	7	91	8	99	8	86	8	–	–
PPO	90	5.5	112	7	100	6	117	5	102	5.5	91	5
BBOT	95	8	123	8	111	10	122	6	112	6	104	6
PBO	120	9	152	7.5	139	10	153	6	144	8	124	6
Butyl-PBD	124	20	153	12	131	20	150	12	128	12	111	24
PBD	120	11	155	12	139	12	148	7	130	11	108	10
BIBUQ	123	16	160	24	147	24	160	20	152	20	127	16

The values of the maximum RPH and optimum concentration c_0 (g/l) for each of the binary solution systems are listed in Table 1.

The susceptibility of the various binary solutions to impurity quenching was compared using carbon tetrachloride as quencher. The addition of a molar concentration [M] of quencher reduces the RPH from V_0 (when [M] = 0) to V, where V is given by the Stern–Volmer relation:

$$V_0/V = 1 + [\mathrm{M}]/[\mathrm{M}]_{0.5} \quad (1)$$

and $[\mathrm{M}]_{0.5}$ is the half-value quencher molar concentration at which $V = 0.5V_0$. Graphs of V_0/V against the carbon tetrachloride molar concentration [M] are presented for binary solutions in benzene, toluene, xylene, *p*-xylene, mesitylene and *p*-dioxan containing 100 g/l naphthalene, of the following primary solutes: 5 g/l TP (excluding the *p*-dioxan/naphthalene solvent) (Fig. 9), 6 g/l PPO (Fig. 10), 8 g/l BBOT (Fig. 11), 8 g/l PBO (Fig. 12), 10 g/l butyl-PBD (Fig. 13), 10 g/l PBD (Fig. 14) and 15 g/l BIBUQ (Fig. 15).

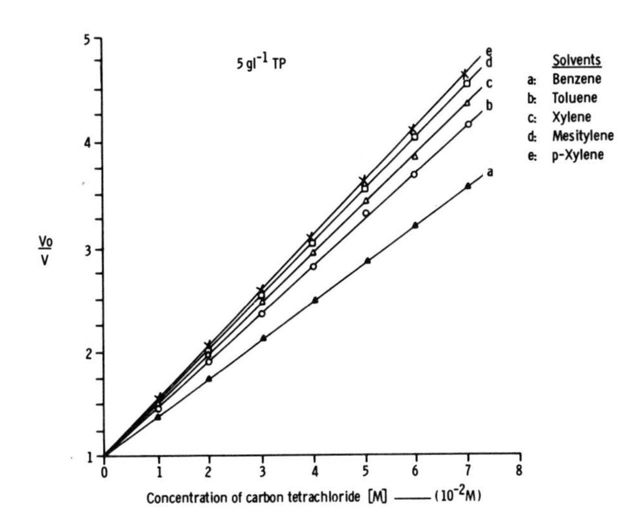

Fig. 9.

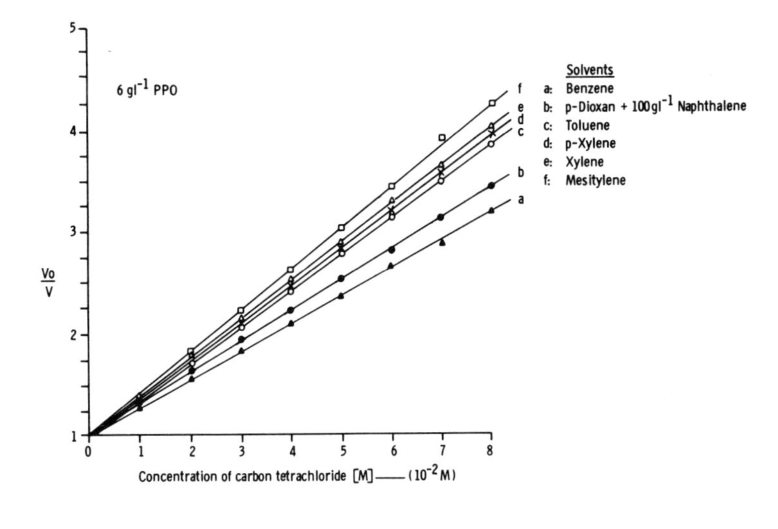

Fig. 10.

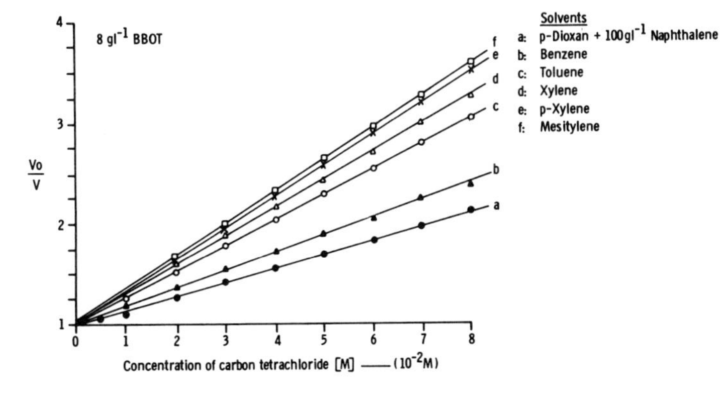

Fig. 11.

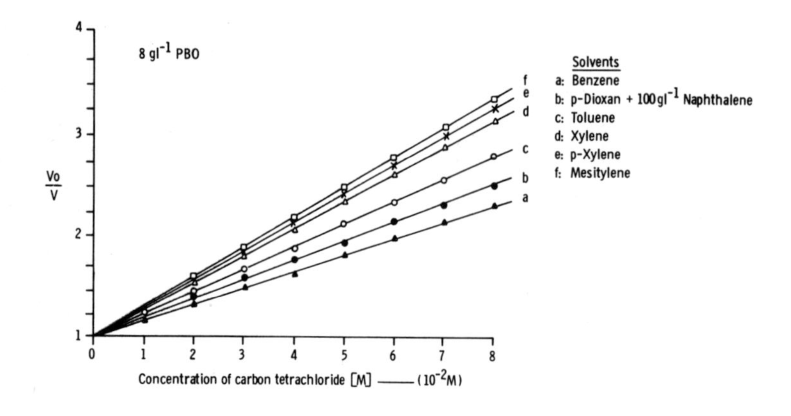

Fig. 12.

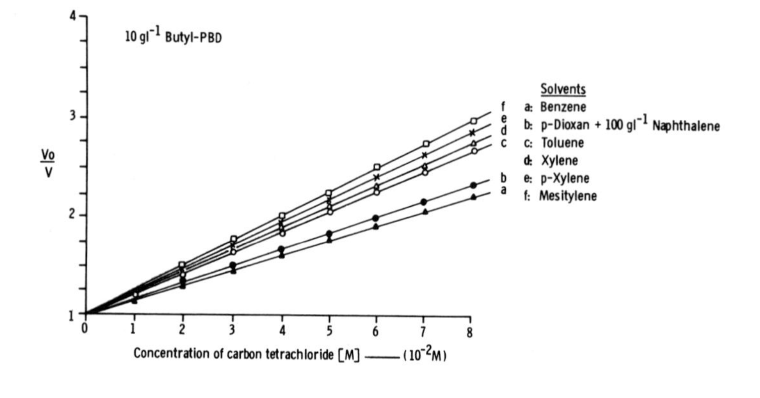

Fig. 13.

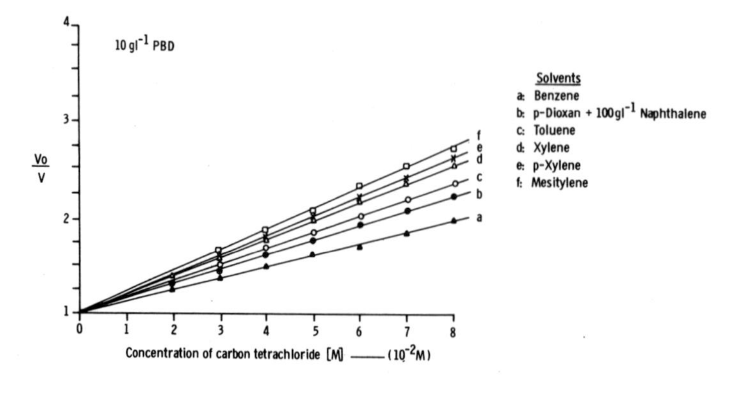

Fig. 14.

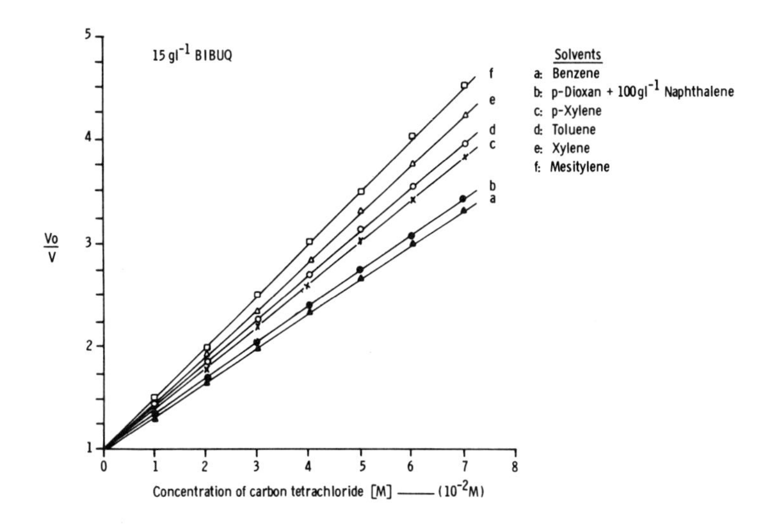

Fig. 15.

Table 2. Binary solutions – quenching by carbon tetrachloride. Half-value quencher concentration $[M]_{0.5}$ (units: 10^{-2} M).

Solute concentration	Solute	Solvent Benzene	Toluene	Xylene	*p*-Xylene	Mesitylene	*p*-Dioxan + 100 g/l naphthalene
5 g/l	TP	2.74	2.22	2.12	1.91	1.97	–
6 g/l	PPO	3.96	2.78	2.64	2.72	2.44	3.25
8 g/l	BBOT	5.64	3.82	3.56	3.17	3.09	7.23
8 g/l	PBO	6.12	4.50	3.71	3.51	3.42	5.25
10 g/l	Butyl-PBD	6.31	4.66	4.60	4.22	3.98	6.03
10 g/l	PBD	8.17	5.79	5.06	4.83	4.50	6.44
15 g/l	BIBUQ	2.98	2.24	2.19	2.46	1.93	2.85

Linear Stern–Volmer plots of V_0/V against [M], consistent with Eqn. (1), are observed in all cases. The values of $[M]_{0.5}$, the half-value quencher molar concentration, are listed in Table 2.

TERNARY SOLUTIONS

The influence of secondary solutes on representative binary solutions was studied. Measurements of RPH against secondary solute concentration (g/l, plotted on a logarithmic scale) were made for the addition of DPH, α-NPO, dimethyl POPOP, BBO, POPOP, bis-MSB, and (in some cases) PBBO, to the following binary solutions: 4 g/l TP in benzene (Fig. 16), 4 g/l TP in toluene (Fig. 17), 4 g/l TP in xylene (Fig. 18), 4 g/l TP in *p*-xylene (Fig. 19), 4 g/l PPO in benzene (Fig. 20), 4 g/l PPO in toluene (Fig. 21), 4 g/l PPO in xylene (Fig. 22), 4 g/l PPO in *p*-xylene (Fig. 23), 5 g/l butyl-PBD in benzene (Fig. 24), 5 g/l butyl-PBD in toluene (Fig. 25), 5 g/l butyl-PBD in xylene (Fig. 26), 5 g/l butyl-PBD in *p*-xylene (Fig. 27).

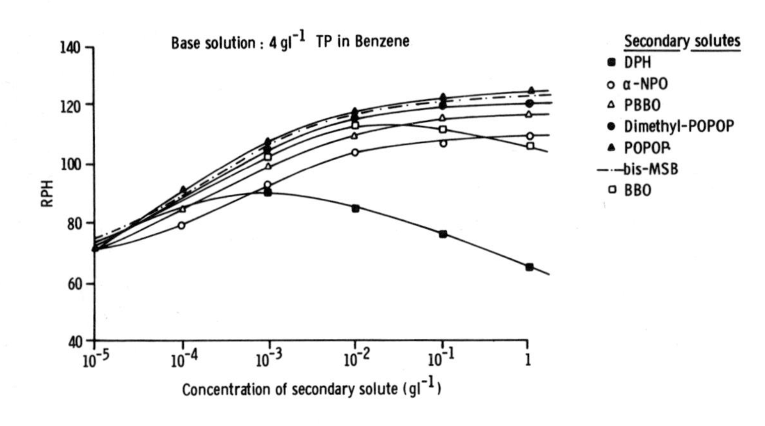

Fig. 16.

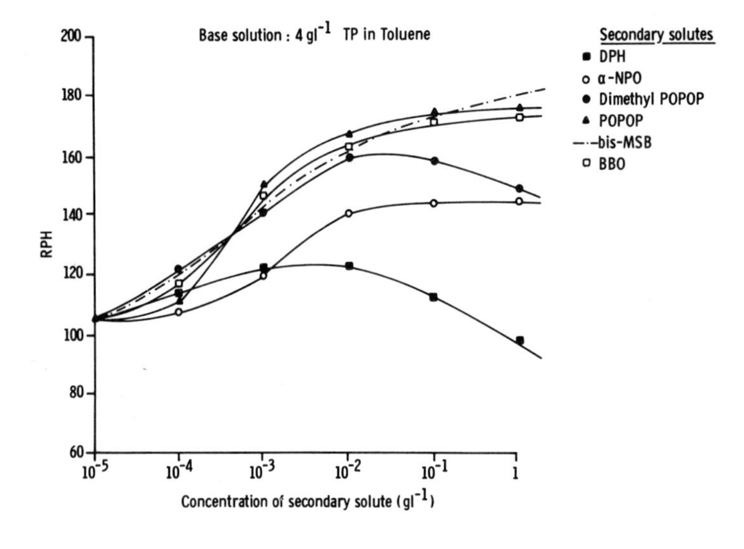

Fig. 17.

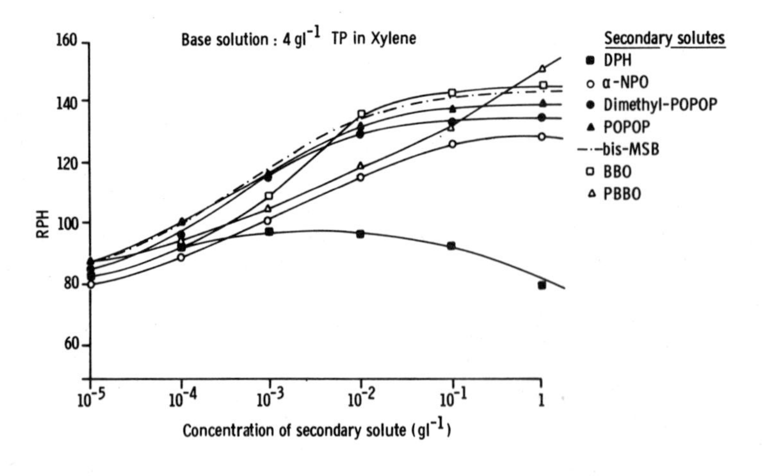

Fig. 18.

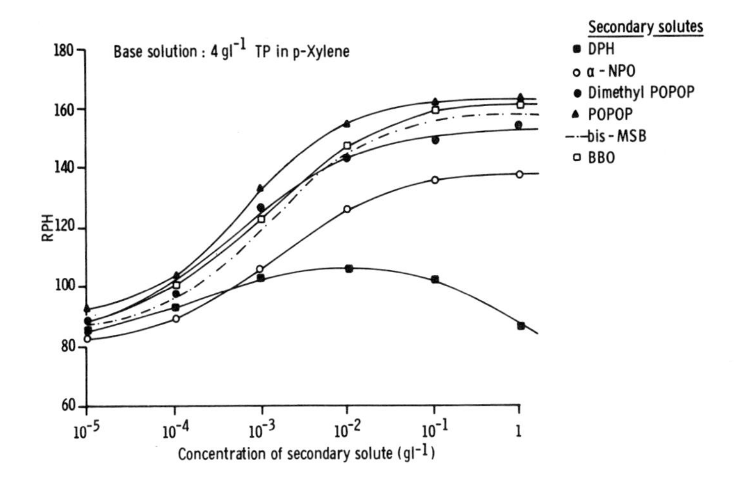

Fig. 19.

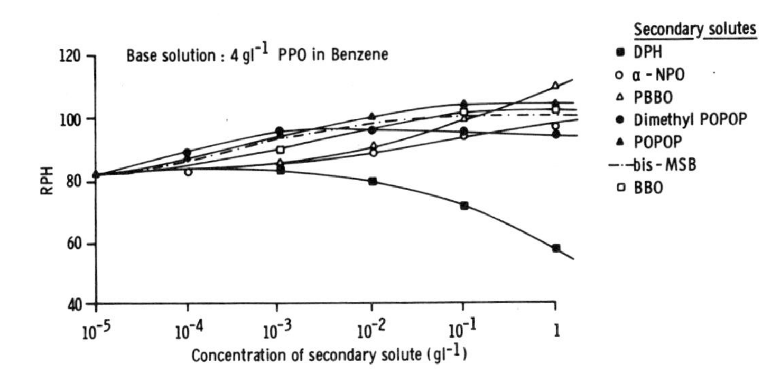

Fig. 20.

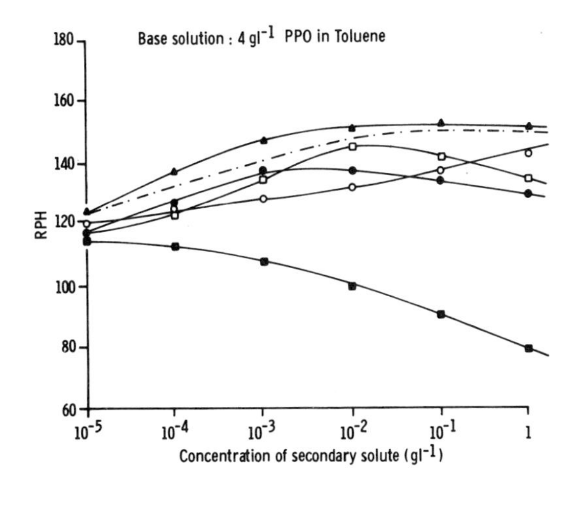

Fig. 21.

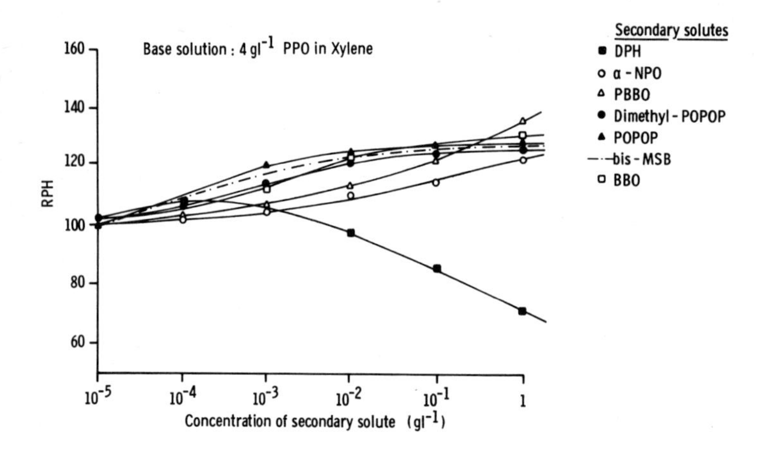

Fig. 22.

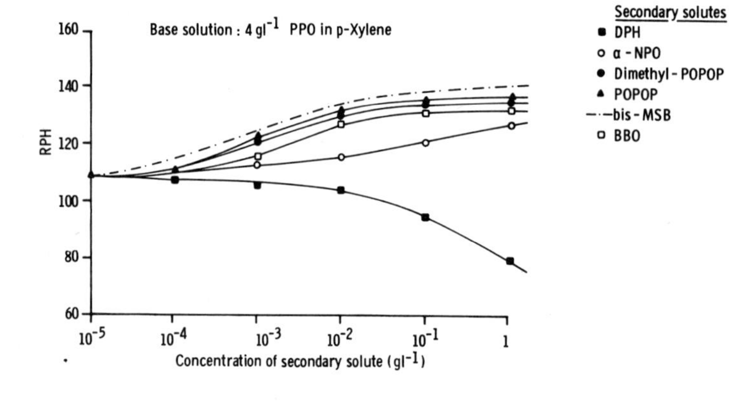

Fig. 23.

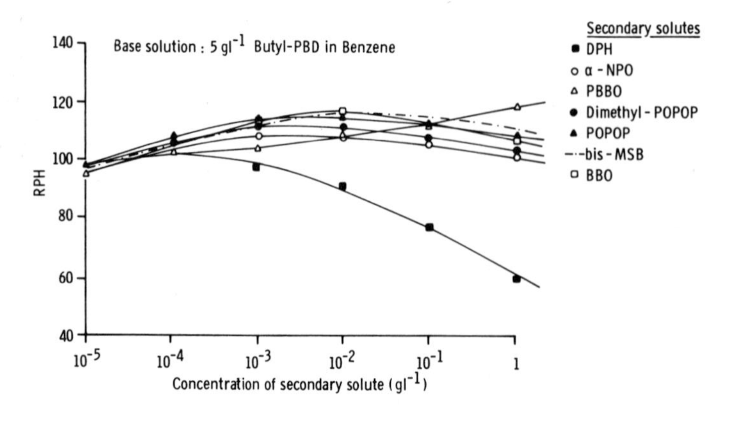

Fig. 24.

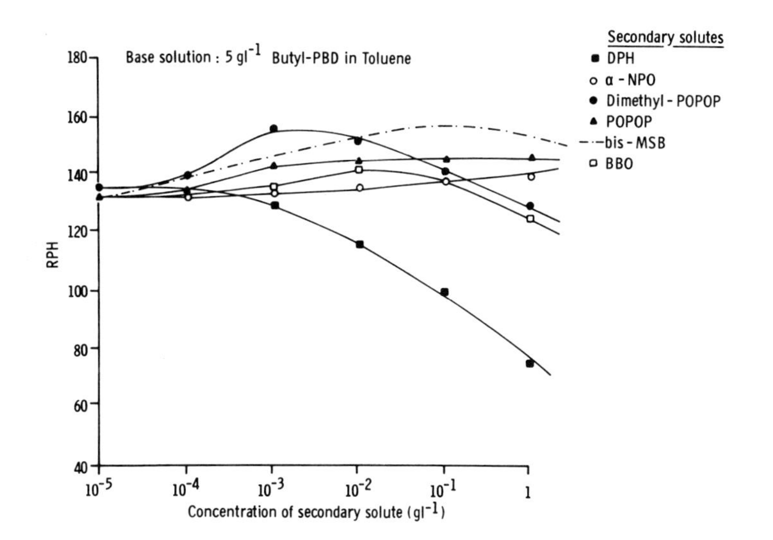

Fig. 25.

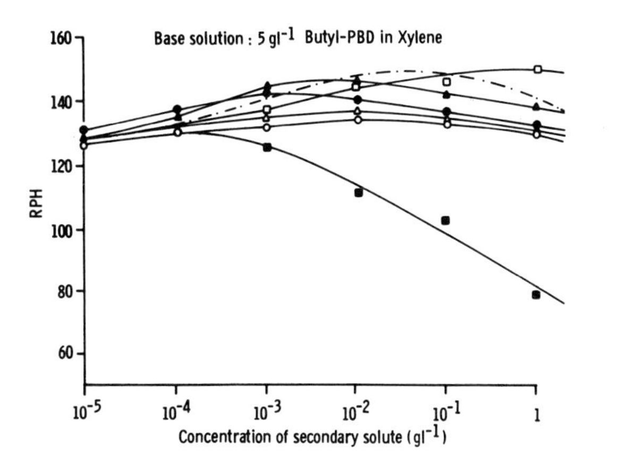

Fig. 26.

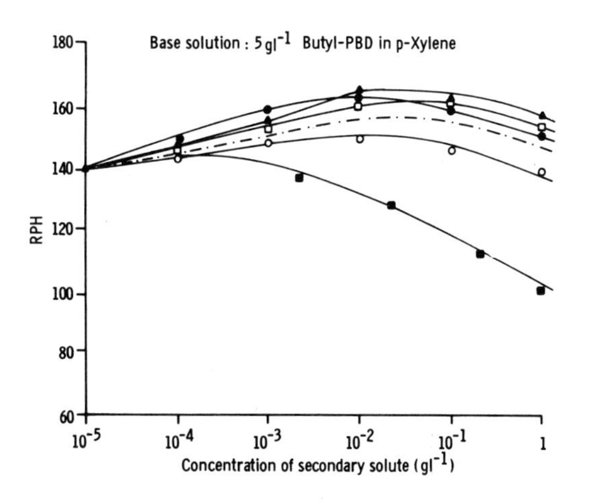

Fig. 27.

Table 3. Ternary solutions containing 4 g/l TP – maximum RPH and optimum secondary solute concentration c_0 (g/l).

Secondary solute	Solvent							
	Benzene		Toluene		Xylene		*p*-Xylene	
	RPH	c_0	RPH	c_0	RPH	c_0	RPH	c_0
DPH	91	10^{-3}	124	3×10^{-3}	101	3×10^{-3}	106	10^{-2}
α-NPO	110	1	145	1	130	1	137	1
BBO	115	2×10^{-2}	173	1	143	1	161	1
bis-MSB	120	0.2	184	1	141	0.1	155	1
dimethyl POPOP	121	1	161	2×10^{-2}	134	1	154	1
POPOP	125	1	175	1	140	1	163	1
PBBO	117	1	–	–	153	1	–	–

Table 4. Ternary solutions containing 4 g/l PPO – maximum RPH and optimum secondary solute concentration c_0 (g/l).

Secondary solute	Solvent							
	Benzene		Toluene		Xylene		*p*-Xylene	
	RPH	c_0	RPH	c_0	RPH	c_0	RPH	c_0
DPH	84	10^{-4}	115	10^{-5}	108	10^{-5}	108	10^{-5}
α-NPO	97	1	142	1	123	1	128	1
BBO	102	1	147	2×10^{-2}	128	1	132	1
bis-MSB	101	1	150	0.2	130	1	141	1
dimethyl POPOP	97	3×10^{-3}	138	10^{-3}	124	1	135	1
POPOP	105	0.2	153	0.1	129	1	138	1
PBBO	110	1	–	–	137	1	–	–

Table 5. Ternary solutions containing 5 g/l butyl-PBD – maximum RPH and optimum secondary solute concentration c_0 (g/l).

Secondary solute	Solvent							
	Benzene		Toluene		Xylene		*p*-Xylene	
	RPH	c_0	RPH	c_0	RPH	c_0	RPH	c_0
DPH	102	10^{-4}	134	5×10^{-5}	130	10^{-5}	144	10^{-4}
α-NPO	108	2×10^{-3}	139	1	134	2×10^{-2}	149	10^{-2}
BBO	118	10^{-2}	142	10^{-2}	150	1	161	7×10^{-2}
bis-MSB	117	0.1	156	0.1	147	5×10^{-2}	156	2×10^{-2}
dimethyl POPOP	111	5×10^{-3}	155	10^{-3}	142	2×10^{-3}	163	10^{-2}
POPOP	115	10^{-2}	146	1	146	5×10^{-3}	165	10^{-2}
PBBO	119	1	–	–	136	2×10^{-2}	–	–

The values of maximum RPH and optimum secondary solute concentration c_0 for each of the ternary solution systems are listed in Tables 3, 4 and 5.

DISCUSSION

The RPH depends on the *solvent* (Table 1). For all the solutes, the lowest values of RPH are for solutions in benzene and in *p*-dioxan + 100 g/l naphthalene, intermediate values of RPH are obtained for solutions in mesitylene and in xylene, and the highest values of RPH are for solutions in toluene and in *p*-xylene. The preferred solvents, in the absence of quenching, are thus toluene and *p*-xylene.

The RPH depends on the *primary solute* (Table 1). In all the solvents the TP solutions have the lowest RPH. (This is due to the low transmission coefficient of the glass vial for the TP emission. The RPH of the TP solutions is increased considerably if a polyethylene or quartz vial is used). The PPO solutions have the next lowest RPH, followed by the BBOT solutions. The PBO, PBD and butyl-PBD solutions have similar high RPH values in all the solvents, but the optimum concentration c_0 is higher for butyl-PBD than for the other solutes. Under the conditions of measurement BIBUQ is the best primary solute. The preferred binary solutions with the maximum RPH, in the absence of quenching, are:

(A) 24 g/l BIBUQ in toluene (RPH = 160)
(B) 20 g/l BIBUQ in *p*-xylene (RPH = 160)

The susceptibility to impurity quenching, which is proportional to $1/[M]_{0.5}$ (Table 2), depends on the *solvent*. The least susceptible solvents are benzene and *p*-dioxan + 100 g/l naphthalene; those of intermediate susceptibility are toluene and xylene; and those most liable to quenching are *p*-xylene and mesitylene. *Toluene* is therefore to be preferred to *p*-xylene under quenching conditions. Under such conditions the relative insensitivity of benzene and *p*-dioxan/naphthalene solutions to quenching may be sufficient to compensate for their lower 'unquenched' RPH.

The susceptibility to impurity quenching depends on the *primary solute* (Table 2). In each of the alkyl benzene solvents the susceptibility to quenching increases in the order PBD, butyl-PBD, PBO, BBOT, PPO, BIBUQ and TP. BIBUQ and TP are 2 to 3 times more prone to quenching than PBD, butyl-PBD and PBO. Hence, in the presence of quenching, the preferred binary solutions with maximum RPH are:

(C) 12 g/l PBD in toluene (RPH = 155)
(D) 12 g/l butyl-PBD in toluene (RPH = 153)
(E) 7.5 g/l PBO in toluene (RPH = 152)

It is difficult to compare the relative merits of *secondary solutes,* since the RPH of a ternary solution depends on several factors as discussed above. A mean order of increasing merit can, however, be derived from the data of Tables 3 to 5, as follows: DPH (1); α-NPO (2.1); dimethyl POPOP (3.7); BBO (4.5); bis-MSB (5.1); POPOP (5.3) amd PBBO (5.9), where the number in parentheses is the average order of increasing RPH from among six or seven secondary solutes. The preferred secondary solutes are therefore PBBO, POPOP, bis-MSB and BBO.

The addition of a secondary solute to a 5 g/l butyl-PBD solution produces only a marginal increase in RPH (Table 5). This is to be expected, since the butyl-PBD emission is already well matched to the photomultiplier spectral response. The addition of a secondary solute to a 4 g/l PPO solution produces a useful increment in RPH (Table 4) due to the improved spectral matching. Three of the ternary solutions:

(F) 4 g/l PPO + 0.1 g/l POPOP in toluene (RPH = 153)
(G) 4 g/l PPO + 0.2 g/l bis-MSB in toluene (RPH = 150)
(H) 4 g/l PPO + 2 × 10^{-2} g/l BBO in toluene (RPH = 147)

have RPH values comparable with the preferred binary solutions (C), (D) and (E).

The addition of a secondary solute to a 4 g/l TP solution produces a dramatic increase in RPH (Table 3), and yields RPH values exceeding those of the best binary solutions (A) and (B). The best of these ternary solutions are:

(I) 4 g/l TP + 1 g/l bis-MSB in toluene (RPH = 184)
(J) 4 g/l TP + 1 g/l POPOP in toluene (RPH = 175)
(K) 4 g/l TP + 1 g/l BBO in toluene (RPH = 173)

It should be noted, however, that the ternary solutions (F)–(H) and (I)–(K) will be subject to impurity quenching, similar to that of the binary PPO toluene solutions and binary TP toluene solutions, respectively. The order of merit of solutions (A)–(K) which, in the absence of quenching is (I), (J), (K), (A), (B), (C), (D), (F), (E), (G), (H), will thus be modified when impurity quenching is present. The 'quenched' RPH values can be determined from the 'unquenched' RPH values and the values of $[M]_{0.5}$ (Table 2) using Eqn. (1).

In the present studies only one impurity quencher, carbon tetrachloride, has been used. Different quenching behaviour is to be expected with other impurities. It is hoped to extend the studies to other quenchers, including amines, carboxylic acids and other molecules which have groups characteristic of biological molecules.

ACKNOWLEDGEMENTS

We wish to thank Philips, Eindhoven, Netherlands for the loan of a Philips automatic liquid scintillation analyser, and Koch-Light Laboratories Ltd., Colnbrook, Bucks. and Nuclear Enterprises Ltd., Edinburgh, for the gift of scintillator chemicals.

REFERENCES

1 M. Ageno, M. Chiozzotto and R. Querzoli, *Accad. Naz. Lincei* **6**, 626 (1949); *Phys. Rev.* **79**, 720 (1950).
2 J. B. Birks, *The Theory and Practice of Scintillation Counting,* Pergamon Press, Oxford, 1964.
3 F. N. Hayes, B. S. Rogers and P. C. Sanders, *Nucleonics* **13**, 46 (1955).
4 F. N. Hayes, D. G. Ott, V. N. Kerr and B. S. Rogers, *Nucleonics* **13**, 38 (1955).
5 F. N. Hayes, D. G. Ott and V. N. Kerr, *Nucleonics* **14**, 42 (1956).
6 J. B. Birks and K. N. Kuchela, *Proc. Phys. Soc.* **77**, 1083 (1961).
7 J. B. Birks, *Photophysics of Aromatic Molecules,* Wiley-Interscience, London and New York, 1970.
8 W. J. McDowell and L. C. Henley, Oak Ridge National Laboratory Report ORNL-TM-3676 (March 1972).

DISCUSSION

B. E. Gordon: Are you planning to study quenchers which interact with the solvent? This would alter the diffusion coefficients of the solvents and so change the order of the effectiveness of the solvent in preventing quenching.

J. B. Birks: Carbon tetrachloride interacts with excited solvent (or solute) molecules to form excited complexes (exciplexes). The competition between the radiationless dissipation of the excitation energy of the exciplex and the dissociation of the exciplex determines the magnitude of the quenching. The exciplex dissociation rate depends on its enthalpy, which is determined by the ionization potential and electron affinity of the solvent (or solute) and quencher molecules. The magnitude of the solvent quenching increases with decrease in ionization potential I of the solvent, e.g. benzene ($I = 9.25$ eV), toluene ($I = 8.8$ eV), *p*-xylene ($I = 8.4$ eV), mesitylene ($I = 8.4$ eV). It is probable that the susceptibility of the solute to quenching similarly increases with decrease in its ionization potential.

As mentioned in the paper it is proposed to extend the studies to other quenchers which have groups characteristic of biological molecules. These quenchers act in a similar manner to carbon tetrachloride, i.e. they form exciplexes with excited solvent or solute molecules. They do not form donor–acceptor complexes with unexcited solvent or solute molecules, which would be undesirable since it would lead to immediate (static) quenching rather than to diffusion-controlled (dynamic) quenching. The distinction between these different types of quenching has been discussed elsewhere.[7]

The diffusion coefficients of benzene and its alkyl derivatives are similar, so that diffusion is not the determining factor in the present studies. It is difficult to envisage any quencher which would appreciably alter the diffusion coefficient of a solvent without altering the nature of the solvent.

F. Battig: According to your figures in a system containing toluene, butyl-PBD and oxygen or quencher there was some positive but small effect of bis-MSB and dimethyl POPOP on pulse height of carbon-14. Would you therefore suggest the general use of secondary solutes together with butyl-PBD in practical measurements?

J. B. Birks: The results referred to are shown in Fig. 25. They show that the addition of 0.1 g/l bis-MSB or 10^{-3} g/l dimethyl POPOP to a 5 g/l butyl-PBD in toluene solution increases the RPH from 135 to 155 or 156. However, reference to Fig. 6 shows that a 12 g/l butyl-PBD solution in toluene has an RPH of 153, which is about the same as that obtained with the solutions containing secondary solutes. Providing the optimum concentration of butyl-PBD is used, there appears to be no advantage in using a secondary solute under our particular experimental conditions.

I would hesitate to lay down any general prescriptions. We have indicated the solutions with the maximum RPH, but we have also presented all the experimental data so that users can estimate the influence of the solvent and the nature and concentration of the primary solute and secondary solute and the effect of quenching for themselves. The ultimate test of any scintillator solution is its performance under the particular operational conditions.

A. R. Ware: You commenced your lecture by confining your remarks to aromatic solvents. In view of some recent success I have had using a long chain hydrocarbon solvent, I would like to hear your comments on the possible use of solvents other than aromatic solvents in scintillation solutions. Would these solutions offer any beneficial properties from the quenching viewpoint?

J. B. Birks: This is a most interesting question. *p*-Dioxan, which is an aliphatic compound, is an efficient scintillator solvent when used in conjunction with 100 g/l naphthalene. Recent work by Hirayama and Lipsky, described elsewhere,[7] has shown that a large number of aliphatic hydrocarbons are fluorescent, with a fluorescence spectrum peaking between 207 and 230 nm. The quantum yields for excitation at 147 nm wavelength increase with molecular size, e.g. hexane (0.0002), decane (0.002), hexadecane (0.0045), cyclohexane (0.0035), methylcyclohexane (0.0055), bicyclohexyl (0.02) and decalin (0.021). The latter values are of similar magnitude to those for liquid benzene excited at 147 nm. It is thus to be expected that the larger aliphatic hydrocarbons should function satisfactorily as scintillator solvents. The choice of scintillator solute may differ, because the solvent fluorescence spectrum differs from that of an alkyl benzene, and the use of a secondary solvent, like the 100 g/l naphthalene added to *p*-dioxan, might prove advantageous. The ionization potential of the aliphatic hydrocarbons is considerably higher than that of benzene, so that the quenching of the solvent excitation should be even less than for benzene. Provided no snags are encountered with solubility or viscosity, liquid scintillators based on aliphatic solvents appear promising alternatives to the existing scintillator solutions.

A. Dyer: Further to the possible use of other solvent systems in liquid scintillation counting, there is a recent publication by Gomez (E. Gomez *et al., Intern. J. Appl. Radiation Isotopes* 22, 243 (1971)) citing the use of various nitriles with apparent success.

J. B. Birks: No comment.

A. Dyer: Would you care to comment on the excitation of scintillator systems by processes other than β-particles, i.e. α-particles, γ-rays, etc.?

J. B. Birks: I have discussed this topic at length elsewhere[2] and will therefore confine myself to a few remarks. Internal liquid scintillation counting of α-particles has been brought to a fine art by McDowell and Henley[8] who have resolved the 4.0, 5.4, 5.7, 6.3 and 6.8 MeV peaks of thorium-232, thorium-228, radium-224, radium-220 and polonium-216 and 4.19 and 4.79 MeV peaks of uranium-238 and uranium-234 in a solution of 5 g/l PPO and 200 g/l naphthalene in toluene using a single photomultiplier and a multichannel analyser. They used organic phase-soluble solvent-extraction reagents to incorporate the radionuclides in the scintillator with minimal quenching. This is a remarkable achievement since quenching increases rapidly with the atomic number of the impurity.
γ-rays interact with matter in a different manner from charged particles: by the photoelectric effect, by the Compton effect, and by pair production. Pair production does not occur at energies below 1.02 MeV, and can usually be neglected. The photoelectric effect is negligible in normal liquid scintillators above 30 keV, but it extends to higher energies in the walls of the vial and in scintillators loaded with heavy elements. Ignoring these points, the principal γ-ray absorption is due to the Compton effect, which yields electrons ranging in energy from zero to about 0.75 of the γ-ray energy. These Compton electrons interact with the scintillator in the same manner as β-particles. If the Compton electron energy exceeds the Cerenkov threshold (electron velocity $>$ velocity of light in scintillator), Cerenkov emission occurs in addition to the scintillation. The detection efficiency for γ-rays is generally much less than for β-rays because of the reduced stopping power of the detection element.

B. W. Fox: Is it ever possible to use the 95% of energy lost in the initial fluorescence process?

J. B. Birks: Not in present liquid scintillation counters. Possibly in the future, the development of solid state systems may enable this lost energy to be utilized.
The 95% energy loss to which you refer is that of the overall scintillation process. The solute fluorescence, which is the final step, has an efficiency of 80 to 90%. The major loss occurs at the first step of the process. Only 14% of the electrons in benzene are π-electrons

(even less in the case of the alkyl benzenes), so that only one in seven of the initial solvent ionizations and excitations can lead to a scintillation. The manner in which the 14% gets reduced to 5% has been discussed previously.[2] Possible methods of increasing the 5% are being explored, but the 14% appears to be inherent to liquid scintillators based on aromatic solvents. The 95% energy loss appears as heat. It could be detected by calorimetry. Calorimetry was used to measure radioactivity over 60 years ago.

Chapter 2

Liquid Scintillation Counting as an Absolute Method

J. A. B. Gibson

Health Physics & Medical Division, UKAEA Research Group, Atomic Energy Research Establishment, Harwell, England.

INTRODUCTION

The liquid scintillation counter is well established as an analytical tool in physics, chemistry, biology and medicine (see for example Gibson and Lally, 1971).[1] The principle of the method for absolute counting is to dissolve a known weight of the isotope in the liquid scintillator and obtain an integral bias curve. Extrapolation of this bias curve to zero bias will give an activity which is less than the absolute activity because of statistical fluctuations which result in a significant probability that no output will be obtained from an event in the scintillator. This *zero probability* described by Gibson and Gale in 1968,[2] can be determined from the overall efficiency of the scintillator and the photomultiplier tube. This efficiency or *figure of merit* (Horrocks and Studier, 1961)[3] is shown by the number of electrons produced at the input of the first stage (dynode 1) of the photomultiplie from the deposition of 1 keV of energy in the scintillator.

This chapter demonstrates the basis of the theory and includes methods of measuring the figure of merit, P, together with calculations of the modifying factors which cause losses in the scintillator and result in a reduction in the number of first stage electrons. The results of measurements for three isotopes will be given and a method of calculating the zero probability will be shown. The proposed use of a high gain photomultiplier is included.

THEORETICAL METHOD

The essential processes in the liquid scintillation counter are shown in Table 1. The figure of merit is defined as:

$$P = \frac{n(E)}{E.F(E)} \tag{1}$$

where $n(E)$ is the average number of electrons at the first stage of the photomultiplier, E is the electron energy in the scintillator and $F(E)$ is the correction for energy losses due to high rates of energy deposition and losses into the walls of the scintillator cell. P is independent of the electron energy and defines the efficiency of the system. Thus if P is

Table 1. Schematic representation of a liquid scintillation counter with typical parameters at 5 keV.

System	Schematic	Process	EMI 9524S Gain	P.M. tube Quanta	RCA C31000D Gain	P.M. tube Quanta
Liquid Scintillator	β-ray (5 keV)	Energy (=425 nm)[a]		1700		1700
	Solvent excitation	dE/dx losses	0.63	1050	0.63	1050
	Solute excitation → Light emission	Quenching and light losses	0.01	10.5	0.01	10.5
Photomultiplier	Photocathode	Conversion efficiency	0.13	1.4[b]	0.30	3.15[c]
	First dynode	Stage gain	7.73	10.8	30.0	94.5
	Second and later dynodes	Stage gain	2.12	23.0	3.16	300.0
	Anode	Overall gain	1.4×10^4	2×10^4	10^7	3×10^7
Electronics	Head amplifier	Conversion to volts	–	100 μV	–	4 mV
	Main amplifier	Gain	2×10^4	2 V	500	2 V

[a] Maximum emission of the scintillator
[b] Zero probability, exp [−1.4] = 0.247
[c] Zero probability, exp [−3.15] = 0.044

measured by one of the methods to be discussed, then n_i can be calculated for any energy E_i such that:

$$n_i = P.E_i.F(E_i) \quad (2)$$

The probability distribution of n_i at the first dynode is assumed to be a Poisson distribution of mean n_i.[4] This Poisson distribution is modified by the photomultiplier to produce an output distribution which can be calculated exactly or be approximated by various functions, e.g. Gaussian, Gamma, etc.[5]

The zero probability for a single photomultiplier tube

The zero probability of a Poisson distribution is $\exp[-n_i]$. This probability is increased by the photomultiplier (see Appendix) such that at E_i:

$$Z_i = \exp[-n_i(1 - Z_s)] \quad (3)$$

where Z_s is the zero probability for a single electron input to the first dynode. Z_s is normally very small when the first stage is arranged to have a gain >5 ($Z_s < 0.01$) and Z_i approximates to $\exp[-n_i]$. Thus it is only necessary to determine Z_s accurately when the first stage gain is less than 5.

The precision of Z_i and hence the value of the absolute activity ($\propto(1 - Z_i)$) depends mainly upon the accuracy of n_i. Thus assuming $Z_s = 0$ in Eqn. (3) and differentiating:

$$\frac{dZ_i}{Z_i} = dn_i$$

or the relative precision of Z_i depends upon the absolute precision of n_i. For example, if $dn_i/n_i = 0.1$, then for an absolute activity, ϵ_i, $d\epsilon_i/\epsilon_i$ is as follows:

n_i	dn_i	Z_i	dZ_i	ϵ_i	$d\epsilon_i/\epsilon_i$
1	0.1	0.368	0.037	0.632	0.059
2	0.2	0.135	0.027	0.865	0.031
3	0.3	0.050	0.015	0.950	0.016
4	0.4	0.018	0.007	0.982	0.007
5	0.5	0.007	0.003	0.993	0.003
6	0.6	0.002	0.001	0.998	0.001

This shows that if the precision of n_i is only ±10% then at least six electrons must be produced at the first dynode to achieve a precision of 0.1% in the absolute activity. Fewer electrons or an improved precision would require a better measurement of n_i.

The total zero probability for a β-ray spectrum is obtained by dividing the spectrum into bands, calculating the probability for each band and summing to give the total probability (see Appendix).

Zero probability for two tubes in coincidence

In this case it is necessary to have an electron at the output to each tube. The probability of obtaining an output for a single tube is given by:

$$\epsilon_i = 1 - Z_i$$

where Z_i is given by Eqn. (3). If both the tubes have the same sensitivity then the probability of obtaining an output from both tubes is given by:

$$\epsilon_c = (1 - Z_i)(1 - Z_i)$$

and the zero probability is:

$$\begin{aligned} Z_c &= 2Z_i - Z_i^2 \\ &= 2.\exp[-n_i(1 - Z_s)] - \exp[-2n_i(1 - Z_s)] \end{aligned} \tag{4}$$

The total zero probability for β-ray spectra is obtained as before except that $[H_i \exp(-n_i)]^2 \neq H_i \exp(-2n_i)$ and the two terms in Eqn. (4) must be calculated separately.

The complete distribution

In absolute counting it is necessary to determine the shape of the bias curve near to zero bias in order to make the appropriate extrapolation. Gale and Gibson[5] have shown that the differential bias curve can be approximated by the gamma function:

$$P(x) = Cx^{\alpha}\exp(-\beta x) \tag{5}$$

where α and β are constants dependent upon the electron energy and the gain of the system respectively. C is a constant for normalizing the spectrum such that:

$$\begin{aligned} C &= (1 - Z_i)\left[\int_0^{\infty} x^{\alpha}\exp(-\beta x)\mathrm{d}x\right]^{-1} \\ &= (1 - Z_i)\frac{\beta^{\alpha+1}}{\alpha!} \end{aligned}$$

where:

$$\begin{aligned} \alpha! &= \Gamma(\alpha + 1) \\ &= \int_0^{\infty} x^{\alpha}\exp(-x)\mathrm{d}x \qquad (\alpha \geqslant -1) \end{aligned}$$

is the tabulated gamma function. The function $P(x)$ has a single peak at $x = \alpha/\beta$ and tends to a Gaussian distribution at large values of α. Z_i, the zero probability, is calculated from Eqn. (3) and α and β are calculated from the mean and variance of the theoretical distribution. Thus with stage gains m_1 and $m_2 = m_3 \ldots\ldots = m_K$:

Mean:

$$\bar{r} = n_i \cdot m_1 \cdot m_2^{K-1}$$

$$= \frac{\alpha + 1}{\beta}(1 - Z_i) \tag{6}$$

Variance:

$$\sigma^2 = n_i \cdot m_1^2 \cdot m_2^{2(K-1)} \left[1 + \frac{m_2}{m_1(m_2 - 1)}\right]$$

$$= \frac{\alpha + 1}{\beta^2}(1 - Z_i)[1 + Z_i(\alpha + 1)] \tag{7}$$

The relative variance is $\sigma^2/\bar{r}^2$ and when Z_i is negligible (n_i large) then:

$$\frac{\sigma^2}{\bar{r}^2} = \frac{1}{n_i}\left[1 + \frac{m_2}{m_1(m_2 - 1)}\right]$$

$$= \frac{1}{\alpha + 1} \tag{8}$$

In a practical system it may not be possible to determine the stage gains precisely enough to predict a value of α from Eqn. (8) and it is then necessary to obtain a numerical approximation by a method which will minimize the sum of the squares between the calculated and measured curves. This will only be necessary where the mean β-ray energy is small (<100 keV) and the precision required is of the order of ±0.1%. Thus for a β-ray spectrum β is a constant which is approximately independent of energy, $(\alpha + 1)$ is proportional to n_i which is calculated from Eqn. (2) and Z_i is calculated from Eqn. (3).

The energy-dependent losses in the scintillator

The scintillation efficiency is assumed to be independent of the β-ray energy but this assumption requires corrections where the specific energy loss ($\mathrm{d}E/\mathrm{d}x$) is large (i.e. at low electron energies) and also where the range of the β-rays is of similar magnitude to the cell dimensions (large electron energies). Both these processes reduce the scintillation efficiency and their effect must be calculated in order to find n_i.

The effects of *ionization quenching* have been described by Birks.[6] He showed that the specific light output ($\mathrm{d}L/\mathrm{d}x$) from any scintillator could be described by the simple formula for the quenching effect, i.e.

$$\frac{\mathrm{d}L}{\mathrm{d}x} = \frac{S\,\mathrm{d}E/\mathrm{d}x}{1 + \mathrm{kB.d}E/\mathrm{d}x} \tag{9}$$

where S is the scintillation efficiency and kB is a constant. The total light output is obtained by numerical integration to give:

$$L = \int \mathrm{d}L = S \int_0^E \frac{\mathrm{d}E}{(1 + \mathrm{kB.d}E/\mathrm{d}x)} = S.E.Q(E) \tag{10}$$

with $Q(E)$ equal to the value of the integral at E divided by E. dE/dx was calculated for the scintillator using the formula given by Nelms.[7] Kolarov, Gallic and Vatin[8] suggest that the specific light output will decrease more rapidly with increasing dE/dx but the evidence which they quote is for protons and heavy ions. The formula in Eqn. (9) should be adequate for electron interactions within the scintillator at energies in excess of a few keV. The value of kB is 9 mg cm^{-2} MeV^{-1} for a liquid scintillator[6] and this value was used in the calculations. The error in kB will be less than ±10% and thus at 5 keV the error in $Q(E)$ will be less than ±3%, and over a complete spectrum this error will be further reduced. If this produces too large an uncertainty in n_i and hence in ϵ_i (the absolute activity) then the scintillation efficiency can be checked by using a source of iron-55 to produce electrons of 5.9 keV.

At high energies Benjamin *et al.*[9] produced a formula for the fraction of energy deposited within a cylinder of radius r and height h, i.e.

$$W(E) = 1 - 0.5\,[1/r + 1/h]\,T(E) \tag{11}$$

where $T(E)$ is the electron range calculated from Nelms' formula.[7] The value of $W(E)$ is stated to have a precision of ±1% (Benjamin *et al.*)[9] but unless very small cells are used this effect is only important above 100 keV and will not affect the zero probability.

The total correction to the efficiency is thus:

$$F(E) = Q(E)W(E) \tag{12}$$

which should be accurate to ±3%. Calculated values of $Q(E)$, $W(E)$ and $F(E)$ are given in Fig. 1.

MEASUREMENT OF THE FIGURE OF MERIT

The figure of merit, P, may be measured by one of four methods:

1. Comparison of the observed spectrum position with the single electron spectrum.
2. Comparison of the total spectrum shape (i.e. resolution) with the theoretical distribution.
3. Measurement of the zero probability.
4. Comparison between the output of the coincidence and single tube counting rates as described by Kolarov *et al.*[8]

The first three methods have been used in preliminary experiments but an assessment of the Kolarov method will be left until the publication of the second part of his paper.

Spectrum position

The output of the system can be expressed in terms of a single electron input to the first stage of the photomultiplier. This is obtained by using a low intensity light source (e.g. the phosphorescence of an empty cell which has been exposed to sunlight) to give an output distribution on a multi-channel analyser.

The output distribution from conversion electrons will have a peak at a position which is proportional to $E.F(E)$ and is quoted as the number of electrons, $n(E)$. Then P can be calculated using Eqn. (1). This method can be used without calculating the full output distribution but can be applied to a β-ray spectrum if computing facilities are available, e.g. Gale and Gibson.[5] This may be necessary with conversion electrons if they are

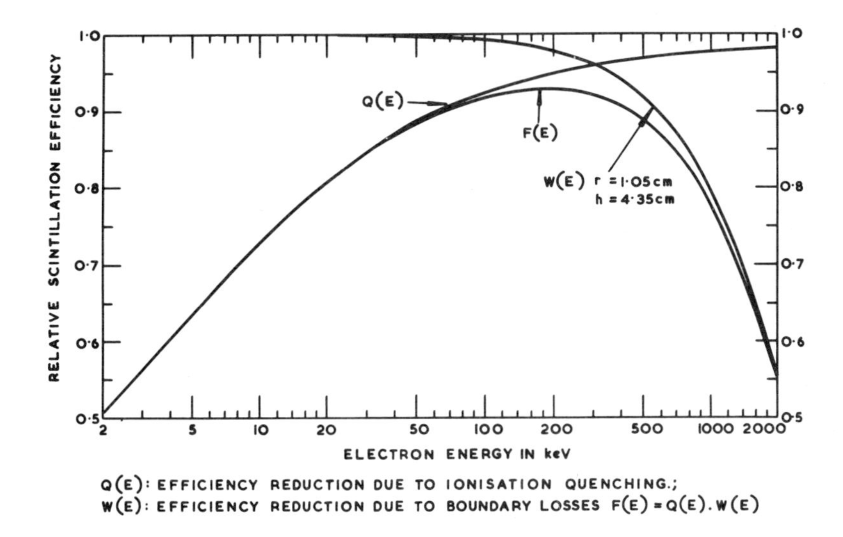

Fig. 1. Variation of scintillation efficiency with electron energy. (Reproduced from *J. Phys. E., Ser. 2,* **1**, 99 (1968), with kind permission of the Institute of Physics).

produced from the K, L and M shells of the atom. This method should produce a value of n which is accurate to better than ±5% with conventional photomultiplier tubes, but this precision can be improved with high gain tubes. The precision of P and hence n at other values of E should be about ±6%.

Spectrum shape and resolution

The peak from conversion electrons is practically a symmetrical distribution with a resolution R defined as the width w at half the maximum height divided by the peak position $\bar{x}$. This is related to the fractional variance such that:

$$R^2 = \frac{w^2}{\bar{x}^2} = \frac{\sigma^2}{\bar{r}^2}\, 8 \times \ln 2$$

and hence using Eqn. (8):

$$n = \frac{8 \times \ln 2}{R^2}\left[1 + \frac{m_2}{m_1(m_2 - 1)}\right]$$

This method is not as precise as the method of measuring the spectrum position since it is difficult to allow for conversion electrons of similar energies but it can be used as a check on the other method. Again it is more accurate to compare the total calculated distribution with the measured spectrum.

Zero probability

This is not really a distinct method of measuring the absolute activity but if the absolute activity is required for higher energy β-emitters e.g. carbon-14, then a low energy β-emitter such as tritium can be used to determine P. The precision with which the activity of the tritium is required could be low. The value of n for carbon-14 is then calculated from the measured zero probability for the tritium [Eqn. (3)]. This method is not strictly absolute but it will give, for example, a precision of better than ±0.1% for carbon-14 from ±5% for tritium.

Kolarov method

This method (Kolarov *et al.*, 1970)[8] is based upon three counts with a coincidence system:

a) with tube 1,
b) with tube 2, and
c) with tubes 1 and 2 in coincidence.

The probability of an output for a single energy is, for the three cases:

a) $C_1/\epsilon_0 = 1 - Z_1$
b) $C_2/\epsilon_0 = 1 - Z_2$
c) $C_c/\epsilon_0 = 1 - (Z_1 + Z_2) + Z_{12}$

where $Z_{12} \neq Z_1 Z_2$, from which Z_1, Z_2 and ϵ_0 can be calculated if C_1, C_2 and C_c are measured. A modification of the method to reduce the noise is discussed below (see p. 31).

PRELIMINARY EXPERIMENTAL MEASUREMENTS

These results have been presented previously[2] and hence will only be discussed very briefly to indicate how a precision of about ±2% can be obtained. Absolute standards were obtained from the Radiochemical Centre at Amersham (RCC) and from the National Bureau of Standards, USA (NBS). The results of a comparison between the liquid scintillation method and the certified activities are given in Table 2. A small correction has been applied to the results for carbon-14 and chlorine-36 to allow for the different method used to calculate the zero probability for a coincidence system.

Table 2. Comparison of the measured activity with RCC and NBS standards.

	Single tube system $P = 0.45$ electron keV^{-1}			Coincidence system $P = 0.36$ electron keV^{-1}		
Isotope	$1 - Z$	Measured activity d.p.s.	RCC standard d.p.s.	$1 - Z$	Measured activity d.p.s.	NBS standard d.p.s.
Tritium	0.629[a]	3400±120	3480±70	0.430	1980±60	1930±19
Carbon-14	0.964	7.15±0.14	7.40±0.22	0.937	850±17	873±9
Chlorine-36	0.996	72.0±0.8	78.8±7.0	0.992	753±8	748±7

[a] 1 ml of solution added: corrected for quenching, i.e. $P = 0.45 \times 0.82$

The single tube system had a very high noise level and it was therefore necessary to use 1 ml of aqueous solution in the tritium determination. This produced a significant quenching effect but Gibson and Gale[10] have shown that quenching is equivalent to a simple gain change which is independent of energy. The *relative quenching factor*, g, is a constant which is less than or equal to unity and hence the figure of merit determined for an unquenched system is reduced from P to gP. The zero probability is then determined from Eqn. (3) by using gn instead of n.

These results were obtained with photomultipliers having a photocathode efficiency of about 13% at 425 nm for the single tube and about 20% for the coincidence systems. The latest efficiencies are in excess of 25% at this wavelength. This means that n will be increased and the precision will be improved (see Table 1). The use of the methods to be discussed below should also reduce the background from the single tube and hence improve the precision of the extrapolation.

PROPOSAL TO USE A HIGH GAIN PHOTOMULTIPLIER TUBE

Photomultiplier tubes with high quantum efficiency (30% at 425 nm) and high gain first dynodes (30 to 50) using gallium phosphide (RCA type C31000D) are now in use in various scintillation systems.[8] They are expensive but could well prove to be very useful in the specific application of absolute counting. The high quantum efficiency means that the figure of merit for most liquid scintillators can be increased to $P \geqslant 1$ electron keV^{-1} for a single tube or $P \sim 0.5$ electron keV^{-1} for two tubes in coincidence. These values could be increased by using more efficient scintillators. However, if values of P of 1.0 and 0.5 electron keV^{-1} are assumed for single and coincidence systems respectively, then an assessment of this tube can be made.

Zero probabilities for single and coincidence systems

In absolute counting it may not be necessary to use the lowest possible background that is attainable with a coincidence system. The background counting rate for a single tube with an equivalent bias level of 1.5 electrons will be of the order of 10 c.p.s. Thus it is possible to compare a single tube counting two and more electrons with a coincidence system counting effectively all electrons. The ratio of the zero probabilities is given by:

$$\frac{Z\,(\text{single})}{Z\,(\text{coincidence})} = \frac{\exp(-n)(1+n)}{2\exp(-n/2)-\exp(-n)}$$

$$= \frac{\exp(-n/2)(1+n)}{2-\exp(-n/2)}$$

This ratio is less than one for all values of $n > 0$ and thus as can be seen in Table 3 and Fig. 2, the single tube is always more efficient than the coincidence system.

Extrapolation of the integral bias curve

The shape of the differential bias curve for a conversion electron is calculated by using the gamma function described in Eqn. (5). The relative variance with a first stage gain of 30 and subsequent gains of 3.16 (overall gain $\sim 10^7$) will be $1.025/n$ and hence from Eqn. (8), $\alpha + 1 = n/1.025$ and $\beta = 1/1.025$ when $n \geqslant 5$. For $n < 5$, Z is significant and α and β can be calculated from Eqns. (6) and (8).

The complete β-ray spectrum is calculated for a series of discrete energy bands as described above. In theory the computed integral bias curve should agree with the measured bias curve but in practice it may be necessary to adjust α and β to obtain the best fit to the data. This can be done using a simple iterative or regression procedure.

Problems to be examined

The most difficult problems will be associated with the counting of tritium where the maximum efficiency for counting two and more electrons is 69% (Table 3). In this

Table 3. Comparison between the efficiency for a single tube counting two and more electrons and a coincidence system with half the figure of merit for each tube.

Isotope	Mean β energy (keV) E_a	Correction factor (Fig. 1) $F(E_a)$	Effective energy (keV) $E_aF(E_a)$	Single tube $P = 1.0$ electron keV^{-1} Electron number n_a[a]	Zero probability Z	Z + single elec. prob. $Z + 1e$	Efficiency	Coincidence $P = 0.5$ electron keV^{-1} Electron number n_a[a]	Zero probability Z	Efficiency
Tritium	5.7	0.652	3.7	3.7	0.175	0.315	0.685	1.85	0.427	0.573
Iron-55	5.9[b]	0.463	2.7	2.7	0.067	0.248	0.752	1.35	0.451	0.549
Radium-228	10.1	0.730	7.4	7.4	0.125	0.217	0.783	3.70	0.291	0.709
Nickel-63	17.5	0.794	13.9	13.9	0.073	0.124	0.876	6.95	0.179	0.821
Thulium-171	25.5	0.831	21.8	21.8	0.050	0.092	0.908	10.8	0.124	0.876
Carbon-14	48.9	0.880	43.0	43.0	0.017	0.034	0.966	21.5	0.047	0.953

[a] $n_a = E_aF(E_a).P$. electrons
[b] X-ray

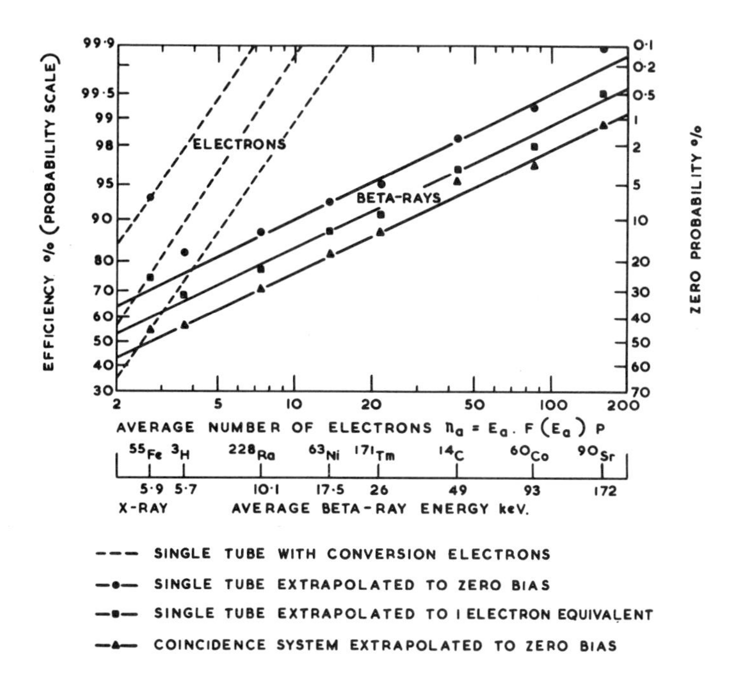

Fig. 2. Efficiency and zero probability for single and coincidence system at 1 electron keV^{-1}.

case it is most important to check the shape of the differential bias curve for tritium against a theoretical curve. This will demonstrate any variations in the system and give an indication as to how good a fit can be obtained with the gamma function. This can be repeated for other isotopes although the fit is less important for the smaller extrapolations.

The stability of the background, which is mainly due to photomultiplier noise in a single tube system, requires some study. The fluctuations can be reduced by cooling and the elimination of most single electron pulses should improve the background stability. Standard errors of ±1 c.p.s. in the background mean that if 10^6 counts are required (±0.1% standard error) then a source of output equivalent to 10^5 c.p.s. must be used to avoid an increased error in the count. The use of a second coincidence tube may well be necessary to avoid such high activities.

There is some evidence for gain reductions or fatiguing at high counting rates and this could well cause a drift in the bias curve during an experiment. The magnitude of this effect can be as high as 10% or higher for some photomultiplier tubes.[11]

After-pulses are known to be a problem with most systems. These appear as an enhanced counting rate at low energies which is associated with the source and may be due to effects in the photomultiplier. These pulses reduce the precision of the extrapolation and should be eliminated if at all possible.

The shape of the β-ray spectrum near to zero β-ray energy is not well known for some isotopes and it is therefore important to confirm that the theoretical fit to the observed differential bias curve is correct.

Ionization quenching which was discussed on p. 27 is obviously important at low energies and the shape of the curve for light output [Eqn. (10)] requires confirmation

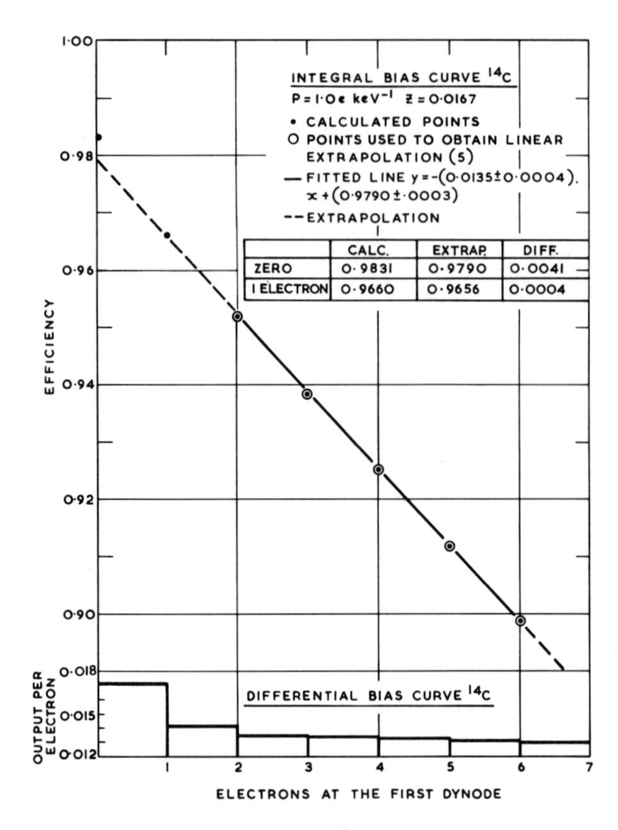

Fig. 3. A linear fit to the theoretical integral bias curve for carbon-14 from two to six electrons.

with the scintillator to be used. This could be examined by using β-excited X-ray sources placed within the scintillator since the establishment of the peak position is the essential requirement for this experiment.

Precision of the method

If a reasonable solution can be found to the points discussed above then the remaining problems are associated with:

1. the precise determination of Z, and
2. the extrapolation of the bias curve.

Z depends upon n, and with 1.85 electrons for tritium (Table 3) and a precision of $dn = \pm 0.10$ then, as was seen on p. 25, the final precision in the absolute activity will be ±2% (0.02). The value of dn could be reduced if an exact shape is determined for the light output curve. The extrapolation for tritium can be restricted to a limited part of the integral bias curve (< 3 electrons) or be fitted over a much wider range but it is unlikely that a precision of better than ±2% will be obtained. Much higher precision is to be expected at higher energies and above a mean energy of 20 keV the absolute activity can be calculated to better than ±0.1%. The error in the extrapolation will then depend on the extent of the after-pulses but for carbon-14, with a mean energy of 48.9 keV, the extrapolation from a bias level of two electron equivalents is only 1% (see Fig. 3) and this should be measurable to 0.1%. The overall error should be about ±0.14%. Attainment of such precision will depend upon solution of the earlier problems.

CONCLUSIONS

It seems feasible that most isotopes which decay by β-ray emission can be counted absolutely in a liquid scintillation counter. A precision of ±5% is attainable for tritium (β mean 5.7 keV) and this could be improved to ±2% with careful investigation of various factors in the system. Errors of ±2% for carbon-14 and ±1% for chlorine-36 have been found but these measurements could well be improved to the 0.1% level if the efficiency of the system is carefully measured.

The problems requiring further investigation are:

1. the stability of the counter background
2. the effects of high counting rates on the gain of the system
3. elimination of after-pulses
4. the shape of the β-ray spectrum near to zero energy
5. the light output as a function of the energy of the incident electron.

Solution of these problems will ensure an accurate measurement of the figure of merit and hence give a precise value of the zero probability.

Fundamental improvements will only come with more efficient scintillators and photocathodes but a more detailed understanding of some of the problems of the liquid scintillator will probably make it very competitive with other forms of absolute standardization.

ACKNOWLEDGEMENT

The early experimental work and theoretical calculations were done with the active collaboration of Mr. H. J. Gale.

REFERENCES

1 J. A. B. Gibson and A. E. Lally, *Analyst,* **96**, 681 (1971).
2 J. A. B. Gibson and H. J. Gale, *J. Sci. Instr. (J. Phys. E.)* **1**, 99 (1968).
3 D. L. Horrocks and M. H. Studier, *Anal. Chem.* **33**, 615 (1961).
4 G. T. Wright, *J. Sci. Instr.* **31**, 377 (1954).
5 H. J. Gale and J. A. B. Gibson, *J. Sci. Instr.* **43**, 224 (1966).
6 J. B. Birks, *The Theory and Practice of Scintillation Counting,* Pergamon Press, Oxford, 1965, p. 185.
7 A. T. Nelms, National Bureau of Standards Circular, 577, 1956.
8 V. Kolarov, Y. Le Gallic and R. Vatin, *Intern. J. Appl. Radiation Isotopes* **21**, 443 (1970).
9 P. W. Benjamin, C. D. Kemshall and D. L. E. Smith, AWRE Report NR/A-2/62, (1962).
10 J. A. B. Gibson and H. J. Gale, *Intern. J. Appl. Radiation Isotopes* **18**, 681 (1967).
11 O. Yourgbluth, *Appl. Opt.* **9**, 321 (1970).

APPENDIX

Calculation of the zero probability for a single photomultiplier tube

It is assumed that the input to the first dynode is Poisson distribution of mean n electrons. The photomultiplier tube consists of K stages with gains, $m_1, m_2, m_3 \ldots\ldots m_K$.

If the stage gains are not all equal then it is necessary to calculate the zero probability for a single electron input backwards from the eleventh stage such that:

$$Z_s(K) = \exp(-m_K)$$

$$Z_s(K-1) = \exp[-m_{K-1}(1 - Z_s(K))]$$

$$Z_s(K-2) = \exp[-m_{K-2}(1 - Z_s(K-1))]$$

$$Z_s(1) = \exp[-m_1(1 - Z_s(2))]$$

The zero probability at the input to the photocathode with an efficiency a is:

$$Z_s(0) = \exp[-a(1 - Z_s(1))]$$

As an example let $a = 0.25$, $m_1 = 7$, m_2 to $m_9 = 2$, m_{10} to $m_{11} = 4$ and then:

Stage, K	m_K	$1 - Z_s(K+1)$	$Z_s(K)$
11	4.00	–	0.0183
10	4.00	0.9817	0.0196
9	2.00	0.9804	0.1406
8	2.00	0.8594	0.1796
7	2.00	0.8204	0.1938
6	2.00	0.8062	0.1996
5	2.00	0.8004	0.2020
4	2.00	0.7980	0.2030
3	2.00	0.7970	0.2032
2	2.00	0.7968	0.2032
1	7.00	0.7968	0.0038
0	0.25	0.9962	0.7795

The total zero probability for a complete β-ray spectrum is obtained by dividing the spectrum into bands, calculating the probability for each band and summing to give the total probability, i.e.

$$Z = \sum_i Z_i = \sum_i H_i \exp[-n(E_i)(1 - Z_s(1))]$$

where H_i is the fraction of the spectrum within band i and $\sum_i H_i = 1$.

DISCUSSION

J. B. Birks: Is the single-tube method superior to the coincidence method for absolute counting when allowance is made for the photomultiplier noise?

J. A. B. Gibson: Yes, provided the photocathode has a high conversion efficiency (~30%) and the single tube is specially selected for low noise. The exceptions are if the amount of available activity is low or the tube shows the effects of fatigue at high counting rates.

J. L. Spratt: I was struck by your comment that people want 0.1% accuracy. How many want, or even deserve, such a degree of accuracy?

J. A. B. Gibson: Probably relatively few. However, the International Bureau of Weights and Measures are interested in using liquid scintillation counters in place of other absolute counting methods if this is possible. For most uses i.e. for biological samples etc., certainly 1% and probably 5% is quite adequate.

B. E. Gordon: The method of Kolarov (*Intern. J. Appl. Radiation Isotopes* **21**, 443 (1970)) was in fact described by Dr. V. P. Guinn in *Liquid Scintillation Counting* (ed. C. G. Bell and F. N. Hayes), 1958, p. 166, about 10 years ago and it became obvious that it was not accurate for tritium but appeared to be an adequate method to obtain the absolute activity of carbon-14 (perhaps ±2%). I suspect that the problem for tritium still remains because of noise in counting samples of weak energy.

J. A. B. Gibson: Yes, I agree but the increased efficiency of photocathodes should improve the accuracy of tritium counting by this method.
The method as described by Guinn is indeed similar to Kolarov's method but contains a fallacy. Guinn states that the counting rates in channels 1 and 2 would be $A\epsilon_1$ and $A\epsilon_2$ with an absolute activity ϵ. He then says that the coincidence counting rate is $A\epsilon_1\epsilon_2$. This is only true for conversion electrons and is not true for a β-ray spectrum (see p. 26).

Chapter 3

Colour Quenching in Liquid Scintillation Coincidence Counters

F. E. L. ten Haaf

N. V. Philips' Gloeilampenfabrieken, Nuclear Applications Laboratory, Eindhoven, The Netherlands

INTRODUCTION

The liquid scintillation counting technique requires the addition of a radioactive sample to a solution of scintillating material prior to measurement. Usually the scintillation properties of the solution are deteriorated due to the addition of the sample and this effect is called *quenching. Chemical quenching* is a common name for the phenomena that reduce the number of photons per MeV primary particle energy, initially produced in a scintillation event. The term *colour quenching* or *optical quenching* is used to denote the subsequent loss of photons along their paths to the photocathodes of the photomultiplier tubes in the detector.

Various authors have discussed the experimental aspects of quenching and, in particular, the difficulties which are sometimes encountered in the determination of the counting efficiency of coloured samples.[1-3] Instrumental methods for determining the efficiency often require different calibration curves for samples of different composition and this seems to imply that the spectral response of the instrument in each case is affected in a different way.

So far, however, little effort has been made to explain these effects in terms of spectrometer response. No doubt the reason for this is the extreme complexity of the problem, where many parameters are involved.

In 1969 Neary and Budd[4] published pulse height analyser spectra obtained from liquid scintillation samples, quenched by different chemicals. They showed that coloured samples produce broader pulse height distributions than uncoloured samples with approximately the same counting efficiency. The authors explain this by suggesting that the light produced in different parts of a sample will take paths with different lengths. For coloured samples this results in a pulse height variance which adds to the normal statistical variance of a scintillation event.

In 1971 Kaczmarczyk[5] discussed a detailed model of the response of a liquid scintillation counter which, however, did not take colour quenching into consideration.

In this chapter a simple mathematical model is set up, which illustrates qualitatively some typical features of chemical and optical quenching. Pulse height spectra generated by the model for different conditions will be compared with experimental data.

A SIMPLE MATHEMATICAL MODEL FOR QUENCHING

Assuming a linear relationship between pulse height and energy, we represent the height $h(E)$ of a pulse caused by a β-particle with the energy E, in the absence of colour quenching as:

$$h(E) = qcE \tag{1}$$

where c is the conversion factor for an unquenched sample and q represents the pulse height reduction factor due to chemical quenching.

The range of low energy β-rays can be considered small in comparison with the sample dimensions. In this case we may assume that the initial absorption of each single β-particle, the transfer of part of its energy to scintillator molecules, and the subsequent conversion of the energy into light, all occur in a small volume element within the sample. By definition the effect of chemical quenching is confined to the same small volume element which, for simplicity, we shall regard as a point.

In the presence of colour quenching, the additional attenuation of the pulse due to light absorption has to be accounted for (see Fig. 1(a)). The position of each scintillation event must now be considered. However, to account for all the parameters involved would lead to a complicated multiple integral, but we simplify matters by considering the system as if it were one-dimensional. In the original sample (Fig. 1(a)), the activity is spread evenly throughout the three-dimensional sample. In the mathematical model the 'sample' is represented by the line joining the photomultiplier photocathodes, which has a length $2d$ (see Fig. 1(b)). The activity is considered to be evenly spread along this line segment.

We now consider one scintillation event with a position given by a point on the line segment with the coordinate x, where $-d > x > d$. There are now only two possible light paths: one with a length $d + x$, towards photomultiplier 1; the other, with a length $d - x$, towards photomultiplier 2. If the optical attenuation constant is μ, the light traversing the first path will be attenuated by a factor $\exp(-\mu(d + x))$. The light traversing the second path will be attenuated by a factor of $\exp(-\mu(d - x))$. Consequently, the pulse heights h_1 and h_2 at the outputs of the photomultipliers will be:

$$h_1(E,x) = \tfrac{1}{2} \exp(-\mu(d + x))\,[qcE] \tag{1a}$$

and

$$h_2(E,x) = \tfrac{1}{2} \exp(-\mu(d - x))\,[qcE] \tag{1b}$$

In coincidence counters it is customary to add these two pulses together in a pulse summation circuit (Fig. 6) prior to analysis. The pulse height at the output of this circuit is then given by:

$$h_s(E,x) = h_1 + h_2 = \frac{\exp(-\mu(x-d)) + \exp(-\mu(x+d))}{2}\,[qcE]$$

This can also be written as:

$$h_s(E,x) = \cosh(\mu x)\,[\exp(-\mu d)qcE] \tag{2}$$

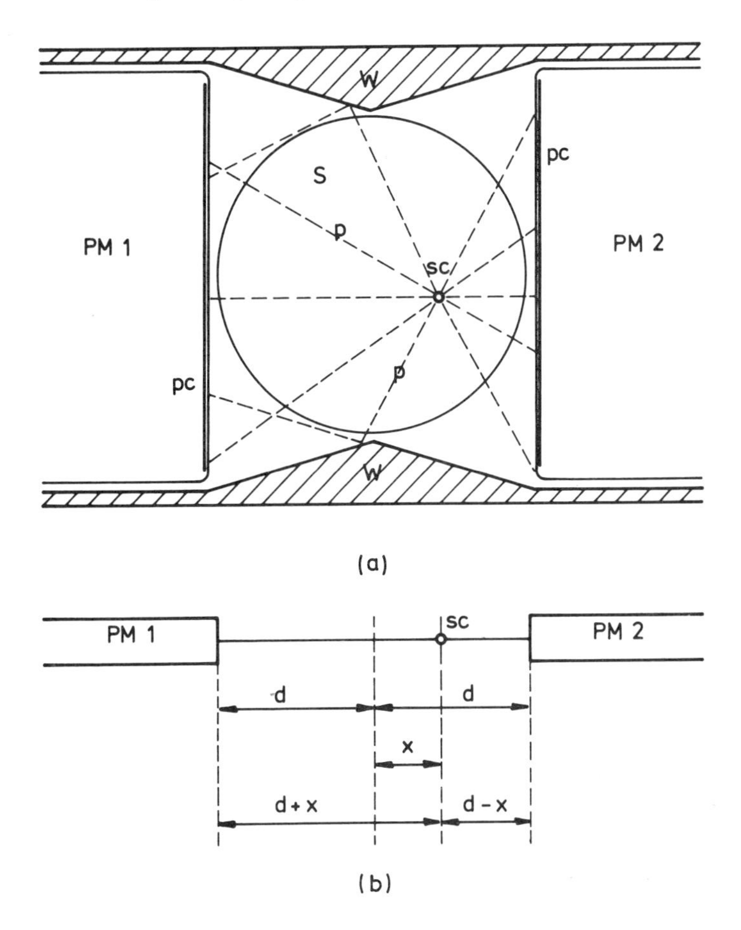

Fig. 1. (a) Schematic cross-section through liquid scintillation detector chamber with photomultipliers (PM) and their photocathodes (pc), reflective chamber walls (W), sample (S), arbitrary scintillation event (sc) and photon paths (p). (b) Simplified one-dimensional representation as discussed in text.

Equation (2) shows that, unless $\mu = 0$, the pulse height for a given energy depends on the position of the scintillation event in the sample.

The functions h_1, h_2 and h_s are represented graphically in Fig. 2 for arbitrary values of E, q and μ. This figure shows that a pulse from an event at x_1 will be counted in a coincidence system, but a larger pulse at x_2 will not be counted. The reason is that in the first case h_1 and h_2 are both above the one photoelectron level, a condition which is required to register the event as a coincidence. In the second case h_1 is below the one photoelectron level, and a coincidence becomes improbable. In fact, there is a low probability that any events will be counted which occur outside the region shown shaded in Fig. 2. This region will be smaller for larger values of μ, and consequently the coincidence losses will increase. If μ is small and E is sufficiently large, the coincidence losses will be insignificant.

It is possible to derive an analytical expression for the single energy response of the model (see also Fig. 3). For this purpose assume that a number of scintillation events, equal to the unit of activity, are distributed evenly in the 'sample', that is, along the line

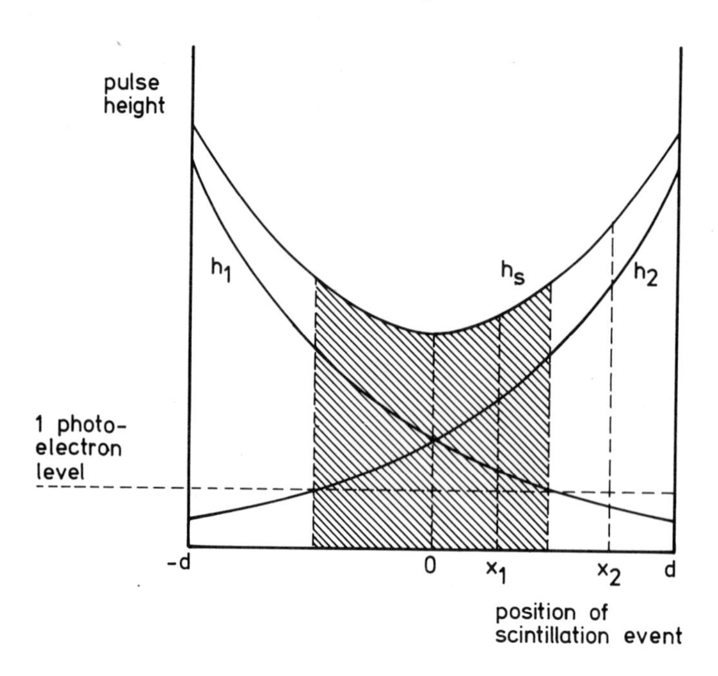

Fig. 2. Graphical presentation of heights of single pulses and summed pulse as a function of x for arbitrary fixed values of E, q and μ.

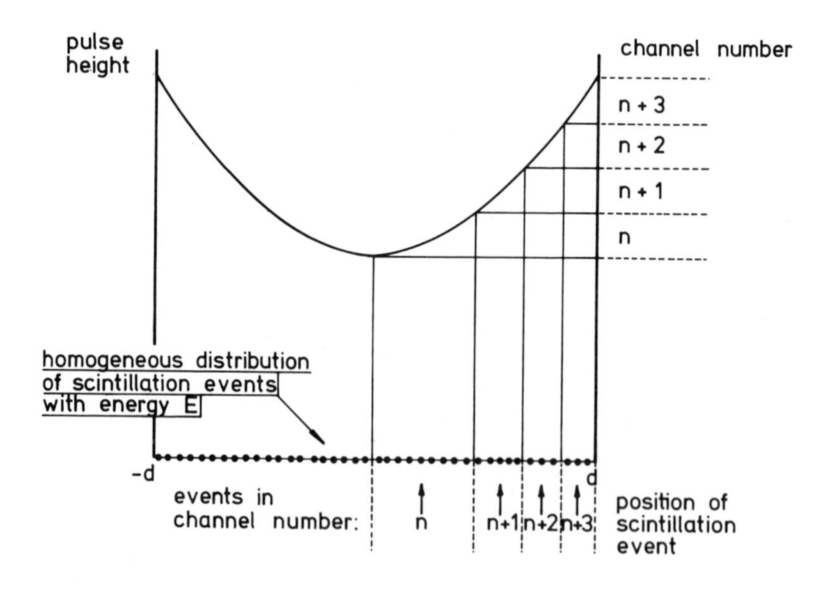

Fig. 3. Illustration of the correlation between the position of events with the same energy in a sample and the channel in which they will be counted, if optical quenching is present.

segment $(-d, d)$. All events have the same energy E. If we call the number of events n, we have:

$$\int_{-d}^{+d} \mathrm{d}n = 1, \text{ while } \int_{-d}^{+d} \mathrm{d}x = 2d$$

Consequently:

$$\mathrm{d}x = 2d.\mathrm{d}n \tag{3}$$

From Eqn. (2) we get, by differentiation with respect to x:

$$\left(\frac{\mathrm{d}h_s}{\mathrm{d}x}\right)_{E,x} = \mu \sinh(\mu x).h_s(E,0) \tag{4}$$

where $h_s(E,0)$ stands for $\exp(-\mu d)qcE$, which is the pulse height for an event in the centre of the sample. After inverting Eqn. (4) and substituting Eqn. (3), we have:

$$\left(\frac{\mathrm{d}n}{\mathrm{d}h_s}\right)_{E,x} = \frac{1}{2d\,\mu \sinh(\mu x) h_s(E,0)} \tag{5}$$

Equation (5) represents the pulse height response for a sample with a homogeneous distribution of scintillation events, all having the same energy E. In Fig. 4 two examples are shown. Figure 4(a) represents the effect of pure chemical quenching, where there is no dispersion at all. This is, of course, due to the fact that the model does not account for statistical pulse height deviations. In reality some dispersion does occur, especially at small pulse heights. Figure 4(b) shows the effect of pure optical quenching. There is appreciable pulse height dispersion and we note that, at the high energy side, there is a region where pulses will be lost if a coincidence condition is imposed, as discussed previously.

In order to construct the pulse height response of the model for a given β-emitter we need an expression for the β-energy distribution. Nuclear theory provides a distribution function which, after some remodelling takes the form:

$$p(E) = \mathrm{k}F(E + 511)(E_{\max} - E)^2 (E^2 + 1022E)^{\frac{1}{2}} \tag{6}$$

where $p(E)$ is the probability that a β-particle has an energy E (in keV). The function F is a correction factor for the influence of the electrical field of the nucleus and depends on E and the nuclear charge Z. For simplicity we will disregard this effect and make $F = 1$. The constant k will serve to normalize the distribution function in such a way that:

$$\int_0^{E_{\max}} p(E)\mathrm{d}E = 1$$

From Eqn. (2) we have:

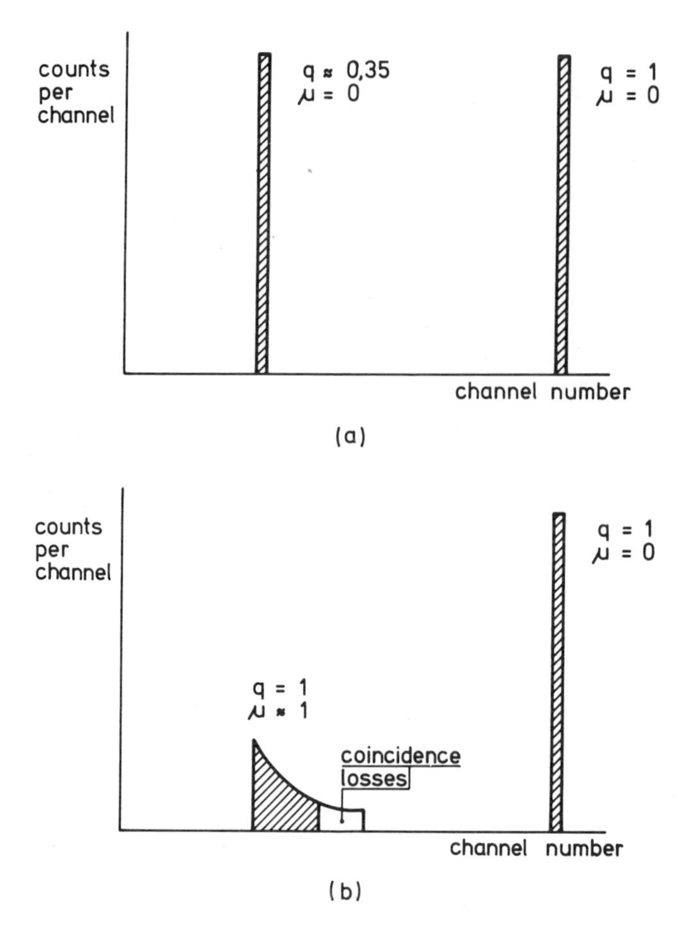

Fig. 4. Single energy response from a chemically quenched sample (a) and an optically quenched sample (b) according to the model.

$$E = \frac{h_s}{\cosh(\mu x)\exp(-\mu d)qc} \quad (7)$$

Assuming values for E_{max}, c, q and μ, one can now find the pulse height distribution $p(h_s)$ for any given value of x, by substituting Eqn. (7) in Eqn. (6) and calculating the value of Eqn. (6) for a sufficiently large number of values of h_s. To find the pulse height response for an evenly distributed activity one has to repeat this procedure for a sufficiently large number of equidistant x values, and to superimpose all the obtained distributions.

Figure 5 shows some results, in the form of integral pulse height distributions, obtained by computer calculation, for different values of q and m, where $m = \mu/\log 2$. As E_{max} was given a value of 158 keV, the spectra can be associated with carbon-14. The constant c was taken as 0.7 photoelectrons/keV, and the distance d as 1 cm.

To obtain pulse height distributions representative of a coincidence counter, the condition that both h_1 and h_2 [Eqns. (1a) and (1b)] have to be at least equal to one for any event in order to be counted, was 'built into' the computer program.

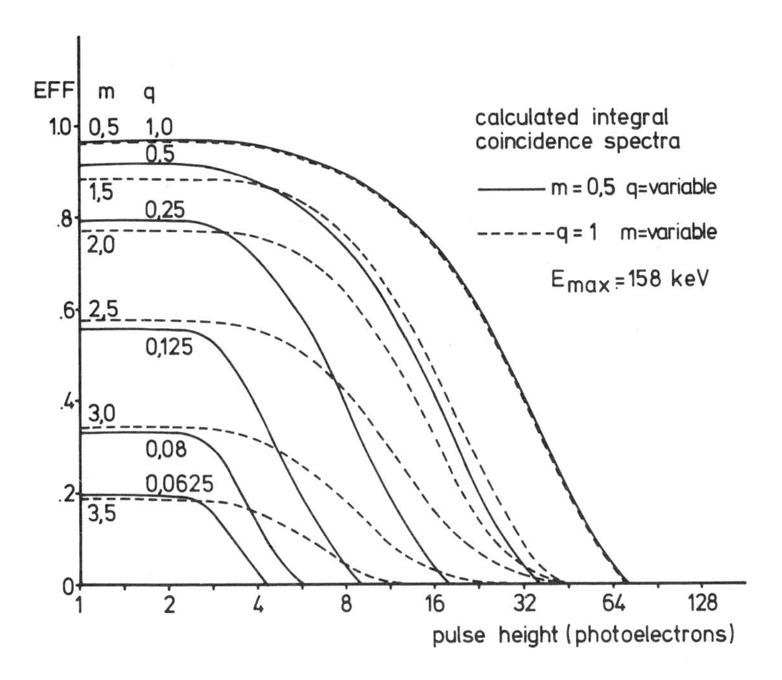

Fig. 5. Pulse height distribution curves, calculated by computer, for values of q and m. (E_{max} = 158).

The pulse height distributions shown in Fig. 5 consist of two series. The full lines represent distributions for different values of q, simulating different degrees of chemical quenching. The parameter m here has a constant low value of 0.5. The reason for making it 0.5, and not 0, is a trick to account for the appreciable geometrical effect in a coincidence counter.

The dotted lines represent pulse height distributions for different values of m, which simulate different degrees of optical quenching. Here q is constant and has the value of 1.

A comparison of the two series shows that the pulse height distributions associated with optical quenching cover a much wider energy range than distributions associated with chemical quenching.

Figure 9 shows calculated integral pulse height spectra with and without the coincidence condition imposed. It demonstrates that optical quenching is associated with large coincidence losses, whereas in the case of chemical quenching the coincidence losses are slight.

COMPARISON WITH EXPERIMENTAL RESULTS

Experimental pulse height distributions were obtained by means of a commercial liquid scintillation counter (Philips model PW 4510). This instrument has pulse-amplifiers of the linear type, but it is provided with a logarithmically calibrated gain control. Spectra were taken in the integral mode with a fixed lower level setting, and the logarithm of the gain was varied in steps.

Under normal conditions coincidence spectra are measured, but if the coincidence circuit output cable is disconnected from the coincidence gate (see Fig. 6), the gate remains open and single spectra can be obtained.

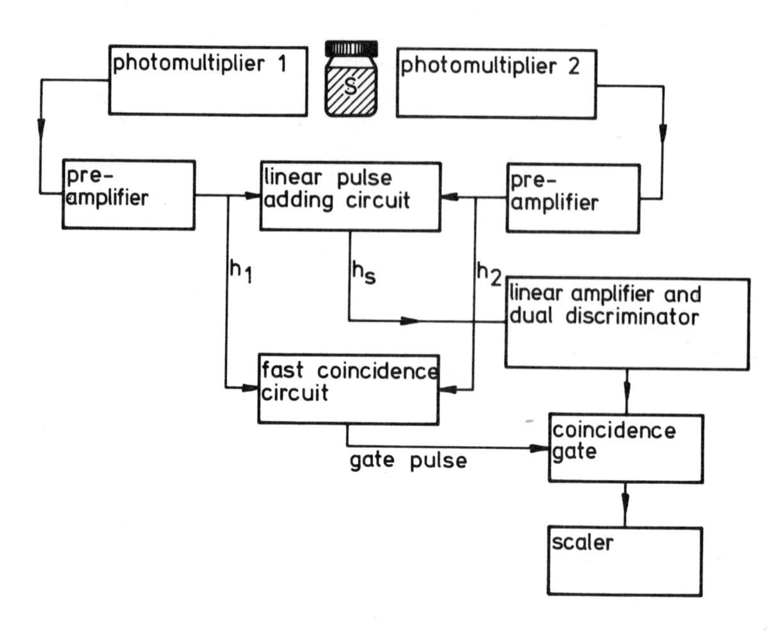

Fig. 6. Simplified block diagram of liquid scintillation counter.

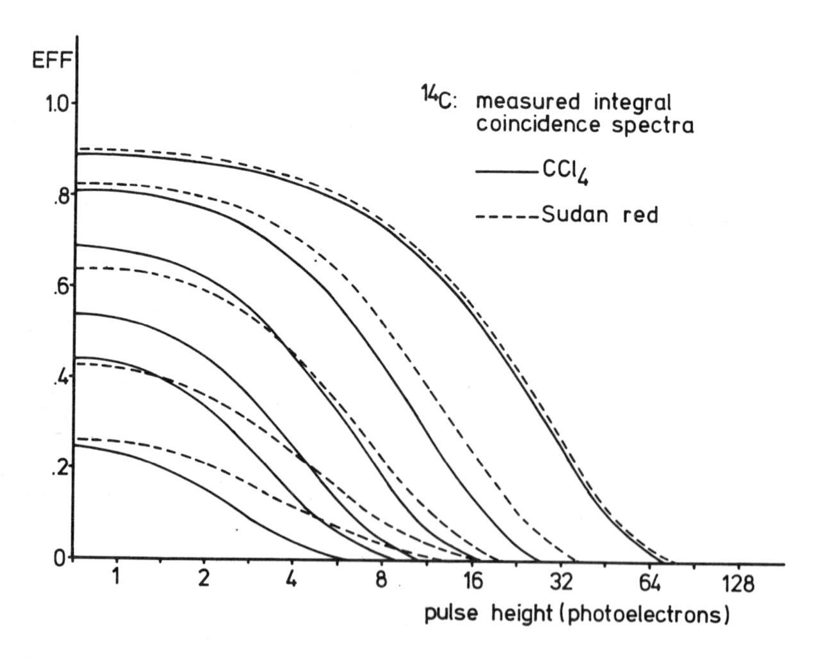

Fig. 7. Coincidence spectra of chemically quenched and optically quenched samples of carbon-14.

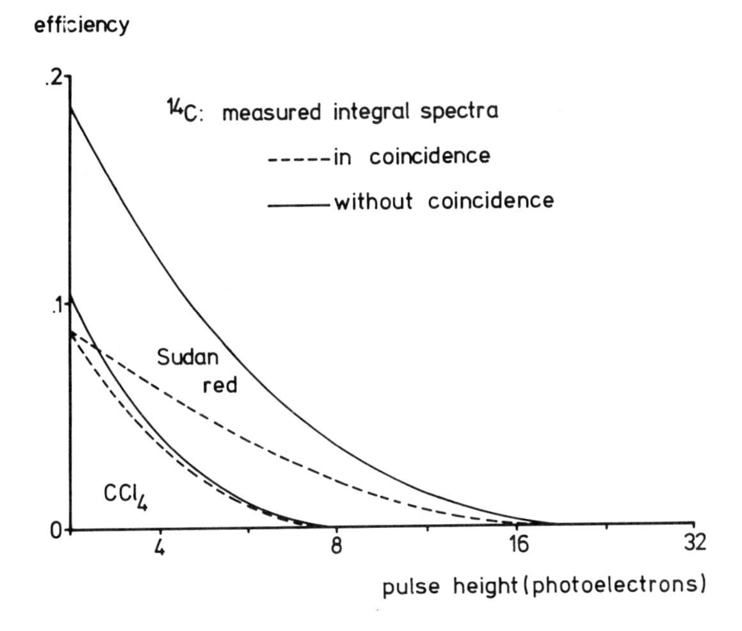

Fig. 8. Coincidence and single spectra of a chemically quenched and an optically quenched sample. The non-coincident spectra could be measured down to the three photo-electron level without disturbance from the thermal noise of the photomultipliers.

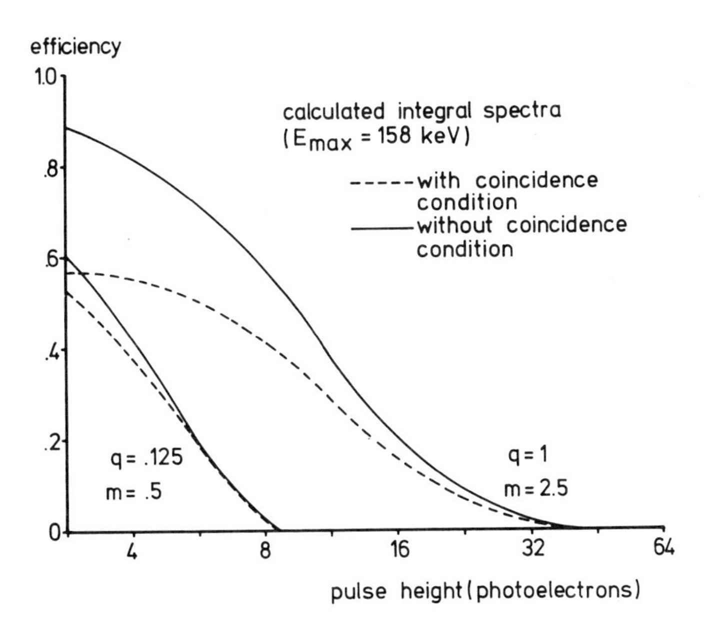

Fig. 9. Calculated integral pulse height spectra, with and without the coincidence condition imposed.

Two series of ^{14}C-spiked samples were measured, all having the same activity and scintillator composition. The scintillator was butyl-PBD (6 g/l toluene). To one series increasing quantities of CCl_4 had been added and to the other increasing quantities of Methyl-red. In Fig. 7 the coincidence spectra obtained from these samples are shown, where the count rate is divided by the sample activity. On the horizontal axis the pulse height is represented by the estimated equivalent number of photoelectrons.

The colour-quenched samples, when compared with chemically quenched samples of approximately equal integral counting efficiency, have the same tendency towards broader pulse height distributions as shown by the calculated spectra in Fig. 5.

Figure 8 represents the coincidence and single spectra of a colour-quenched sample and a chemically quenched sample, having approximately the same integral counting efficiency in coincidence. The figure shows that the colour-quenched sample suffers much larger coincidence losses than the chemically quenched sample. This is in agreement with the calculated results shown in Fig. 9.

CONCLUSION

Colour-quenched liquid scintillation samples show broader energy distributions and larger coincidence losses than chemically quenched samples which have approximately the same counting efficiency. These effects can be explained qualitatively by means of a simple mathematical model.

REFERENCES

1 E. T. Bush, *Anal. Chem.* **35**, 1024 (1963).

2 H. H. Ross, *Anal. Chem.* **37**, 621 (1965).

3 H. H. Ross, *Radioisotope Sample Measurement Techniques in Medicine and Biology*, IAEA, Vienna, 1965, p. 409.

4 M. P. Neary and A. L. Budd, *MIT Conference on Current Status of Liquid Scintillation Counting, April 1969*, Grune and Watson, New York.

5 N. Kaczmarczyk, *Organic Scintillators and Liquid Scintillation Counting*, Academic Press, London and New York, 1971, p. 977.

DISCUSSION

J. B. Birks: Some instrument manufacturers have claimed that quench correction methods using channel ratios and/or external standards are applicable to both chemical quenching and colour quenching. The present analysis shows that this is not valid. An internal standard appears necessary to correct properly for colour quenching.

F. E. L. ten Haaf: I quite agree that this is true for the general case where both chemical and optical quenching vary strongly within a given series of samples. However, if the quenching is mainly due to one or the other type, the external standard and/or channels ratio method can be used successfully, provided of course the proper calibration curve is used for each type. At very low degrees of quenching one might use one and the same curve without making large errors.

J. A. B. Gibson: How good is the agreement between the theoretical and practical distributions?

F. E. L. ten Haaf: Not very good, since the theoretical model is a very simplified one. I look on this as a mathematical exercise, which has shown that our simplified model does bear considerable relation to reality.

Chapter 4

Application of Solid Scintillators in High Pressure Radio Column Chromatography

G. B. Sieswerda and H. L. Polak

Laboratory of Analytical Chemistry, University of Amsterdam, Amsterdam, The Netherlands

INTRODUCTION

Column chromatography of compounds labelled with a weak β-emitter is applied in two different ways. The most common method is the discontinuous measurement of the radioactivity in the individual samples from a fraction collector. The samples are counted with the appropriate scintillators and solvents in a non-aqueous environment. On the other hand, continuous measurement of radioactivity in the effluent of a chromatographic column by means of a flow cell may be applied. Before enumerating the advantages and disadvantages of both systems, it is better to carry out some calculations in order to get an objective insight into the theoretical possibilities of both detecting systems.

Table 1 shows the results of the discontinuous counting of 10 nC. The efficiency is supposed to be 80%, the background 100 c.p.m. One minute counting time produces a resulting signal of 20 000 counts. The signal to noise ratio is 2000.

Table 2 applies to the continuous measurement of the same amount of radioactivity. Let the efficiency be 50%. An eluent flow rate of 1 ml/min and a cell volume of 0.1 ml results in 0.1 min residence time of each particle in the detector. So the counting time of the radioactivity is 0.1 min. The signal amounts to 1110 counts. The elution time of the total amount of radioactivity is 5 min, so the background contribution becomes 500 counts and the signal to noise ratio is decreased to 50. This means a decrease by a factor of 40 in comparison with the discontinuous measurement. Moreover one should remember that the signal to noise ratio cannot be improved in the case of the continuous measurement unless the counting time is increased. This can only be achieved by a decrease in the flow rate, which is in contradiction with the demand for a rapid separation of the compounds under investigation. On the other hand, the signal to noise ratio in the case of discontinuous measurement can be improved very simply by the choice of a longer counting time of the individual samples. This can be done without violating the separating conditions.

From the foregoing example it is quite clear that flow counting can never compete with discontinuous measurement. Consequently it may be wondered why continuous measurement is applied. The reasons for this are merely practical and are not difficult

Table 1. Discontinuous measurement.

Radioactivity	10 nC
Efficiency	80%
Counting time	1 min
Signal	~ 20000 counts
Background	100 c.p.m.
Noise = $\sqrt{\text{Background}}$	10 c.p.m.
Signal to noise ratio	20000/10 = 2000

Table 2. Continuous measurement.

Radioactivity	10 nC
Efficiency	50%
Flow rate (F)	1 ml/min
Cell volume (V)	0.1 ml
Counting time (V/F)	0.1 min
Signal	1110 counts
Background	100 c.p.m.
Elution time	5 min
Background contribution	5 X 100 = 500 counts
Noise = $\sqrt{\text{Background contribution}}$	~ 22 counts
Signal to noise ratio	1110/22 ≙ 50

to understand:

1. In contrast to the discontinuous method the continuous method is non-destructive in the case of weak β-emitters.
2. The method is fast. In the case of discontinuous counting all fractions have to be examined for radioactivity. The preparation of the samples is time-consuming.
3. The method is fully automatic. The ratemeter/spectrometer signal as a function of time can be registered by means of a strip chart recorder.

Weighing the advantages and disadvantages of both methods against one another, it was decided to combine both techniques in order to profit by the accuracy of the discontinuous method as well as by the gain of time of the continuous counting.

With this end in view it is our intention to construct a chromatographic system in which the flow cell serves as a level detector and steers as such a fraction collector. In other words, as soon as the background rises to an adjustable level, the effluent is collected until the background returns to its original level. The collected fraction can then be prepared for discontinuous measurement.

CONSTRUCTION OF THE FLOW CELL

For a good operation of this combined system it is necessary to know the properties of the flow detector. We constructed a flow cell suitable for continuous measurement of weak β-radiation in conjunction with a high pressure chromatographic column. With the construction attention has been paid to the possibility of changing the scin-

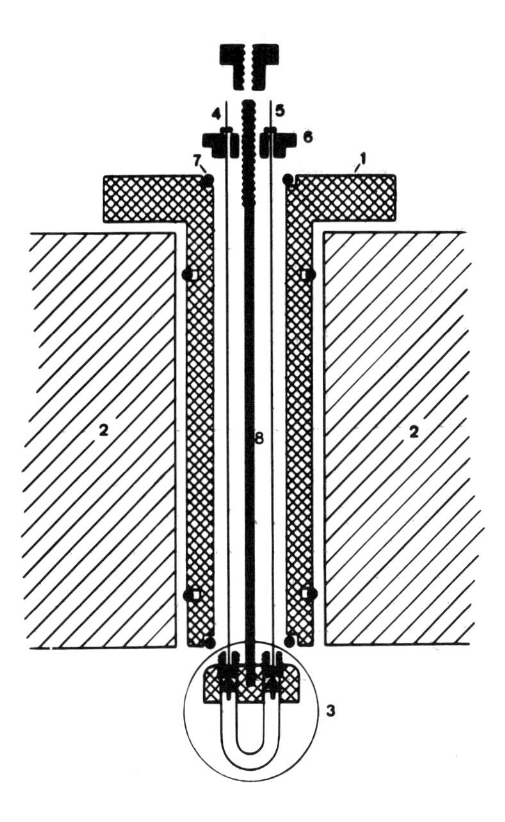

Fig. 1. Cell holder with flow cell. 1 = aluminium cylinder, 2 = photomultiplier housing, 3 = flow cell (part encircled Fig. 2), 4 and 5 = inlet and outlet capillaries, 6 = lid, 7 = 'O' ring, 8 = screw.

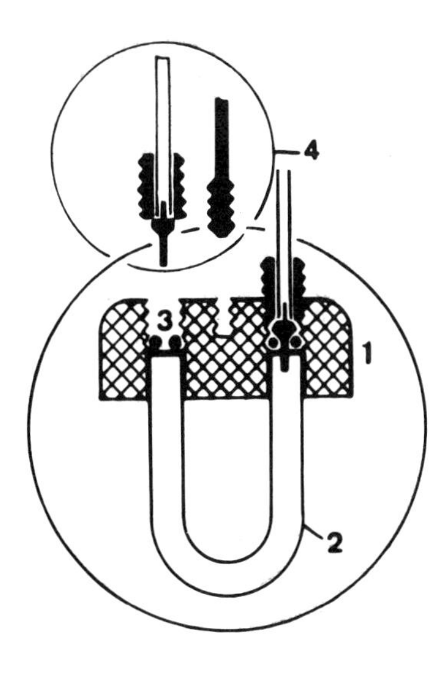

Fig. 2. Flow cell. 1 = socket, 2 = quartz U-pipe, 3 = 'O'-ring, 4 = joint and screw.

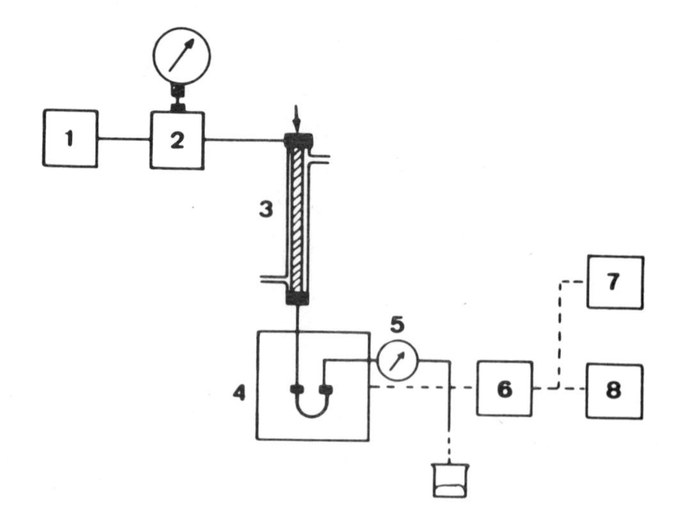

Fig. 3. Scheme of the apparatus. 1 = eluent reservoir, 2 = pump, 3 = thermostated column, 4 = detector cell mounted in counting chamber of the photomultiplier assembly, 5 = flow meter, 6 = ratemeter/spectrometer, 7 = multi-channel recorder 8 = recorder.

tillating material, thus enabling a free choice of scintillator and a rapid renewal in case of contamination.

Figure 1 shows the cell holder and cell constructed in our laboratory. The cell holder consists of a hollow cylinder. The actual cell is a quartz U-pipe mounted on a socket which is attached to the lower part of the holder.

Figure 2 shows the actual cell and the joints for the inlet and outlet tubing. The cell is filled with solid scintillating material. Several commercially available scintillators were used.

Figure 3 shows the chromatographic and electronic system. The eluent is pumped through by a pulsating pump. The pump and the injection port are connected by a teflon tube. The connector at the bottom of the column is joined to the detector cell in the same way. The scintillations caused by the interaction of the radioactivity and the scintillating material are detected by the photomultiplier assembly. The recorder signal of the ratemeter/spectrometer is fed into a strip chart recorder. The digital output signal is fed into a multi-channel analyser used as a multi-scaler.

EXPERIMENTS

Relative static efficiencies

By measuring the relative static efficiency, the scintillator with the best properties was chosen for further experiments. The results are shown in Table 3. It is obvious that PPO and PTP are the most efficient scintillators. However, POPOP was preferred for further experiments because it shows better cell packing properties and maintains a constant flow rate. The flow cell filled with POPOP possesses an effective fluid volume of 0.1 ml.

Table 3. Relative static efficiency of several scintillators for ^{14}C-labelled ethanol.

Scintillator	Relative static efficiency (S_{rel}) (%)
PPO	100
PTP	78.6
POPOP	54.3
DPS	46.7
trans-Stilbene	43.5
Anthracene	34.9
ANPO	34.5
LiCl	24.6
Naphthalene	17.0

Table 4. Absolute static efficiency of POPOP for ^{14}C-labelled amino acids and HTO.

Compound	Absolute static efficiency (S_{abs}) (%)
Aspartic acid	50.9
Threonine	53.4
Glutamic acid	53.2
Serine	50.7
Proline	53.6
HTO	5.4

Absolute static efficiencies

The results of the determination of the absolute efficiency of POPOP for some ^{14}C-labelled amino acids and tritiated water are shown in Table 4. The results are in good agreement with one another. The high efficiency values found are the more remarkable in view of the small effective volume of the detector.

Calibration curve

For quantitative measurements a calibration curve has to be realized. Figure 4 shows the calibration curves for carbon-14 and tritium radioactivity. The input signal is represented by the amount of radioactivity injected, the output signal by the number of counts recorded on the multi-channel analyser.

The output signal of the detector as a function of the input signal shows good linearity. This figure also shows the dependence of the output signal on the β-radiation energy as can be concluded from the different slopes of the calibration curves for carbon-14 and tritium.

Peak broadening

The most important characteristic of a flow cell is its peak broadening effect at

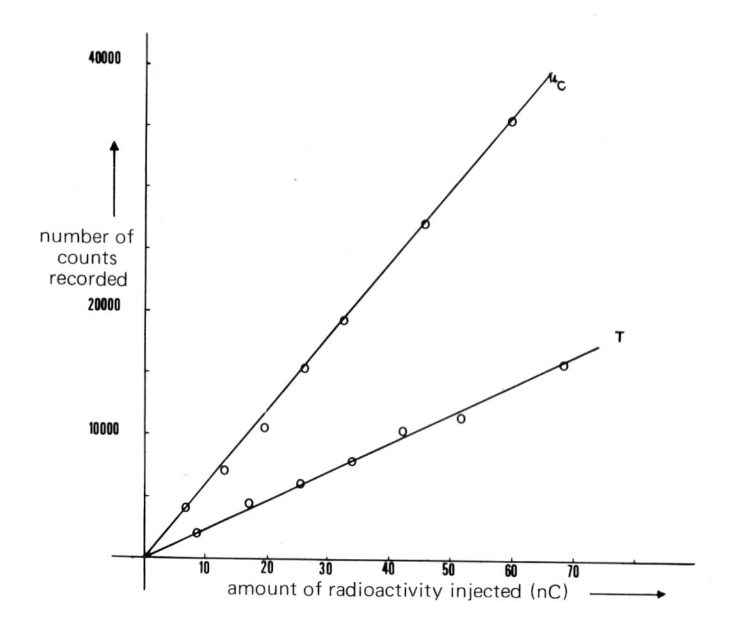

Fig. 4. Calibration curve for ^{14}C-labelled aspartic acid and for ^{3}H-labelled water.

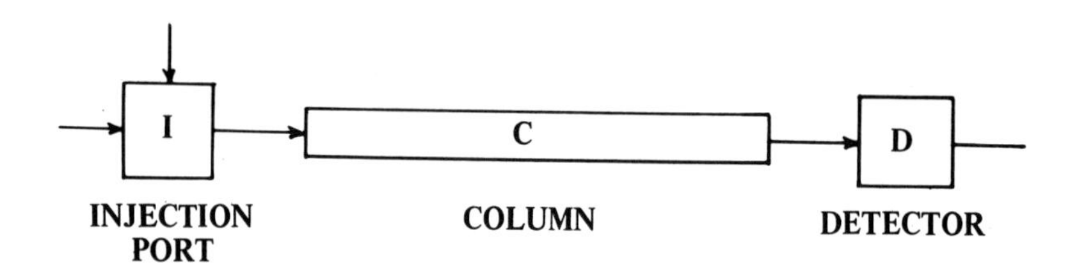

Fig. 5. Chromatographic system.

a certain flow rate. The detector is expected to deliver a signal time function which is a true representation of the elution function.

Figure 5 shows the chromatographic system: the injection port, the feed line to the column, the column, the feed line to the detector and the detector itself. The signal time function of a component, the elution function, has a Gaussian shape and can be characterized by the standard deviation. A concentration peak leaving the injection port will disperse further in the feed line to the column and subsequently in the column, the feed line to the detector and the detector. These peak broadening effects caused by the individual components of the chromatographic system are independent of one another. Consequently the total dispersion can be expressed in terms of variances and summarized as shown in Fig. 6.

As mentioned above the detector is expected to deliver a true representation of the elution function. That means that the contribution of the detector to the overall peak broadening should be small. The determination of the detector contribution is

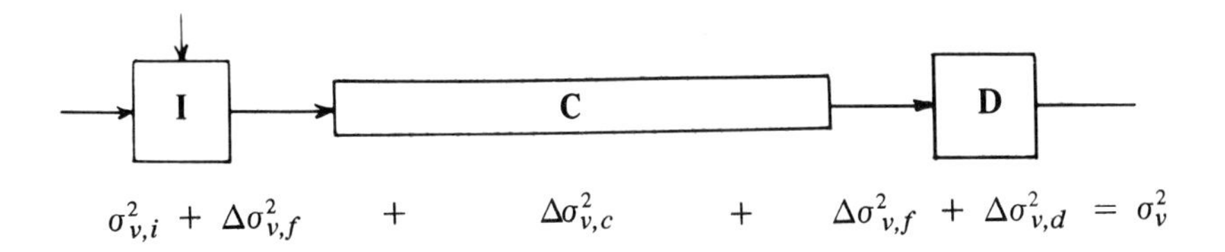

Fig. 6. Peak broadening contributions in the chromatographic system expressed in terms of variances.

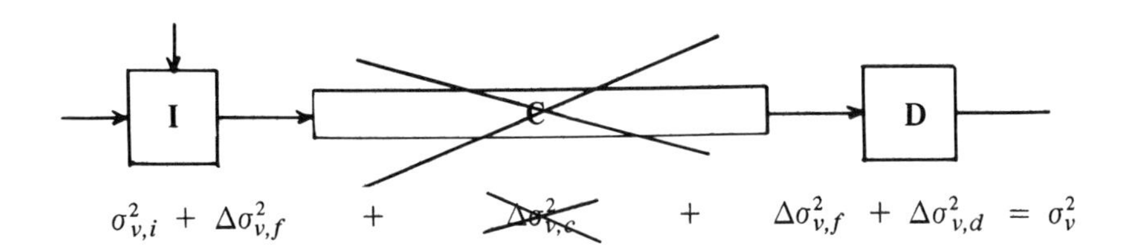

Fig. 7. Measurement of the peak broadening effect of the flow cell.

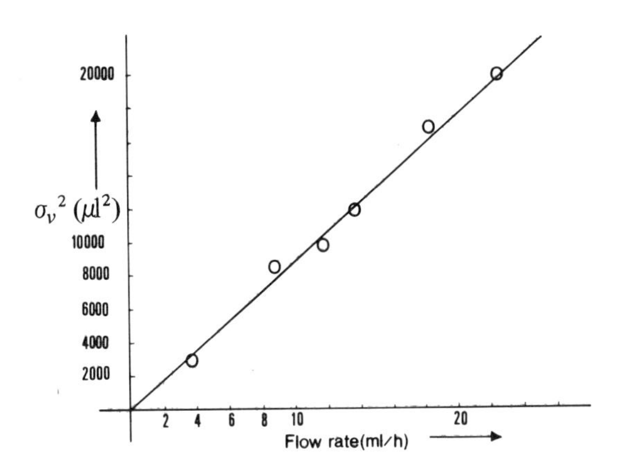

Fig. 8. Peak broadening caused by the flow cell as function of the flow rate.

possible by experimenting without a column. This is made clear in Fig. 7. The radioactivity is injected directly into the detector. By measuring the peak broadening in this way we can calculate the peak broadening effect of the flow cell, since the contributions of the other part of the system are constant.

Figure 8 shows the results. The relationship between the flow rate and the variance is an important property of the flow cell. Experiments including a column provide the overall peak broadening. Consequently one can decide on an acceptable contribution of the flow rate to the overall peak broadening.

Contamination

In view of the peak broadening effect, another aspect has to be considered – namely the contamination by radioactive material. A longer residence time of the activity in the detector and consequently an increase of the peak broadening can be expected in cases of contamination.

In our experiments tritium-labelled water did not contaminate the scintillator at all. However, the amino acids contaminate POPOP to a high degree in the absence of a decontaminator in the eluent. The output signals have not the normal Gaussian shapes but show tailing effects. The presence of a detergent in the eluent, however, eliminates the contamination and consequently a constant background was measured before and after the passing of an amount of radioactivity through the detector.

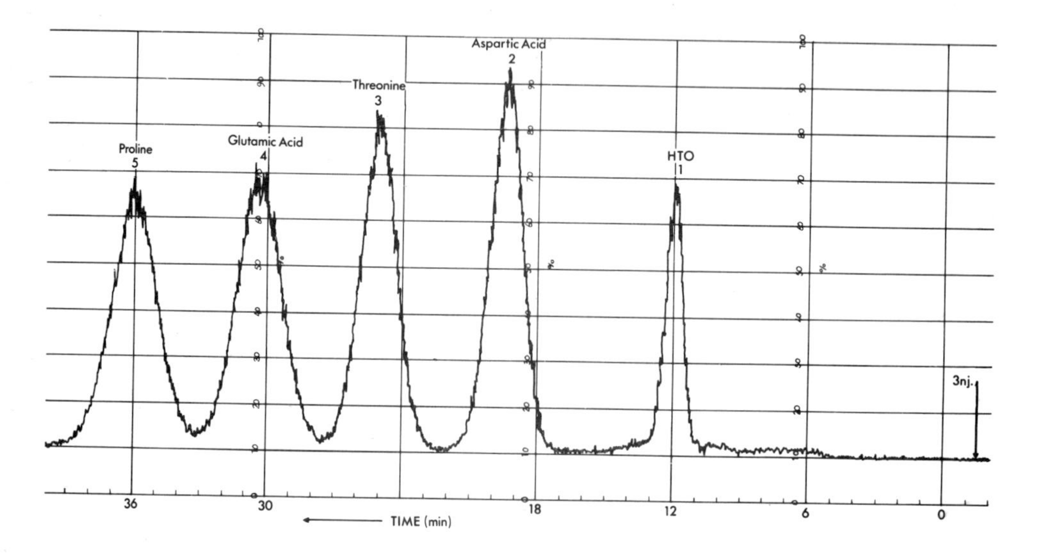

Fig. 9. Chromatogram of a mixture of (1) tritiated water, (2) aspartic acid, (3) threonine, (4) glutamic acid and (5) proline. Separation on Aminex Q-150S with citrate buffer. Counting range 25 kcpm; time constant 1.5 s; flow rate 12.4 ml/h; recorder full scale 10 mV; chart speed 1 cm/min; temperature 60°C; column dimension 50 × 0.5 cm.

Separation of a mixture of amino acids

In practice the monitor is used in conjunction with a column filled with cation exchanger. A mixture of ^{14}C-labelled aspartic acid, threonine, glutamic acid, proline and tritiated water was used as the separation example. The result of the separation is shown in Fig. 9. At first sight one may be satisfied. The separation is completed within 40 min and the components of the mixture are separated almost entirely. However, one should also remember the contribution of the flow cell to the overall peak broadening. At a flow rate of 12.5 ml/h the contribution of the flow cell to the peak broadening is 100 μl. The function of tritiated water in the mixture is now clear. Assuming interaction between ion exchanger and water, the column will disperse the tritiated water peak to a very small extent. The quantity of peak broadening of more than 100 μl will be caused by mixing in the column. In the same way every peak in the chromatogram possesses a

Table 5. Detector contribution to total peak broadening.

Compound	σ_v (μl)	Detector contribution (μl)		Detector contribution (%)
HTO	105	100	=	95
Aspartic acid	210	100	=	48
Threonine	235	100	=	43
Glutamic acid	260	100	=	38
Proline	287	100	=	35

detector peak broadening contribution of 100 μl. Table 5 shows the results of these simple calculations. One can see that 95% of the dispersion of the tritiated water peak is caused by the detector cell. So only 5% is due to mixing. The detector peak broadening effect on the second component is also large, namely 48%, and decreases in the elution sequence down to proline (35%).

CONCLUSION

It is obvious that the result is affected to a great extent by the detector. Satisfaction is changed to dissatisfaction when one considers the applicability of the detector. One can assume two things:

1. The column is too efficient for the detector, so the separating system will have to be adjusted.
2. The detector is insufficient and will have to be corrected.

The first recommendation is in flagrant contradiction to the chromatographic demands. The detector should not ruin the separation. It is clear that the flow cell will have to be adjusted. This can be done by reducing the detector volume. However, this means a reduction of the signal since the counting time is proportional to the cell volume. Likewise a reduction of the efficiency means an increase of the detection limit. A solution to this difficulty is the so-called peak tracer method. If the amount of radioactive-labelled compound in a mixture is too small for flow detection one can inject a large amount of this particular compound before the actual separation of the mixture. The flow detection then indicates the place in the chromatogram for the compound under investigation. Next the separation of the mixture is carried out under the same conditions. The effluent is collected at the place in the chromatogram indicated by the peak tracer method and counted in the discontinuous way until a significant signal is recorded.

One can conclude that the combination of a flow cell, as a level detector or a peak tracer detector, with a fraction collector provides an improvement in the sensitivity of the determination and decreases the time of analysis.

BIBLIOGRAPHY

E. Rapkin, *Lab. Scintillator* (Picker Nuclear Publication), **11**, 6L (1967).
E. Schram, *Anal. Chim. Acta* **17**, 417 (1957).
E. Schram, *Anal. Biochem.* **3**, 68 (1962).
K. H. Clifford, *J. Chromatog.* **40**, 377 (1969).
D. H. Spackman, *Anal. Chem.* **30**, 1190 (1958).
A. M. van Urk-Schoen and J. F. K. Huber, *Anal. Chim. Acta* **52**, 519 (1970).

DISCUSSION

B. E. Gordon: I don't clearly understand the peak tracer method.

G. B. Sieswerda: Since the intensity of signal is dependent on the cell volume and the flow rate, some small amounts of radioactivity may give no significant signal under certain separation conditions. The only possibility for detection then is to use the peak tracer method. In this method, a much larger and measurable amount of radioactivity is injected under the same separating conditions and in this way the retention time of this particular compound is determined. During the second run, in which the unknown mixture is separated, the effluent is collected at the retention time indicated by the tracer, and counted by the more conventional discontinuous liquid scintillation counting procedure.

J. Murray: Which detergent did you use for reducing contamination of the scintillator by ^{14}C-labelled amino acids?

G. B. Sieswerda: We used a polyethylene lauryl ether (Bry-35) which is also a component of the eluent necessary for the chromatographic separation.

D. Schram: How were the several solid scintillators calibrated for their mesh size, prior to the determination of their relative efficiencies?

G. B. Sieswerda: We did not calibrate for mesh size, we just used the available products. However, depending on the chromatographic requirements, it may be necessary to recrystallize and sieve the scintillators.

B. W. Fox: Do I understand that the only reason for using POPOP as against PPO is because of its better packing properties?

G. B. Sieswerda: Yes.

B. W. Fox: Have you looked into the possibility of altering the crystal habit of PPO to utilize its much greater efficiency?

G. B. Sieswerda: No, we have not done this.

P. Johnson: Dr. Sieswerda, can I ask if you have used any mass detector in addition to your flow detector for radioactivity? The reason I ask is that one of the well accepted advantages of using a flow detector system for radioactivity in parallel with a mass detector, rather than a discrete off-line counting method, is that with discontinuous detection of radioactivity it is possible to miss peaks of high specific activity but insufficient mass to activate the mass detector. It would therefore be of interest to use your system for instance, on-line to one of the commercially available ultraviolet or other mass detector systems.

G. B. Sieswerda: This would be of interest but we have not yet used any detector other than the radioactivity flow cell.

B. Scales: High pressure liquid chromatography is often carried out using mixed solvents, and gradient elution techniques. When organic solvents are incorporated into the system, the increased solubility of the scintillator can be a serious problem. Have you considered the use of plastic scintillators in these situations, or are the counting efficiencies too low?

G. B. Sieswerda: Yes, plastic or glass scintillators can be used to overcome the problems of dissolution of conventional organic scintillators but the counting efficiencies are unacceptably low for weak β-emitters.

P. Stanley: Have you attempted to flow-count macromolecules of biological interest, for instance proteins or nucleic acids? Does PPO affect the biological activities of these macromolecules?

G. B. Sieswerda: The answer to the first question is no. I did not use the flow cell for the detection of macromolecules. In this respect, however, one can expect contamination of the scintillating material. Macromolecules, like proteins and nucleic acids, can complex with the scintillator and can be retained on its surface. So, in answer to your second question, I must say that there is a risk of affecting the biological activities of these macromolecules.

SECTION II

HISTORICAL DEVELOPMENTS AND MISCELLANEOUS PROBLEMS

Chapter 5

A History of the Development of the Modern Liquid Scintillation Counter

E. Rapkin

Short-Hills,
New Jersey, U.S.A.

INTRODUCTION

Following the discovery of the liquid scintillation process and the construction of working instrumentation in university and government laboratories, further progress in equipment design has come about almost exclusively from industry. At one time or another about twenty firms have manufactured liquid scintillation counters with about ten being active today (see Table 1). However, significant new contributions have come from relatively few, with the remainder content to develop and perfect concepts originating with others. There have been, of course, cosmetic changes; these will be ignored. Of more real interest and utility are changes leading to improved counting performance hence greater sensitivity and/or reduced counting time, those providing new convenience in instrument operation, sample handling, or data manipulation and presentation, and those which enhance instrument reliability. In one way or another they all can result in time-saving to the investigator and should therefore be given equal consideration.

These improvements will be considered chronologically; usually they have followed logically from one to the next. It is hoped that knowledge of how and why they have occurred will make for a better understanding of the potentials and limitations of today's equipment and perhaps point out new directions for the still further improved instrumentation which will surely come in the future.

1950–1957: Beginnings

The years between 1950 when the first liquid scintillation counters were built in the laboratory and 1957 when the first automatic counter was announced must be considered a formative period. The few users devoted a large part of their efforts to demonstrating that the liquid scintillation method had a rather broad applicability to various α- and β-containing test substances; little routine use was made of the technique which was then not well-known. Also, because there were so few units, and because the equipment permitted only manual operation, it was really not possible for there to be the build-up of that body of published literature which might have led others to make the large investment ($4500–7500) which liquid scintillation counting then required. Further, it was by no means obvious that liquid scintillation counting was preferable to Geiger or proportional

Table 1. Details of liquid scintillation counter manufacturers.

Company	Country	Entered field	Discontinued	Single tube	Number[a]	Coincidence	Number[a]
ACEC	Belgium	1967	?	no	—	yes	a
Aloka	Japan	1965	active	no	—	yes	a
ANS	U.S.	1964	1969	no	—	yes	b
Baird Atomic	U.S.	1960	1962	yes	a	no	—
Beckman	U.S.	1965	active	no	—	yes	c
Belin	France	1959	1964	no	—	yes	a
Ekco	Great Britain	195?	?	yes	b	no	—
Friesecke and Hoepfner	Germany	1966	active	no	—	yes	a
Intertechnique	France	1968	active	no	—	yes	b
Isotope Dev. Ltd.	Great Britain	1961	1966	no	—	yes	a
Kobe Kogyo	Japan	1965	?	no	—	yes	a
Nuclear Chicago	U.S.	1961	active	no	—	yes	c
Nuclear Enterprises	Great Britain	1960	active	yes	b	yes	a
Packard	U.S.	1954	active	no	—	yes	c
Panax	Great Britain	195?	?	yes	b	no	—
Philips	Holland	1963	?	yes	a	yes	a
Picker	U.S.	1965	1969	no	—	yes	a
Shimadzu	Japan	1965	?	no	—	yes	a
Tech. Measurement	U.S.	1955	1959	no	—	yes	a
Tracerlab	U.S./Belgium	1954	active	no	—	yes	b
Vanguard	U.S.	1962	1965	no	—	yes	a
Wallac	Finland	1968	active	no	—	yes	a

[a] Number in field: a = less than 100; b = 100 to 1000; c = more than 1000.

counting, especially since automatic equipment did exist for handling planchets.

Liquid scintillation counting equipment, other than that used in the earliest experimentation, began with a single, vertically-standing, flat-faced photomultiplier encased in lead shielding. The counting solution was placed in a beaker or vial which was optically coupled to the phototube faceplate with silicone fluid. To maximize light collection a reflector surrounded the sample container; in some later designs the vial had only a transparent bottom with all other surfaces having reflective properties. Electronics consisted of the necessary high voltage supply, a preamplifier at the base of the 1 in photomultiplier followed by a linear amplifier, a threshold discriminator or, in more sophisticated units, a single channel analyser, a scaler and a mechanical timer. With the best counting samples then available, tritium counting efficiency fell in the range of 15 to 20% with backgrounds of 600 to 1200 c.p.m.

Most of the 'background' arose from thermal noise generated within the photomultiplier. While some of this was soon eliminated by refrigerating the detector, even at −20° C, a figure well below the limits of practicality in that neither test substance nor scintillator is adequately soluble and the inevitable frost build-up is a continual nuisance, the noise levels were still excessive. Phototube noise is also high voltage-dependent and can be reduced by lowering the voltage. Unfortunately, this reduces phototube gain thereby reducing the amplitude of the signal available to the preamplifier and then the amplifier. If that signal is still of sufficient amplitude so as not to be lost in the noise generated within the amplification stages, it might be thought that increased electronic gain would overcome reduced input pulse size. However, this was not possible; increasing the gains of preamplifier and amplifier increased both the rates and amplitudes of their noise which, even under normal circumstances, tended to border on the excessive. In fact, the main reason for having the threshold discriminator was not for pulse height analysis but rather to reject low level preamplifier and amplifier noise.

Phosphorescence and chemiluminescence were significant non-electronic problems of single tube counters. Light acting on either the sample container or its contents, or on the photomultiplier faceplate, may cause the slow release of uncountable numbers of single photons. Even more single photons are sometimes released due to chemical processes (usually oxidation) within the vial. Though the photomultiplier photocathode has a relatively low efficiency for the conversion of incident light into photoelectrons – in the 1950 to 1957 period it tended to be 10 to 15% and even now it is but 25 to 30% – when the number of single photons is so large, some counts are recorded. As they do not originate with radioactive decay events they tend to superimpose upon the reproducible background a variable component whose value depends upon the nature of the sample preparation and the manner in which it has been handled.

With strong cooling not possible, with it not possible to lower high voltage, and with the potential problem of spurious light generation ever present, the single tube counter has never proved to be attractive; it will not be considered further. In 1953 Hiebert and Watts[1] described their coincidence counter. Two 2 in diameter photomultipliers viewed the sample. A count could be recorded only if both phototubes pulsed within the coincidence time – about 1 μs in this first coincidence counter. The presumption was that legitimate decay events produce a sufficient number of photons for both tubes to respond whereas randomly generated noise is likely to arise in one tube or the other at any time, but with small probability of occurring in both within the short coincidence time.

With two photomultipliers each having noise levels of 10^4 pulses/min, the number

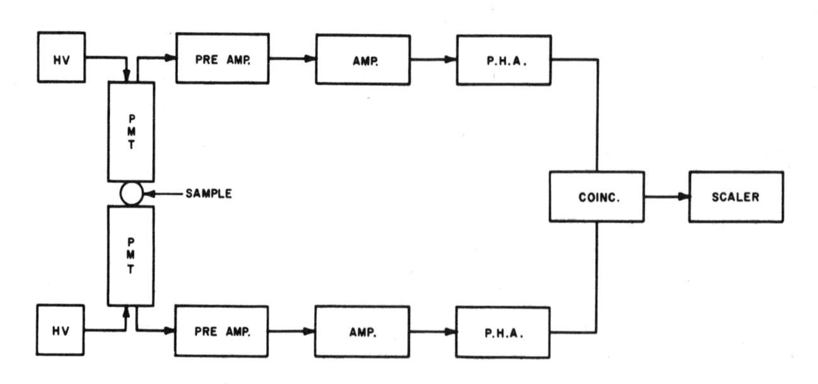

Fig. 1. Simplified block diagram, Los Alamos coincidence counter (1953).

of false coincidences ('accidentals') from simultaneous noise pulsing was 3 to 5 c.p.m., a noise reduction of about 2500 when compared with either tube alone. Continuing the comparison, the legitimate signal would seem to be cut in half since each photomultiplier receives only about half the available light. However, this last is of less concern; the noise rejection potential of the coincidence circuitry allows the phototubes to be operated at higher voltage, hence higher gain, even though under these conditions they have noise levels which would be absolutely intolerable for single tube operation. A block diagram of an early coincidence counter is shown in Fig. 1. The two photomultipliers, having different noise and gain characteristics, were operated at different high voltages chosen to provide about the same amplitude pulse from each when the same number of photons was incident upon the photocathode. The phototube output pulses were subjected to linear amplification and then to pulse height analysis. The lower levels of each single channel analyser were primarily used to reject preamplifier and amplifier noise, its frequency being so great as to exceed the capability of the subsequent coincidence circuitry. The upper level discriminators served to reject high energy background, coincidences where the light distribution to the phototubes was grossly unequal, and conceivably decay events from a second more energetic isotope though these first counters did not really permit dual isotope counting. If the amplified output pulses from each photomultiplier fell between the discrimination levels, the two single channel analysers fired and output pulses were directed to the coincidence circuit; for each coincidence a pulse was directed to the scaler where it was recorded.

With the coincidence system counting efficiency is inherently less than in a single tube counter, especially for low energy isotopes. This is so, even though it is possible to obtain more gain from phototubes operated in coincidence than could be obtained in single tube operation, because with the few available photons from a legitimate low energy decay event divided in two, there is a real probability that one of the tubes may not pulse at all. In contrast to efficiency, which was therefore lowered, but not significantly, there was a marked reduction in background since its principal component in these first counters – photomultiplier noise – was almost completely eliminated. With 15 to 20% tritium integral counting efficiency, background of these early coincidence counters was perhaps 150 c.p.m. Further, the effects of phosphorescence and chemiluminescence were largely overcome since these are single photon phenomena and are not counted in coincidence unless their rates are truly enormous.

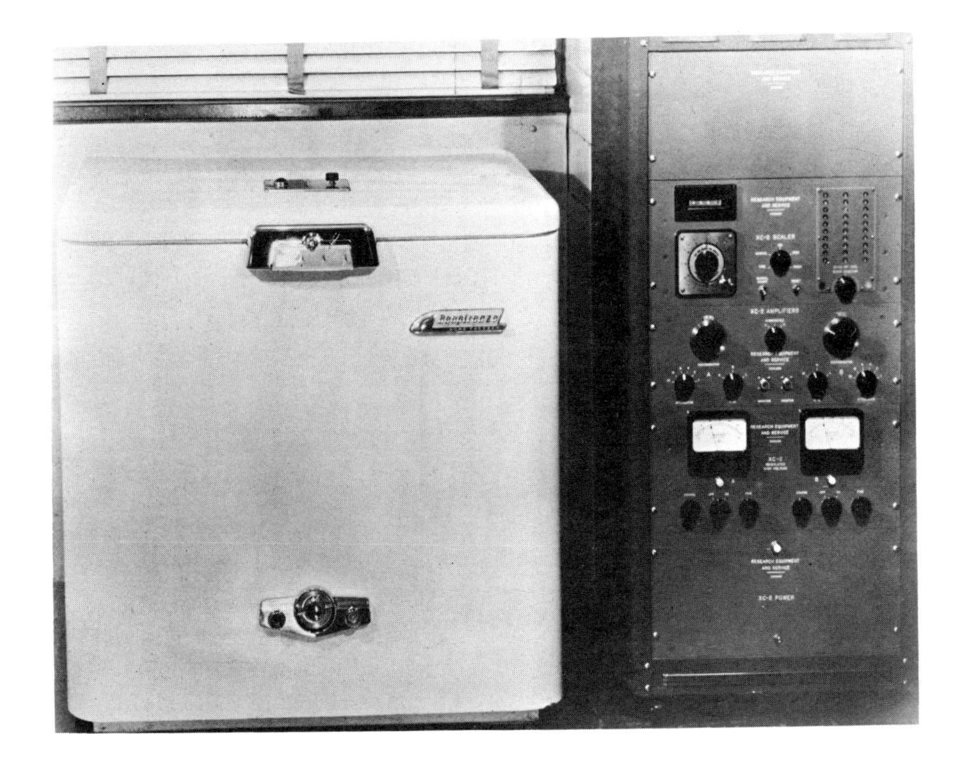

Fig. 2. Packard prototype (1953). Note the separate high voltage supplies, twin amplifier controls, and three decade electronic scaler followed by a mechanical register.

In 1953 to 1954 a predecessor of the present Packard Instrument Company custom built the first commercial coincidence counter for Dr. George Leroy of the University of Chicago (Fig. 2). Sample containers were 60 ml ink bottles. Two horizontally opposed Dumont Type 6292 phototubes were used with a bath of silicone optical coupling fluid between them (Fig. 3). Each photomultiplier had its own high voltage supply but the block diagram differed from that of Fig. 1 in that pulse height analysis was performed on the output of only one phototube (termed the 'analyser') whereas there was only a threshold discriminator examining the output of the other phototube ('monitor').

A somewhat revised production counter was offered by Packard in 1954. The Model 314 (Fig. 4) had a modified pulse height analysis section with a mid-level discriminator in addition to upper and lower, and two six-decade electronic scalers. A single high voltage supply was used; photomultipliers were selected such that, of any pair, the monitor had about twice the gain of the analyser (see Fig. 5). At about the same time Tracerlab also offered a coincidence system and not long afterwards one became available from Technical Measurement Corporation (TMC). After the first few production units Packard abandoned both the optical coupling fluid and the 60 ml ink bottle in favour of air coupling and the 20 ml low potassium glass vial still so widely used; the other manufacturers followed.

In principle the use of coupling fluid results in higher counting efficiencies (and slightly higher background due to Cerenkov events). In practice, however, the fluid tends

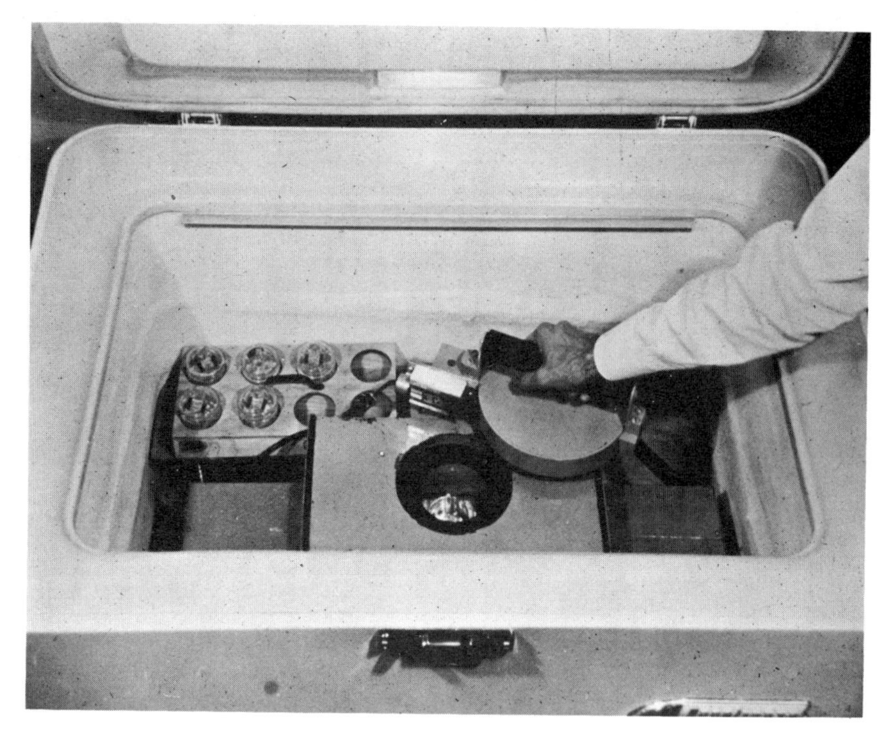

Fig. 3. Packard prototype – shield and detector assembly (1953).

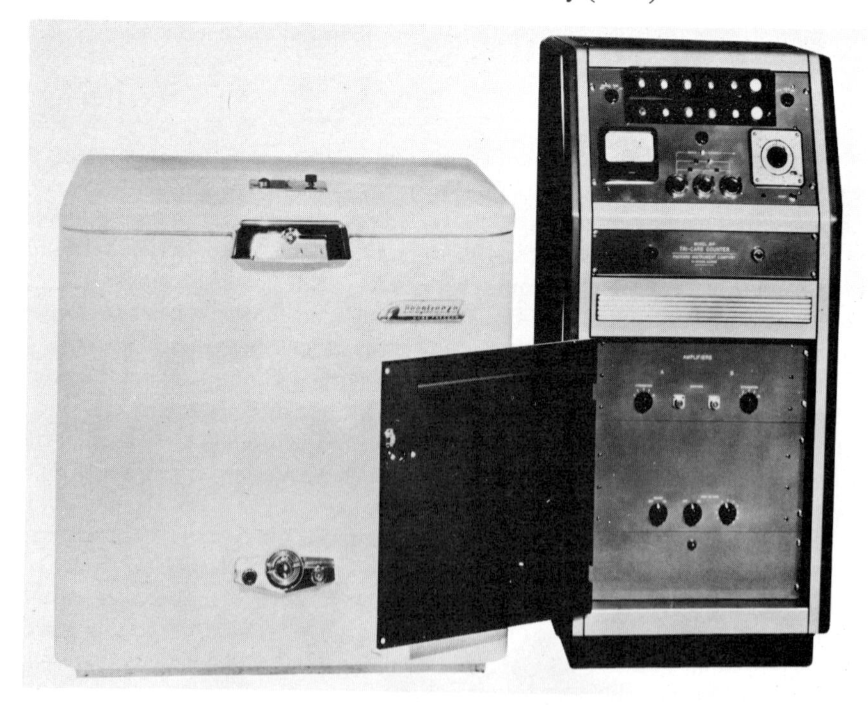

Fig. 4. Packard Model 314 – the first widely sold coincidence counter (1954).

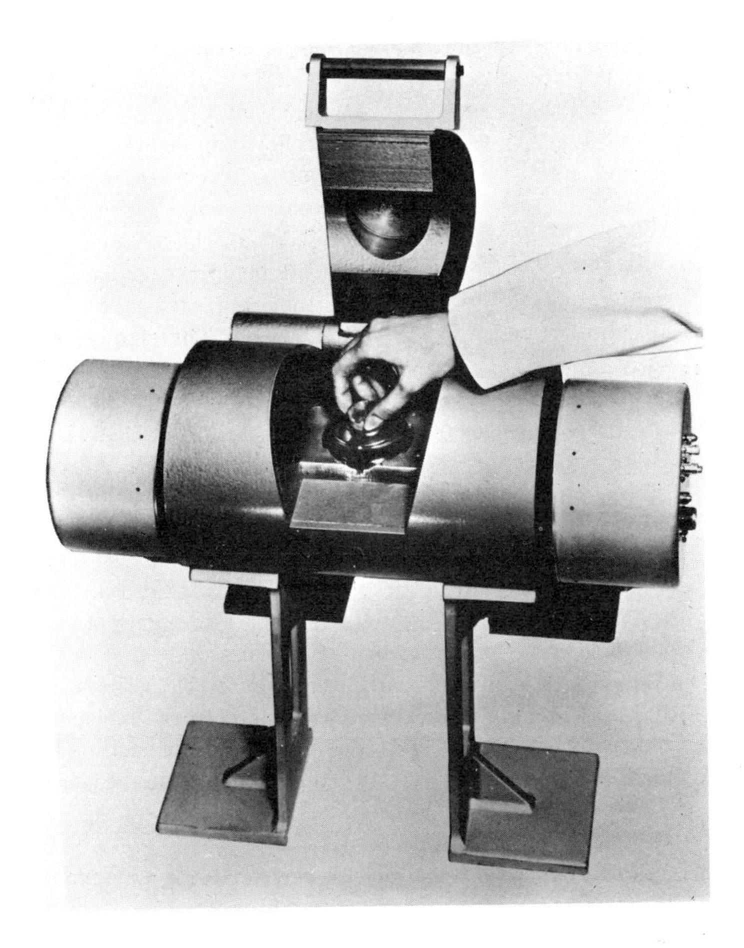

Fig. 5. Packard Model 314 – shield and detector assembly (1954). Note the preamplifiers at either end of the lead shield.

to accumulate dirt as well as traces of activity and scintillators which always seem to be present on the outside of the vial. The first causes reduced light transmission and therefore decreased efficiency while the latter two cause elevated and unpredictable backgrounds. Also, just the fluid itself, wetting each vial and requiring make-up or replacement, is a decided bother.

Reducing the vial size had advantages. With less glass present, the background due to potassium-40 was lowered. Since the vial had a smaller diameter, the light path within the sample was reduced. The phototubes could also be brought closer together. Light collection was improved. In sum, when compared with the Packard prototype, these changes resulted in a moderate increase in integral tritium efficiency – to perhaps 25% – together with a substantial decrease in background from about 150 to 75 to 90 c.p.m.

The first Tracerlab unit was unsuccessful. Whereas the Packard had an almost light-tight manually operated shutter which was closed during sample loading to prevent ambient light from falling upon the photomultipliers (the high voltage was still turned off as an added precaution), the Tracerlab instrument had nothing comparable. Therefore, it was not good practice to open the freezer more often than was absolutely necessary. Even with high voltage off, ambient light can cause phosphorescence of the phototube face and elevated background; with the high voltage on, ambient light may cause permanent damage. In advance of counting, all samples were placed in a rack within the refrigerated compartment and allowed to cool and dark-adapt. The high voltage was turned on and samples were manipulated into the counting chamber by the operator who, though unable to see into the closed freezer, had his arms inserted in heavy opaque gloves fastened into the side walls. The opportunities for error and the inconvenience involved need not be catalogued.

A more interesting aspect of the Tracerlab instrument was the use of phototubes at right angles rather than horizontally opposed. Among the contributors to background in these early counters, and an even more significant contributor today (as a percentage of the total) now that other problems have been minimized, is photomultiplier 'crosstalk' – spurious coincidences arising from phenomena within the phototube which cause light generation, usually in the region of the anode. This light is then seen to varying degrees by both photocathodes. With the phototubes not directly facing one another, it was hoped that these interactions would be reduced. Experimentation did show reduction in crosstalk with phototubes at right angles, the amount being partly a function of the quality and cleanliness of the reflector. However, decrease in background was small, especially since in this period crosstalk was one of the lesser contributors; this minor advantage of the Tracerlab counter was largely ignored in the face of its many disadvantages.

The TMC unit, though of good design and having comparable performance characteristics to the Packard, was apparently not adequately promoted as the Technical Measurement Corporation directed its efforts to the development of instrumentation for nuclear physics. Few systems were sold and manufacture was discontinued in the late 1950's.

1957–1961: Automatic instrumentation

In 1957, with the introduction of the Packard Model 314X, a 100-sample automatic counter, liquid scintillation became serious competition for automatic planchet counting as a method suitable for routine use in biomedical research (Figs. 6 and 7). Samples could now be automatically introduced into the counting chamber, counted to preset time or preset count, and data recorded, all unattended. The automatic mechanism was designed for light-tight sample loading without the need for a shutter directly in front of the photomultipliers. Performance was improved. Efficiencies were higher since, without shutters, the phototubes could be brought closer together. Also, backgrounds were lower, both as a result of heavier shielding and especially because light-tight sample loading eliminated the trauma of turning on and off the high voltage with each sample. But, basically, electronics of the automatic Tri-Carb counter were almost no different from those of its manual predecessor, except for the addition of those elements necessary to control sample changer function and printout. Also, an electronic timer supplanted the mechanical timer of the earlier unit. Older units could be converted to automatic operation and many were. In fact, it was this capability which so quickly established a broad base of users and almost overnight changed liquid scintillation counting from a desirable but often impractical method to one which was soon recognized as the best available technique for the measurement of soft β-emitters.

During this period the Packard company was almost without competition. Tracerlab, which had abandoned its first counter, probably as early as 1956, re-entered the market with a 40-sample automatic counter in 1960. However, for the second time they did not

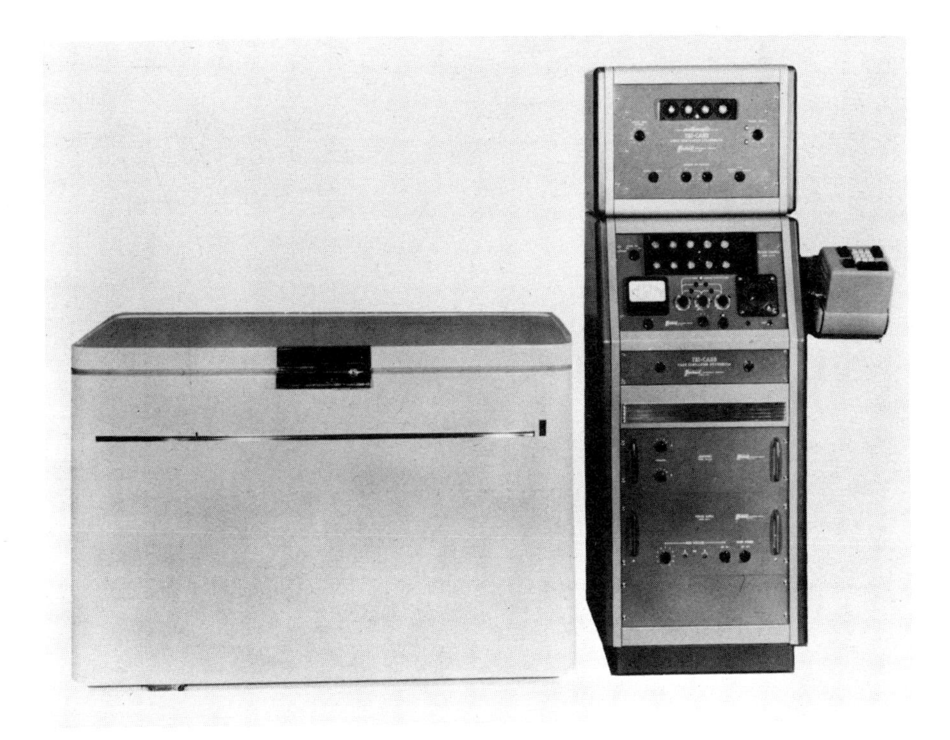

Fig. 6. Packard 314X – the first automatic liquid scintillation counter (1957).

offer a light-tight sample handling mechanism; each time the refrigerator was opened the high voltage had to be turned off. If it were not, at best false counts were generated by the ambient light and, at worst, the photomultipliers were irreparably damaged. In view of this problem and of the inadequate sample capacity, few units were sold and, by 1962, the Tracerlab instrument was no longer a contender in the market.

Though the Tracerlab instrument had disadvantages which made it unacceptable, it did have some interesting aspects. EMI Type 9536S photomultipliers were used. These photomultipliers were markedly superior to the Dumont 6292. Photocathode sensitivity was higher and thermal noise lower, making for higher counting efficiencies and reduced background. Further, the gain of the 9536S was higher than that of the 6292 thereby allowing operation at reduced high voltage (again leading to lowered noise) and/or reducing the requirement for electronic amplification. This, in turn, offered the potential of less preamplifier and amplifier noise, lower threshold settings, and higher efficiency for low energy isotopes. This potential went unrealized in the Tracerlab instrument. Almost the entire design was based upon vacuum tube circuitry (as was largely true of the Tri-Carb as well) except that Tracerlab employed transistorized preamplifiers. At that time such preamplifiers, though probably more reliable than vacuum tube preamplifiers were also appreciably more noisy. Thus, thresholds could not be lowered; the designers apparently accepted less than the maximum potential performance in trade for decreased field service requirements.

Conceptually, the Tracerlab instrument was a single channel design. For the few

Fig. 7. Packard 314X – 100-sample mechanism (1957).

users who felt that they needed two counting channels, a second complete spectrometer was added. While this arrangement was capable of providing more flexibility than the Packard system in which the two channels had at least one common discriminator setting, the advantage that might be gained with fully independent counting channels went unrecognized until the introduction of the Packard 314E series in 1961. Perhaps this at least partly came about because the addition of a second spectrometer was a costly proposition, as well as a complicated one, since the original design did not allow for same. Therefore, the understandable tendency on the part of the Tracerlab sales organization would have been to discourage sales of two channel systems, both to maintain price advantage and to avoid sale of specially modified equipment.

In April 1960, not long after Tracerlab had announced its automatic counter, Packard introduced the 314A series of Tri-Carb counters. These were transistorized versions of the earlier 314 series, unchanged in logic or in features and with the same 100-sample mechanism. One element of the 314 electronics which was retained however, was the vacuum tube preamplifier which was quieter than would have been a transistor design. The 314A series remained in production for only one year. After initial difficulties were overcome, it proved to be a satisfactory instrument and was discontinued so soon only because continuing improvements in transistor technology allowed for the development of a completely transistorized instrument in 1961. From a commercial standpoint, the timing of the 314A was superb; the availability of transistorized equipment, with the enhanced reliability which it certainly offered, secured Packard's market dominance in the face of competition not only from Tracerlab but also from Belin in France and IDL in England, two manufacturers who appeared on the scene late in this period with vacuum tube systems.

1961–1964: Competition

1961 saw the entry of Nuclear Chicago into the liquid scintillation field as well as the introduction by Packard of the aforementioned 314E Tri-Carb series. The Nuclear Chicago unit, a refrigerated counter with a circular array of 50 samples, had a transistorized front end but vacuum tube scalers and a vacuum tube timer. System logic was quite similar to that of the Packard 314 and 314A series, i.e. there were two counting channels with a common discriminator. However, the performance of the Nuclear Chicago instrument was an improvement over the past. 60 ns coincidence time (versus 200 ns in the Packard instruments) made photomultiplier noise less of a concern than ever before. Superior shielding, based on lead, then stainless steel, then copper provided lower backgrounds than could be attained with the iron shield used in the Packard automatic sample changer. All present-day counters employ such graded shielding, resulting in lower backgrounds than can be achieved with an equal thickness of lead by itself.

The Nuclear Chicago instrument also had higher counting efficiencies than had older Packard counters by virtue of the use of the 11-stage EMI Type 6097 phototubes. With the extra dynode providing higher gain, electronic amplification could be reduced with the benefits previously discussed. In these first Nuclear Chicago units tritium efficiency within a counting window was perhaps 30% and the background was under 50 c.p.m. Nevertheless, despite these improvements the first Nuclear Chicago system was unsuccessful and relatively few units were sold.

Probably the principal reason for the poor reception to the Nuclear Chicago instrument was its 50-sample capacity which was of little more interest to prospective buyers than had been Tracerlab's 40 samples. Further, by 1961, it was recognized that transistorized circuitry was more reliable than its vacuum tube equivalent; about half the circuitry of the Nuclear Chicago unit was still based on vacuum tubes. Also, the Packard 314E series, introduced at the same time, offered the only significant advance in system logic since the very first commercial counter delivered by Packard in 1954. Finally, Packard was able to offer comparable counting efficiency to that available from Nuclear Chicago, but with backgrounds unchanged from those of the 314A series, by taking full advantage of the potential of the EMI 9536S for providing higher gain and coupling this with an improved optical chamber which gave greater light collection and therefore larger phototube signals.

In two-channel liquid scintillation counters which preceded the 314E, the output of a single photomultiplier was directed through an amplifier to three discriminators. Several operating modes were switch-selectable but they all involved one discriminator being common to both channels (Fig. 8), the most frequently used arrangement being that of Fig. 8(a). This meant that it was not really possible to count a dual-labelled sample with both isotopes falling within normal windows. Resolution of the system, beginning with the

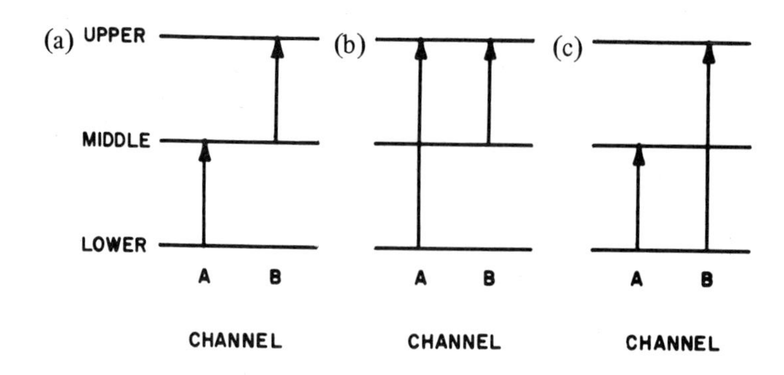

Fig. 8. Adjacent channel operating logic (1954–61).

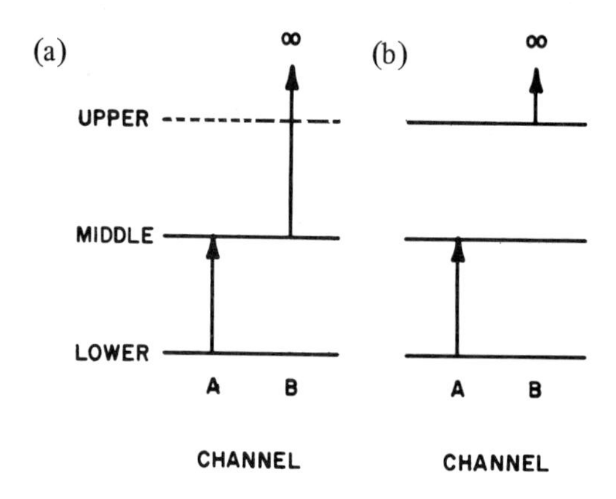

Fig. 9. Dual isotope operating logic (1954–61).

sample and proceeding through the photomultipliers and amplification stages, was not good enough to separate β-emitters of similar energies. For dual isotope counting it was usually thought necessary that energy maxima differ by a factor of at least four. With such differences, if the spectrum of the less energetic isotope was amplified to the extent that it covered half the pulse height analysis range – a minimum figure since, with any less amplification, too much of the spectrum fell below the threshold discriminator – then almost all of the more energetic isotope exceeded the working range of the pulse height analyser. At first, this problem could only be handled by switching off the upper discriminator and counting all pulses more energetic than the mid-level setting; the more energetic isotope was thus counted under integral conditions and with integral background (Fig. 9(a)). A later improvement, first incorporated in the TMC instrument and subsequently in the Packard 314A and Nuclear Chicago units, enabled 'split channel' operation (Fig. 9(b)). Those pulses which exceeded the mid-level discriminator but which were less than the upper-level setting were not counted; the second channel stored only events more energetic than those defined by the upper-level discriminator. This was still integral counting with its disadvantages but it did allow improved dual isotope separation in that the energy

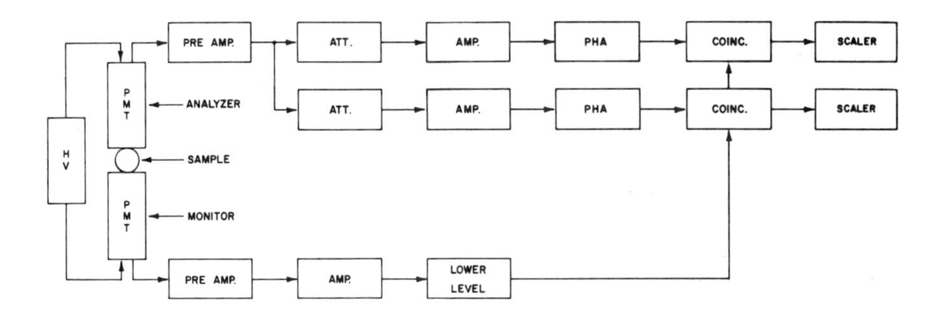

Fig. 10. Packard 314E series – improved two channel system logic (1961).

region of maximum spectral overlap could be ignored. But, to do this and to gain the improvement from it, it was necessary that system gain (in effect, high voltage) be adjusted so that this overlap region fell within the pulse height analysis range, thereby further assuring that a great part of the spectrum of the lower energy isotope fell below the threshold and was lost.

Most of these problems were eliminated by the logic of the 314E. There were no common amplifiers. Each counting channel was provided with its own linear amplifier and single channel analyser and each amplifier was preceded by its own linear attenuator (see Fig. 10). Pulses arising from the analyser phototube, after suitable preamplification, were directed to both channels and later, when three-channel instruments came into vogue, into all three channels. For multiple isotope counting, the phototubes were operated at a high voltage which allowed the least energetic isotope to be counted over the full range of pulse height analysis, the operator generally employing minimal attenuation at the channel input. More energetic isotopes in the same sample were counted, also over the full range of pulse height analysis, with the aid of the channel input attenuators which reduced their over-amplified signals from the preamplifier to a size which enabled them to be handled within a counting window. Counts in the zone of maximum overlap could easily be eliminated by raising the lower discriminator; this had no effect on the less energetic isotope in its own channel. Thus, two or more isotopes could be counted under fairly ideal conditions with higher efficiency for the less energetic, lower background for the more energetic, and better separation than had previously been possible.

In 1962 Nuclear Chicago discontinued its first automatic counter, even though it had only been on the market for one year, and replaced it with a much improved, fully transistorized, 150-sample, three-channel counter (Model 725) which also had computational capabilities in advance of competitive equipment. For the first time the Packard organization had serious competition. Increased capacity in the same refrigerated space that could previously accommodate only fifty samples was achieved by abandoning the circular array. A serpentine arrangement of links was adopted, each link holding an individual sample (Fig. 11); such systems have since been widely used by many manufacturers and must be considered the most popular method even today.

In addition to increased capacity, the Nuclear Chicago sample handling system offered new flexibility in its 'group counting' feature and its ability to detect and bypass empty positions. Previously there had been no sensor to indicate the presence of a sample con-

Fig. 11. Nuclear Chicago 150-sample serpentine, predecessor of all modern serpentine transports (1962).

tainer; at each position the mechanism went through its complete sequence. An empty position was counted until preset time was reached, results were printed, and the mechanism advanced to the next position. With the Model 725 this could not happen; the transport stopped only for valid samples. In the group mode there was even more selectivity. Only samples following a marker, which the operator could place anywhere on the transport, and in any number, were counted. An empty position denoted the end of each group and once that was detected the transport advanced rapidly to the first position following the next marker. This system allowed an operator to place his samples anywhere on the transport and count them to the exclusion of those that someone else may have left.

Three-channel system logic was a hybrid between that of the Packard 314E and the previous Nuclear Chicago instrument in which two channels had shared a common amplifier. The earlier system was retained but an additional attenuator, linear amplifier, single channel analyser, and scaler were added. Thus, in a dual isotope situation, the added channel could be operated without attenuation for the less energetic isotope while the more energetic was counted in one of the two associated channels, making use of the input attenuator to those channels to reduce signal size. The third channel was not intended for a third isotope; even today, few investigators count three isotopes in the same sample. Rather, the third channel was intended for channels ratio quench correction of dual isotope samples. For this purpose there was almost no disadvantage to the constraints imposed by having two channels so closely associated, the channels ratio method involving all of the spectrum of an isotope in one channel and a part of the same spectrum in another.

The Nuclear Chicago Model 725 was also the first instrument to do more than merely print accumulated data. In the first automatic liquid scintillation counters, data were output by a solenoid actuated adding machine; later the adding machine was modified to eliminate its keyboard and its ability to add and subtract. The 'lister' was made for lower cost and enhanced reliability; it removed the temptation to use the adding machine for its originally intended purpose, a pastime which generated numerous unwarranted service calls when the liquid scintillation counter began to print while the adder was being manually operated. With the Model 725 a solenoid actuated printing calculator was substituted for the lister. It could divide counts in one channel by time to give c.p.m.; it could also divide the counts in one channel by those in another to give the channels ratio.

To compete with the increased sample capacity of the Nuclear Chicago unit, Packard, also in 1962, introduced a 200-sample handling mechanism. It too was based on a serpentine chain but with a different linking method than that used by Nuclear Chicago. Group counting and empty position bypass were incorporated in the design. In some units there was visual indication of the sample number at the counting position by means of a Nixie display and a method was provided to manually advance the transport to a chosen sample number in advance of counting the first sample. Otherwise, there was no difference from the original 314E design.

These improvements helped maintain Packard's sales leadership but they were not enough to prevent Nuclear Chicago from establishing a firm foothold in the market. While the Packard unit again had greater sample capacity, the difference between 150 and 200 samples was apparently less important than had been the difference between 50 and 100. Also, there were advantages on the side of Nuclear Chicago – somewhat better performance, an extra channel, and greater computational capability. In late 1962, certainly in response to inroads being made by Nuclear Chicago, Packard began a total redesign of its automatic counter.

In France, beginning in 1958 to 1959, Belin had developed their Carbotrimetre, a refrigerated manual vacuum tube counter, which appeared to be very much a copy of the original Packard 314. A few were sold domestically. After several years an automatic version was also developed but apparently even fewer were sold. Manufacture was discontinued, probably as early as 1964.

During this same period, Isotope Developments Ltd., now joined together with Nuclear Enterprises, also manufactured a room temperature coincidence counter. An automatic instrument – the Tritomat – was offered for some years. Among the more interesting aspects of this instrumentation was the use of 13-dynode EMI photomultipliers with S-type photocathodes. These photocathodes were somewhat less sensitive than the B-type, also available from EMI, and therefore did not give the highest counting efficiencies. However, their noise levels were much lower, thereby permitting acceptable room temperature operation. The phototubes were mounted on a mechanism which spread them apart for sample introduction and brought them together and almost to the counting vial once it was in place – the only such mechanism known to the author.

Also, in 1962, the Vanguard Instrument Company announced a novel 100-sample refrigerated counter which was never put into production (Fig. 12). It did, however, have several new and obviously useful features, later liberally copied by manufacturers then in the field as well as those who came later. The Vanguard unit, which was shown in proto-

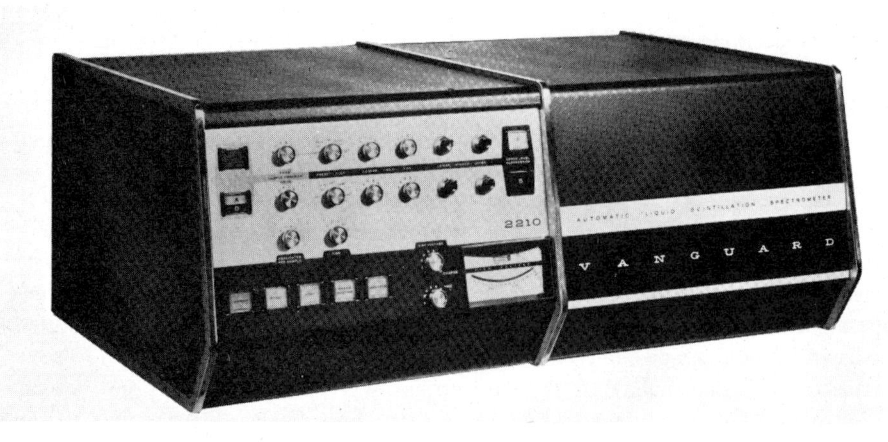

Fig. 12. Vanguard 100-sample counter. An advanced design which never went into production (1962).

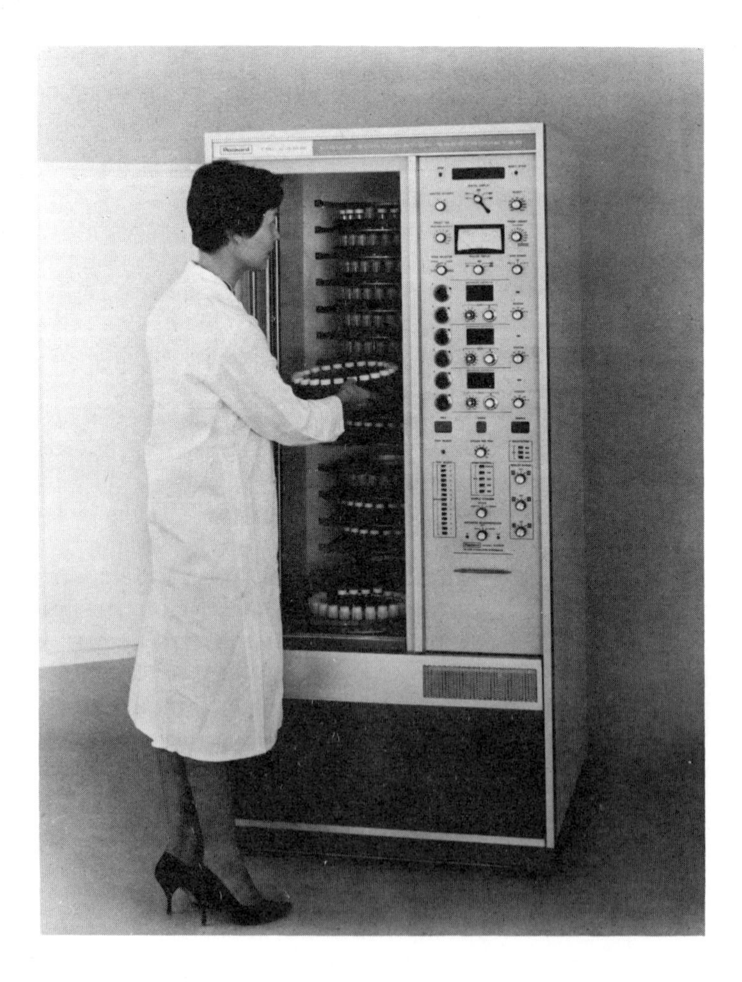

Fig. 13. Packard 4000 series (1963–65).

type form, is believed to be the first unit to combine sample changer and electronics in a single cabinet, the first with a removable sample tray, the first to use a typewriter for data printout, the first to include provision for replicate counting, and the first to include means for automatic rejection of a sample whose count rate did not come up to some predetermined level. To effect economy, the Vanguard counter had no digital display, one concept which has not found favour with many manufacturers.

In April 1963 prototype versions of two new Packard series (termed 3000 and 4000) were on display at the FASEB meeting in Atlantic City, New Jersey; production units were not shipped until approximately six and eighteen months later, respectively. Both series had similar front ends, much revised and improved over the corresponding sections of the 314E. Also, 13-stage EMI Type 9514 photomultipliers replaced the previously used 10-stage tubes and a new and improved graded shield was substituted for the iron shield of the 314E. Three counting channels were provided. In fact, the only similarities between the new 3000 series and the last 314E were the cabinet, the use of the same 200-sample transport, and the same data lister. Even here, a solenoid actuated mechanical calculator was offered as an option. The 4000 series had even greater changes, having a new one-piece cabinet, a tray-type sample changer with 360 sample capacity (15 trays, each with 24 samples) and a ratemeter to facilitate adjustment of counting windows (see Fig. 13). Other common features of these two instrument series, not previously available from Packard, were the ability to make replicate counts, low count rejection, and a Nixie digital display which replaced the glow transfer tubes of earlier units.

Each system also included a crude form of background subtraction in which a predetermined background value for the entire count to preset time was subtracted from the scaler setting before counting began, i.e. the scalers reset to the complement of the anticipated background. The first counts brought the scaler to 000000 and subsequent counts gave a positive number which had no background component. Accurate results maybe had to preset time but the technique does not permit operation to preset count, essentially eliminating the possibility of counting the samples in a series to the same statistics. Further, it does not permit the use of the low level rejection system since, at the end of the first minute of measurement, when an assessment is usually made of whether or not a sufficient number of counts has accumulated to justify continuation, the scalers may very well indicate a value less than zero.

In previous instruments overall gain could be controlled by varying the high voltage, whereas the gain for each channel was independently variable by means of attenuators. It was recognized by the designers of the 3000 and 4000 series that this was an unnecessary redundancy and further that frequent high voltage change did not allow the photomultipliers to settle to their lowest noise levels. Therefore, high voltage was fixed at a value which allowed the least energetic isotope anticipated to be counted with little or no attenuation at the channel inputs. More energetic isotopes could then be handled with increased attenuation. Operation was simplified, not only because the high voltage controls were eliminated, but also because there was no way, as there had been previously, to alter the gain in one channel (by means of high voltage change) and inadvertently affect another. Fixed high voltage has become standard throughout the liquid scintillation industry.

Certainly the most important improvement of these then new instruments was in their front end logic, now used in all modern coincidence counters. In preceding designs pulse height analysis had been performed only on the output of one photomultiplier while the other served only to verify coincidences. If the number of photons reaching both photomultipliers was always equal, and assuming the phototubes and electronics to give truly linear performance, such an arrangement should give a pulse height distribution representative of the energy of the decay events. However, with small numbers of photons – the mean energy tritium event produces about 30 photons even under the best conditions – and with decay events obviously not centred within the vial, there is a high probability for any event that photon distribution will be considerably unequal. For the next decay event producing the same number of photons there is both a small probability of equal distribution or of the same distribution of the preceding event. There is, on the contrary, a high probability that the output pulse of the analyser phototube will not have the same amplitude as that of the previous or following pulses from decay events of comparable energy. System resolution is appreciably poorer than it otherwise might be.

In the Packard 3000 and 4000 series, and in all new instruments to appear since, the concept of analyser and monitor photomultipliers has been eliminated. Rather, the two tubes are as alike as possible, have identical functions, and their outputs are summed. A second output is examined for coincidence (Fig. 14). Coincidence times tend to be of the order of 20 to 50 ns, further minimizing the consequences of phototube noise. Operating with multiplication of 3 to 6 per stage, the 12- and 13-dynode photomultipliers in current use have gains from 10 to 100 times those of the 10- and 11-dynode tubes previously employed. Again, the larger signal reduces the requirement for electronic amplification, and with it electronic noise.

Though new phototubes and faster coincidence led to improved performance, the real advance was in the organization of the front end. Pulse summation provides two benefits: a) improved resolution and b) the pulse arising from a true coincidence and leaving the summation circuit has twice the amplitude of those pulses which would have been fed into the amplifiers of the previous systems. Also, even the smallest pulse from a coincidence has at least twice the amplitude of a typical phototube noise pulse unmatched by a similar noise pulse from the opposed photomultiplier, i.e. the signal to noise ratio has been doubled. That the resolution is improved should be obvious; summing the pulses overcomes a great part of the problem of unequal light distribution, reducing the variation in amplitude of pulses leaving the summation circuit to that due to statistical effects within the photomultipliers. Doubling pulse amplitude prior to electronic amplification further reduces the problem of electronic noise, allowing examination of even lower regions of the spectra of non-energetic isotopes. Performance improvements were quite remarkable. Tritium efficiency was increased to 40 to 45% while the improved shielding and reduced noise resulted in backgrounds of under 30 c.p.m. Dual isotope separation was improved to an even greater extent. Whereas with the old 'unsymmetrical' counters approximately 40% of all carbon-14 events in an ideal tritium/carbon-14 sample could be detected above all of the tritium spectrum with the remainder below, 'symmetrical' logic reversed the situation; slightly more than 60% of the carbon-14 was above and 40% below the tritium end point.

Pulse summation, together with the several other improvements, reinforced Packard's position in the liquid scintillation market. However, the good experience which many investigators had with the Model 725, coupled with the rather long period between announcement of the 3000 and 4000 series and first deliveries, enabled Nuclear Chicago to maintain its share of the market while its engineering department updated their product line. Though ultimately pulse summation, low count rejection, and completely independent

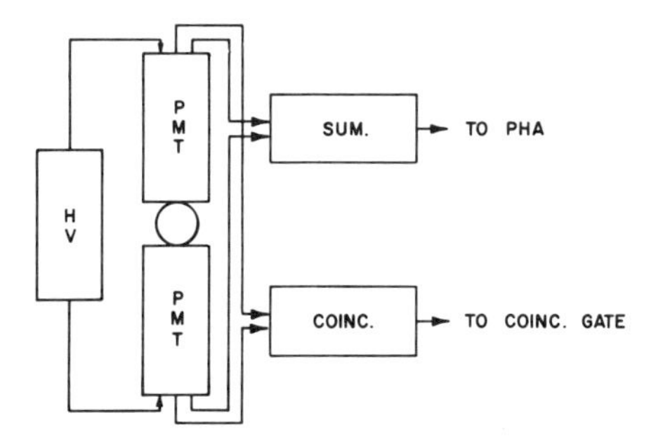

Fig. 14. Pulse summation, a concept used in all modern liquid scintillation counters (1963).

channels were incorporated into the Model 725, the first step was an improved system of background subtraction.

A valid objection to the first channels ratio computation practised by Nuclear Chicago was that the value printed was obtained by dividing the gross counts accumulated in one channel by those accumulated in the second. At high and even moderate count rates the printed result surely approximated the correct value. For low counting samples though, a significant part of the gross count was background and the printed value was in error. To correct this problem, a system for dynamic background subtraction was announced in early 1963; it could be retrofitted to all Model 725's. For each channel a separate variable frequency oscillator generated pulses at the background rate. These were fed to a blocking circuit located at the output of each single channel analyser. Each oscillator pulse was capable of blocking one input pulse to its respective scaler; in effect, background was subtracted. Dynamic background subtraction would seem to be more attractive than the system employed by Packard. Operation to preset count, as well as preset time, is possible. Simultaneous low count rejection is also possible. The digital display correctly shows the counts accumulated at any time less the total background to that time. It is not necessary to change the background subtraction settings at each instance that counting time is altered. However, there was a subtle error in this first system for dynamic background subtraction which was not immediately recognized, but which was later corrected. Nevertheless, dynamic background subtraction, even with this fault which is discussed later (see p. 93), corrected the difficulties with the channels ratio computation and so found favourable reception.

The ANSitron liquid scintillation counter (Fig. 15) appeared on the market in 1964. Though the manufacturer, ANS Inc. was ultimately unsuccessful, the ANSitron counter must be judged a technical success when the number and significance of its features, later incorporated into competitive units, is considered. Of them, three – external standardization, logarithmic amplification and preadjusted counting windows – particularly stand out. Quenching has always been a problem of liquid scintillation counting. Often it is real, but perhaps just as often it has been a negative effect on samples generated by the promotional departments of the various instrument manufacturers. With the first counters, quenching was best assessed with internal standards but the several disadvantages – cost of standards and the nuisance of their addition, destruction of the original sample, possibility of pipetting error, second sample count, etc. – discouraged their use. The channels ratio technique made matters much simpler, eliminating all of these problems, but it

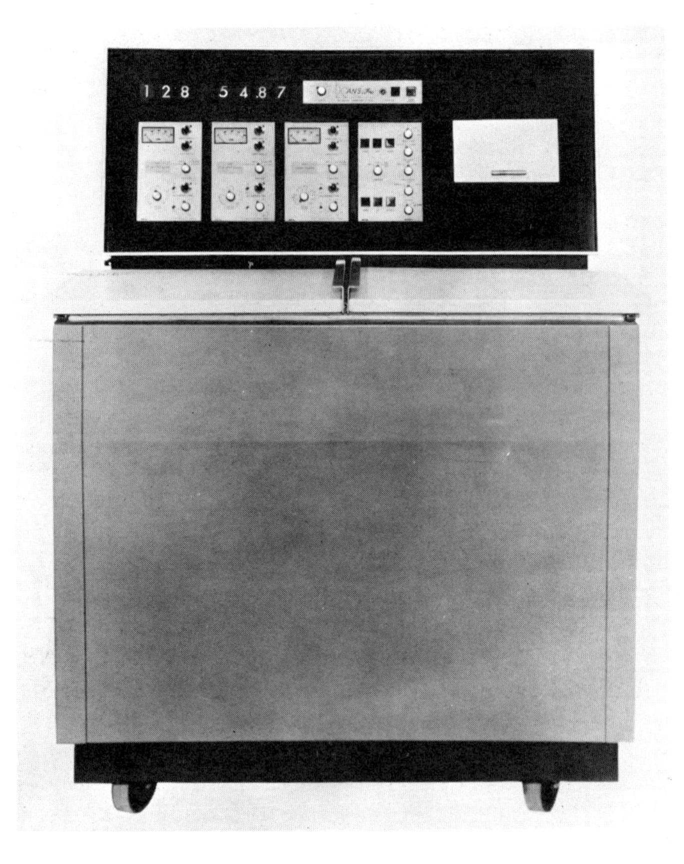

Fig. 15. ANSitron – the first counter to employ automatic external standardization and logarithmic amplification (1964).

required that samples be counted for longer periods than previously in order to obtain satisfactory accuracy in that channel being used to examine the smaller part of the energy spectrum. Such extended counting reduces the number of samples that can be processed in what is certainly rather expensive instrumentation.

The external standard method overcomes these difficulties, making it possible to examine all samples for quenching at no expenditure for standards and with a time expenditure of about 1 min/sample. In the ANSitron, samples were counted to preset time, accumulated data were recorded – an elementary system of electronic computation enabled printout of c.p.m. – and then a gamma source, in this instance caesium-137, previously stored within a lead shield, was moved into position adjacent to the counting chamber. The sample was allowed to count in the presence of the source for 1 min with counts being recorded only in that one of the three channels which had been specifically set aside for that purpose. Any contribution of the sample to that channel, as observed in the counting period prior to standardization, was automatically subtracted from the counts accumulated during standardization. Therefore, the remainder was a direct reflection of the count rate induced into the sample by the external source. Counting efficiency for unknown test substances could be established by comparison of the induced count rate to a plot of counting efficiency in any channel versus induced rate in the standard channel

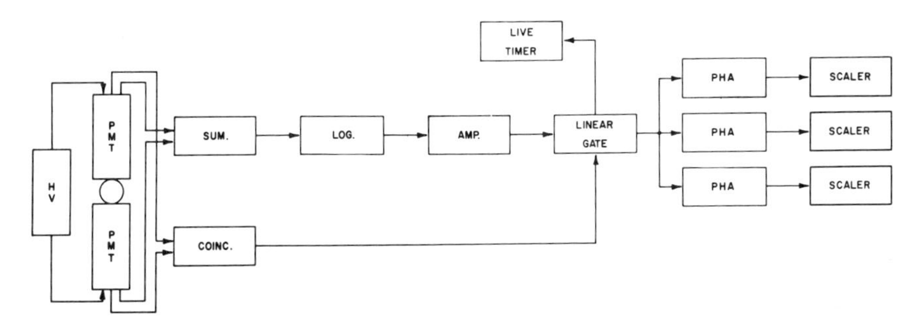

Fig. 16. ANSitron block diagram (1964). A similar arrangement is employed in modern Intertechnique counters.

for a series of samples of known activity, but deliberately quenched and counted with the same window settings. It was desirable that the sample chemistry of the standard series approximated that of the unknown samples and necessary that the volumes of unknowns and standards be the same.

For the user, external standardization satisfied many requirements and imposed few. Initially, it was sensitive to small changes in sample volume but this difficulty has been overcome by later improvements. It also used one counting channel, thereby preventing the operator from using the channel for his samples; this too has been changed in recent designs. What it did accomplish was to facilitate the determination of whether or not a sample was quenched, and if so, by how much. The minute of additional counting for each sample was inconsequential when contrasted with the time consumed either by internal standardization or by the channels ratio method. There are no automatic counters now being manufactured that do not have some form of external standardization.

Logarithmic amplification has met with almost comparable acceptance. While pulse summation and prompt coincidence were treated in the ANSitron as they had been in the 3000 series Packard and the revised Nuclear Chicago unit, output pulses from the summation circuit were fed to a logarithmic attenuator which preceded a single linear amplifier (see Fig. 16). Pulses leaving the amplifier, matched by a pulse from the coincidence trigger, passed through the linear gate and were subjected to pulse height analysis. One, two, or three plug-in single channel analysers were provided. Separate attenuators were unnecessary, the logarithmic element having had minimal effect on small pulses but having caused compression of larger ones. This system offered the maximum in simplicity without any sacrifice in performance. The logarithmic attenuator had a range of more than three decades and was fixed, as was the high voltage. Isotopes from tritium or iron-55 to those more energetic than phosphorus-32 could be counted without any adjustments other than changing discrimination levels.

This simplicity led to the last of the three major advances incorporated in the ANSitron, preadjusted counting windows. With the logarithmic system, the energy limits for counting were defined only by discriminator level settings and these, in turn, were made by adjustment of 10-turn helical potentiometers. Since system amplification was fixed, it was recognized that the window for each isotope could also be fixed with resistors equal to any chosen potentiometer setting. A pair of such resistors – one for the lower level and one for the upper – were mounted in a plug-in termed a β-Set. β-Sets were made available for common single isotope samples and pairs were also provided for dual isotope mixtures. This same concept, either with plug-ins or with switch mounted resistors, has become a standard feature in the product lines of most liquid scintillation manufacturers though due to cost considerations not always in all instruments. It has been incorporated into linear systems as well as those with logarithmic amplifiers. Settings such as these are satisfactory for routine use but, because they encompass a wide energy range, are unlikely to be optimum for low-level counting or for badly quenched samples.

A less novel but also widely adopted aspect of the ANSitron was the routine use of the EMI Type 6255B photomultiplier. The 6255B, a quartz-faced equivalent of the 9514, gave slightly improved counting efficiency due to the somewhat greater transparency of quartz than glass at lower wavelengths. However, its principal advantage was in reduced backgrounds, there being rather less natural radioactivity in quartz than in glass, hence less Cerenkov effect within the phototube faceplate. With the ANSitron, a tritium sealed reference standard could be counted with 45% efficiency with a background of 18 to 20 c.p.m. Once other manufacturers switched to the 6255, their instruments achieved comparable performance.

Other features of the ANSitron that should be noted were its improved system for background subtraction and the flexibility of sample changer operation. Background subtraction in the ANSitron was based upon the dynamic approach first used by Nuclear Chicago. However, it was recognized that the Nuclear Chicago system had a shortcoming, felt only at very low count rates, in that it attempted to match periodic blocking pulses with randomly appearing pulses from background and sample radioactivity. At low rates it might very well be that there would not be a random pulse between two blocking pulses; the first would establish the blocking condition but the second would be lost. In effect, one background count would not be 'subtracted'. In the ANSitron, and later in the Nuclear Chicago counter, circuitry was added to store unused blocking pulses until the blocking position was free to accept them. In that way, none were lost and background subtraction was accurate, even at very low count rates.

In later designs dynamic background subtraction was superseded by less costly and equally accurate methods, generally involving actual mathematical subtraction of the background rate after c.p.m. had been calculated. In adopting this approach, manufacturers and their customers have apparently lost sight of the fact that, when samples are measured to preset count, dynamic subtraction ensures that the measurement is terminated when the desired number of sample counts have been accumulated. With other methods the measurement ends when the total number of counts – sample plus background – reaches the preset number. For low counting samples, a large part of the total counts may be background; the statistical accuracy of the count will be much poorer than the investigator may expect on the basis of his instrument settings.

Automatic systems previous to the ANSitron were not really designed to permit counting of a single sample. Loading one sample usually meant placing that sample in the position adjacent to the loading station and going through an automatic sample change sequence; there was no manual chain advance. In the ANSitron, however, manual pushbutton controls were provided to enable the operator to step through the sample change sequence. Individual samples could be loaded, counted, unloaded, and the chain advanced or not, all under the user's control. Today, the more advanced counters have this capability; in some the transport chain can be made to reverse as well. The ability to interrupt automatic counting in order to manually insert and count one or a few samples is felt to be a necessity by many operators who use this facility to guide experiments in progress.

By the autumn of 1964 Packard had begun to use quartz-faced phototubes and had incorporated a simplified form of automatic external standardization in the 3000 and 4000 series, the latter then being delivered to customers for the first time. In the Packard version, the third channel was used to examine the spectrum induced by a radium-226 source pellet which was pneumatically moved into place or withdrawn to its shield. There was no provision for subtraction of sample contribution from the external standard counts, advantage being taken of the fact that much of the induced radium-226 spectrum could be observed above all of the tritium and carbon-14 spectra. For isotopes more energetic than carbon-14, the operator was required to perform manual subtraction.

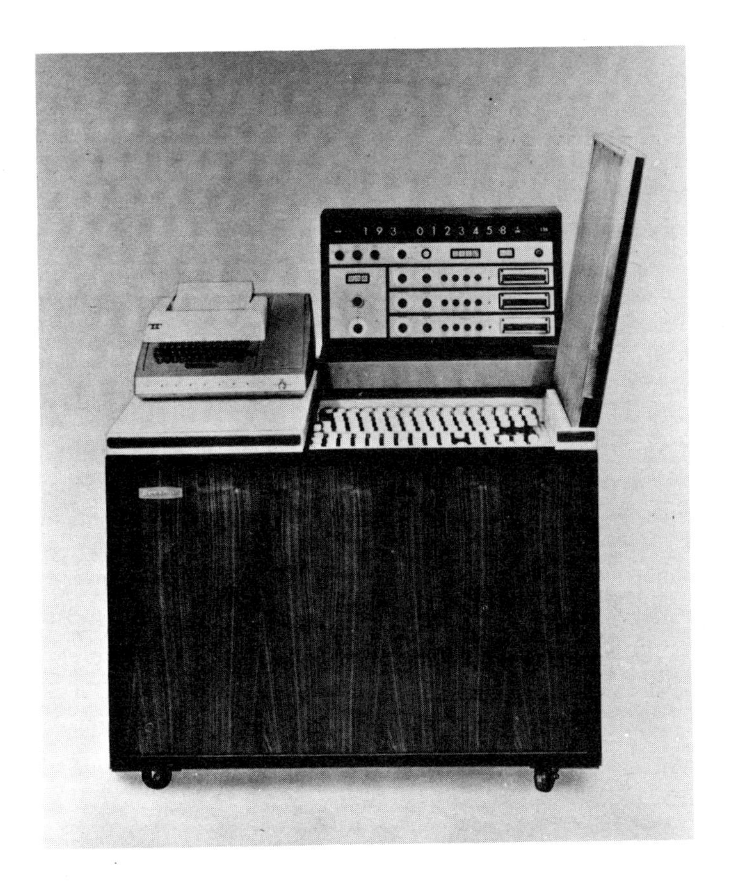

Fig. 17. Beckman LS-200 – 200-sample counter with external standard ratio, electronic calculation and Teletype (1965).

1965–1967: Increasing competition

In 1965 the Beckman Instrument Company entered the liquid scintillation field with the LS-200, an automatic 200-sample counter, in many ways patterned after the ANSitron. The LS-200 (Fig. 17) was modular and could be obtained with one, two, or three channels. The front end had a pseudo-logarithmic characteristic, achieved by the manner in which the photomultipliers were operated, rather than with logarithmic circuitry. The photomultipliers themselves were the recently introduced 12-dynode RCA Type 8575 which had low noise levels yet appreciably higher photocathode efficiencies than previous phototubes. Their low noise, coupled with 20 to 30 ns coincidence, enabled Beckman to offer systems for room temperature operation; a cooling unit was available but was treated as an extra-cost option. Despite low noise, backgrounds were relatively high since these phototubes had severe crosstalk problems. The first instruments in the field exhibited tritium counting efficiencies above 50% but backgrounds were from 35 to 60 c.p.m. High photocathode efficiencies also made for improved tritium/carbon-14 separations, especially in the tritium counting channel where the added tritium counts were largely in a region where there are few carbon-14 decay events. Whereas contemporary competitive instruments which had maximum tritium efficiencies of about 45% gave about 8 to 10% carbon-14 efficiency in the lowermost 30% tritium, the Type 8575 photomultiplier enabled the operator to obtain over 40% tritium efficiency for the same carbon-14 cross contribution.

The Type 8575 photomultiplier gave new stimulus to photomultiplier development which, in turn, has led to modern standards of performance. Though initially having excessive crosstalk and erratic noise levels, the 8575 has been much improved. Not long after the announcement of the 8575, EMI introduced new phototubes with comparable performance. Today, all manufacturers of liquid scintillation counters use one or the other of these bialkali phototubes; a few use both interchangeably. Tritium counting efficiencies of about 60% are now routine with backgrounds in the 20 to 25 c.p.m. range. Separations reflect this improved performance with perhaps only 10% carbon-14 cross contribution in a channel in which tritium is counted with 50% efficiency; more than 60% carbon-14 above all tritium has been the industry standard for some years.

The LS-200 and the Nuclear Chicago Mark I (Fig. 18), introduced at the same time, had improved systems for automatic external standardization, as well as more complete electronic computation. A disadvantage of the external standardization method employed first in the ANSitron and shortly thereafter by Packard was the aforementioned sensitivity to variations in sample volume; increased volume provided increased stopping power for gammas and therefore an increased number of induced counts. The number of induced counts might also change with vial wall thickness or as the source decayed or changed position with respect to the sample. These last two problems were more theoretical than real, it being quite possible to select an external standard with sufficiently long half life and to reproducibly position it with accuracy.

Though the number of induced counts might change with sample volume, the shape of the induced spectrum would be expected to remain fairly constant. Therefore, change in spectral shape, and more important, change in position, rather than the absolute number of induced counts, was used as an index of quenching. The channels ratio method was applied to the external standard spectrum. After the sample count had been completed, the sample was subjected to external standardization. In the Beckman instrument a caesium-137 source was used and was counted in two preadjusted windows specifically reserved for that purpose; in the Mark I a barium-133 source was chosen because its low energy gammas superimposed upon the tritium and carbon-14 spectral regions, and perhaps because shielding was simplified; it was counted in the channels used for tritium and carbon-14. After subtraction of the sample contribution the ratio of external standard counts in the two channels was printed. With the aid of calibration curves, prepared from samples of known quality, it was possible to know the counting efficiency for one or more isotopes within the test samples. The external standard ratio method has gained general

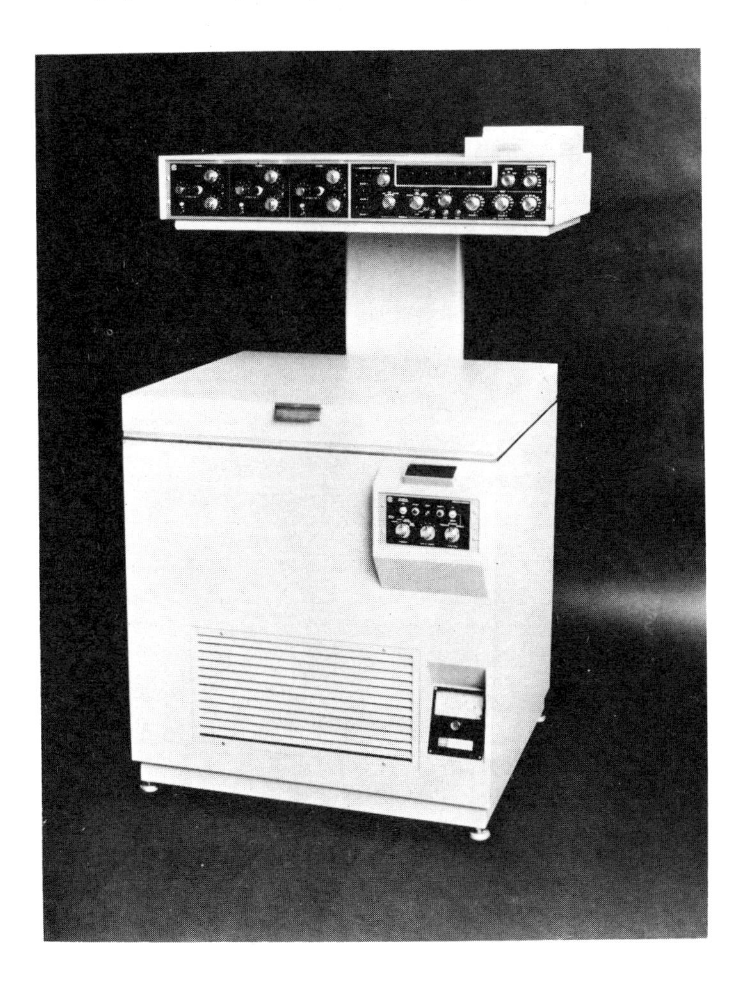

Fig. 18. Nuclear Chicago Mark I – 150-sample counter with external standard ratio and electronic calculation. The first counter to make extensive use of integrated circuitry (1965).

acceptance; most modern counters either print the ratio or, in those which have limited data processing capability, provide the necessary counting information to allow the investigator to calculate it.

Both the Beckman and Nuclear Chicago counters had comparable electronic calculating capabilities which were more extensive than those of the ANSitron and provided far more reliability than could be obtained from solenoid actuated mechanical calculators. In addition to the external standard ratio, both instruments provided c.p.m. whether the sample was counted to preset count or preset time, and performed arithmetic background subtraction and channels ratio computation. The Mark I output data through an adding machine lister; the Beckman was the first counter to make routine use of the Model 33 Teletype page printer.

Until the appearance of the LS-200, it would seem that there was little concern with the data format presented to the instrument user. An occasional counter had been fur-

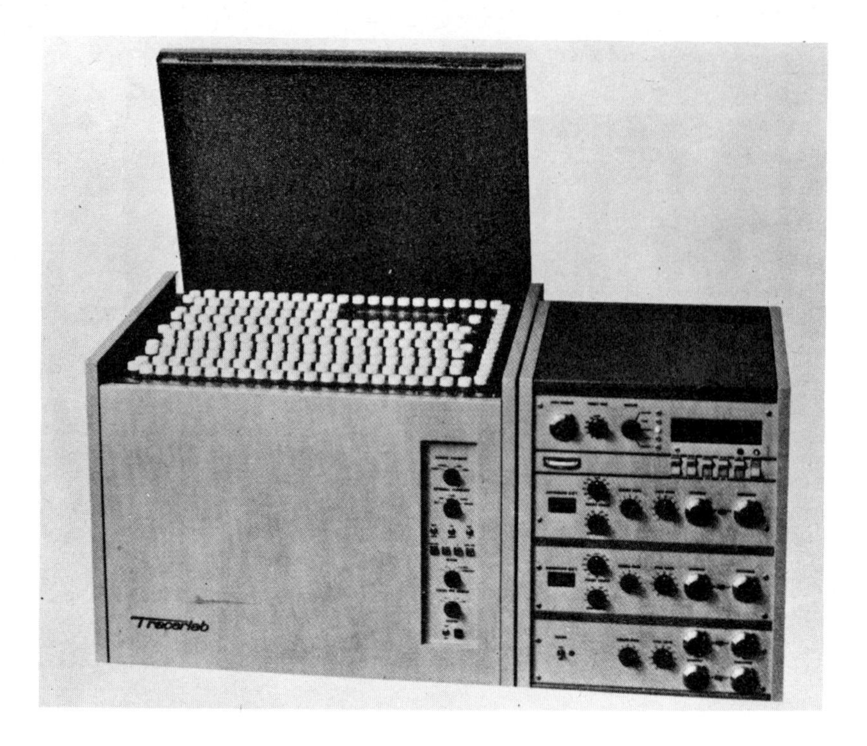

Fig. 19. Tracerlab Corumatic II (1970).

nished with a typewriter, a Teletype, or even a card punch or paper tape punch but these were unusual. The solenoid actuated lister was on almost every counter. However, once the LS-200 appeared, the convenience of the page format was widely appreciated and most subsequent counter designs incorporated either a Teletype or a typewriter, with the Teletype seemingly preferred because it can also provide computer compatible punched tape.

The Mark I was the first liquid scintillation counter to make extensive use of integrated circuitry. As such its electronics package was appreciably smaller than that of all previous automatic units. However, the main objective in turning to integrated circuitry was for enhanced reliability. Today, extensive use of integrated circuitry is made by all manufacturers, not only because of its reliability and compactness, but principally because it is more economical than is conventional transistor circuitry.

The Picker Liquimat, a 200-sample refrigerated counter was also announced in April 1965 but the first production units were not delivered until long after. As a liquid scintillation counter it too was a rather close copy of the ANSitron, the most obvious difference being that it was intended for gamma counting as well as for beta. A shielded sodium iodide detector was horizontally positioned above the liquid scintillation phototubes in such a way that the counting vial passed through a hole perpendicular to the axis of the crystal. Samples could be made to stop in the crystal detector or to descend further to the liquid scintillation phototubes. A separate channel was provided for gamma counting and three channels were available for liquid scintillation. Though the economy of utilizing the same mechanism for counting both betas and gammas has obvious attraction, the

Fig. 20. Frieseke and Hoepfner three phototube coincidence counter (1965–67).

Liquimat apparently did neither in outstanding fashion and relatively few units were sold prior to the discontinuance of manufacture, probably in 1968.

Also in 1965, Tracerlab made a third attempt to enter the liquid scintillation market with their Corumatic, a low cost 50-sample mechanism and detector, intending it for use with existing digital equipment which a customer might have. Few units were sold; later an improved version with 100 samples was offered (Corumatic II) and most recently a 200-sample ambient temperature instrument (Corumatic 200) has been made available. The latter (Fig. 19) is furnished as a complete system with two or three spectrometers and two separate channels for automatic external standardization using a radium-226 source.

The Packard 3375 was introduced in 1966. It is a 200-sample three-channel refrigerated counter which has retained the detector, front end and sample handling mechanism of the previous model. In addition to the three counting channels, two fully independent channels with fixed counting windows are provided for external standardization. Push-button selection allows a choice of preadjusted counting windows. Digital circuitry was completely revised; the mechanical calculator of the previous model was suppressed and replaced by an electronic calculator which gave c.p.m., external standard ratio, and channels ratio which were printed on a solenoid actuated IBM Selectric typewriter.

Integrated circuitry was used throughout.

In the period 1965–67 the German firm Frieseke and Hoepfner offered their Model FHT 770 G2 Liquiszint, probably the strangest liquid scintillation counter ever marketed (see Fig. 20). The sample transport held 17 trays, each with 24 vials for a total of 408; there were three counting channels, each with rather conventional features. What was different about the Liquiszint was the use of three photomultipliers!

It has long been recognized that the coincidence counting efficiency approximates the product of the two single tube efficiencies. On this basis d.p.m. for a single isotope may be estimated by counting the sample both in coincidence and with the two phototubes operating independently:

$$\text{c.p.m.}_{\text{coinc}} = \text{d.p.m.} \times \text{Eff}_{\text{coinc}}$$

$$\text{c.p.m.}_{\text{tube A}} = \text{d.p.m.} \times \text{Eff}_{\text{tube A}}$$

$$\text{c.p.m.}_{\text{tube B}} = \text{d.p.m.} \times \text{Eff}_{\text{tube B}}$$

$$\text{Eff}_{\text{coinc}} = \text{Eff}_{\text{tube A}} \times \text{Eff}_{\text{tube B}}$$

$$\therefore \text{d.p.m.} = \frac{\text{c.p.m.}_{\text{tube A}} \times \text{c.p.m.}_{\text{tube B}}}{\text{c.p.m.}_{\text{coinc}}}$$

However, these expressions do not consider phototube noise which, if large relative to the count rate, as is likely to be the case or there would be no need for coincidence counters, makes the method unsuitable. Frieseke and Hoepfner therefore attempted to employ the third tube to overcome the noise problem by an arrangement in which ABC triple coincidences were used together with AB double coincidences to enable direct d.p.m. calculation. Unfortunately, the author knows of no experimental data to indicate the accuracy of the results obtained. What is obvious however, is that counting efficiency was much reduced since the triple coincidence efficiency is the product of the double coincidence efficiency multiplied by the counting efficiency of the third tube. With the available light divided among three phototubes rather than two, the tritium efficiency of each tube must have been less than the approximately 70% which would have been attainable in normal two tube operation and the overall efficiency was therefore less than the three tube product of about 35% (against perhaps 50% for two tubes), a high price to pay for simplified but only approximate calculation.

By 1967 several Japanese manufacturers were also in the liquid scintillation market. Kobe Kogyo, Horiba, Shimadzu and Japan Radiation & Medical Electronics (Aloka) were all offering three-channel refrigerated or room temperature systems with 200 to 250 sample capacity. External standardization was available at least in the Shimadzu and Aloka units; in the latter, two extra channels were specifically provided for that purpose. Published performance figures for the Kobe Kogyo and Shimadzu units of that time were not up to modern standards but those for the Aloka counter certainly were. None of these instruments had features not already discussed. It is not known whether many of them have been sold outside of Japan.

Beginning in 1965, Nuclear Chicago, then Beckman, and later Packard and Picker introduced low cost counters which were either stripped down top-of-the-line instruments or rejuvenated versions of older equipment which could be sold at attractive prices since the design costs had been largely amortized. Since these instruments, though technically sound, had no features of note, they will not be considered here.

1967–1971: Data handling

By 1967 much of the design of liquid scintillation counters had become quite standardized. With there being only two sources of suitable photomultipliers, and with both producing comparable product, performance – efficiencies, backgrounds, separation – is quite the same for just about all counters. For the most part counters are still refrigerated, no longer for the sake of the phototubes, but rather because of the requirements of certain types of samples; most manufacturers however, supply systems to operate at ambient temperature as well. Front end design is equally divided between linear and logarithmic amplification; preadjusted counting windows are common and are sometimes offered to the exclusion of variable windows. Three channels are the norm and some provision is almost always made to count an external standard in two counting windows.

All of this does not mean that there has been nothing new in the liquid scintillation field since 1967. Rather, what it does mean is that those new developments which have been introduced are mainly outside the areas which, in the past, have received principal emphasis. True, there have been some new sample changer designs – a 420-sample tray type changer from Philips in 1968, a 400-sample tray type unit for β- and/or γ-counting from Nuclear Enterprises, and a 150-sample tray type unit from Packard – but none of these has yet had an impact on the market. Of far more interest have been those counters in which there has been an effort to extend data processing beyond traditional limits.

The first instrument to fall into this category was the Packard 3380/544 (Fig. 21) designed not only to measure quenching by the external standard ratio method but also to correct for it. Initially shown at the 1967 FASEB meeting, deliveries were apparently not made for more than a year. The Model 3380 liquid scintillation counter, a 3375 modified to accept the accessory Model 544 AAA (Absolute Activity Analyzer) (Fig. 22), forms the base of the system. Heart of the 544 is a coil placed around the outside of each photomultiplier in the region between the photocathode and the first dynode. Passage of a current through the coil generates a field which can deflect some of the photoelectrons generated at the cathode to the tube wall where they are lost. If current is increased and the field becomes stronger, the number of electrons diverted from the first dynode is increased with consequent reduction in output pulse amplitude. Thus, an electronic equivalent of quenching is achieved, the effect being quite the same as if one of the more normal factors within the sample caused diminished photon output.

With the 544 the external standard ratio is determined in a brief cycle as soon as the sample enters the counting chamber. The measured value is not likely to be one of ten predetermined ratios and so a small current related to the difference between the observed ratio and the next lower fixed ratio is passed through the coils; the sample is then restandardized and a second ratio is obtained. If that value still does not equal the fixed value the current to the coils is further increased and the process is repeated. From five to fifteen standardizations may be necessary until the sample has been electronically quenched to the predetermined external standard ratio next below the original value for the sample. The sample is then counted. D.p.m. values are calculated by dividing the observed c.p.m. minus background by the counting efficiency at the particular ratio, efficiency having previously been determined for all ten ratios and then stored in banks of switches. The Model 544 is capable of providing calculated d.p.m. for both single and dual isotope samples.

One of the more interesting features of the Model 544 is its system for improving dual isotope separation of highly quenched samples. Ten pairs of preadjusted counting

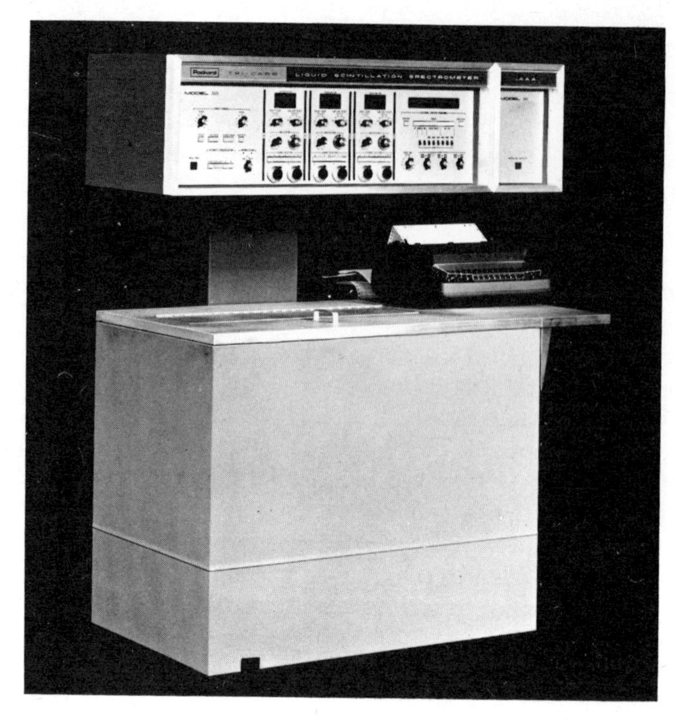

Fig. 21. Packard 3380 with AAA accessory (1967–68).

Fig. 22. Packard Model 544 AAA (1968). Counting efficiencies are stored on banks of selector switches.

windows, one for each degree of quenching, are provided for tritium and carbon-14. After external standard ratio adjustment, appropriate windows are automatically switched into service before each sample count. A useful appraisal of the errors associated with this method of d.p.m. determination has been given by Herberg.[2]

In the Beckman LS-250, successor to the original LS-200, a somewhat different approach has been taken toward improving dual isotope separation. The AQC (Automatic Quench Compensation) feature of the LS-250 controls system gain in response to the difference between the external standard ratio of a reference sample, stored on thumbwheel switches, and that of the unknown. Originally this difference was converted to a voltage which was superimposed upon the reference voltage to the high voltage supply; in more recent instruments the electronic amplification has been varied. The reference sample must be of better quality than unknown test samples. Increasing the high voltage or the amplification in response to quenching causes upward spectral shift. For quenched carbon-14 and more energetic isotopes it may be possible, depending upon the degree of quenching, to 'restore' to the counting window some or all of the counts that would have been there had there been no quenching but which, due to quenching, would normally fall below the window if the gain was not altered. It is also possible to shift the tritium spectrum but not to restore many counts; in tritium counting most of the efficiency loss from quenching results from the fact that light output is so reduced that one or both of the phototubes do not pulse. No amount of gain change can compensate for such loss.

The AQC feature, which has been evaluated by Wang,[3] can be thought of in much the same way as the Packard AAA. In the AQC the goal is to move a peak into a counting window; in the AAA it is to move the window to the peak. Once the sample has been counted the AAA does more with the data since it can provide d.p.m. The LS-250 with its AQC can only provide c.p.m. but operation is very much simpler, almost no time is lost in pre-processing, and the cost of the instrument is appreciably less than is the 3380/544.

In December 1967 Intertechnique, Philips, and the Finnish firm Wallac Oy showed prototypes of new liquid scintillation counters at an exhibition in Stockholm. The Philips counter (Fig. 23), refrigerated, and with 420-sample capacity (21 trays, 20 samples each) was capable of calculating d.p.m. for quenched single and dual isotope samples using either the channels ratio or external standard ratio methods. Quench correction curves are considered to follow quadratic equations, an acceptable approximation over a restricted range of quenching though perhaps less so over a broader range where quench curves exhibit an inflection point. Parameters for the best quadratic fit for experimental data from a quenched calibration series are stored on switches directly mounted onto readily accessible printed circuit cards. Once the sample has been counted the measured count rate, minus background, is converted to d.p.m. which is then printed on a Model 33 Teletype.

The Philips counter has four fully independent counting channels but it is not really intended that they all be used at once. Rather, each sample tray has provision for mounting coding strips which are capable of turning channels on and off, denoting the number of repeat counts, calling for different preset times, and establishing the calculating routine desired. Therefore, sample processing, including the combination of channels used, may vary with each tray. The channel settings are intended to be different and the several instrument users may call out their own requirements automatically as their samples come to the counting position, without the necessity to be physically present to alter control settings.

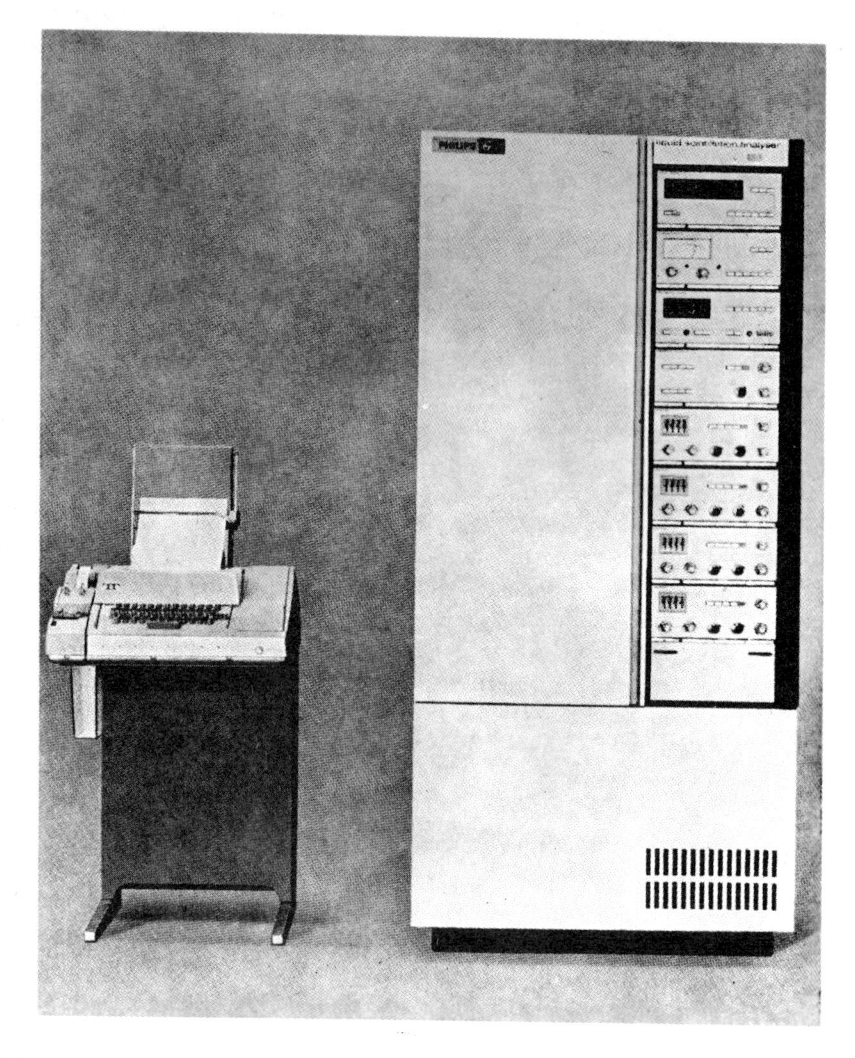

Fig. 23. Philips 420-sample liquid scintillation counter (1968).

The Intertechnique Model SL 40 (Fig. 24) differs from other counters in that it is the only one to incorporate within its electronics a programmable core memory computer. As such, it has computational capabilities in excess of those of other counters though surely not equivalent to those of more costly systems now being delivered, or which have been proposed, which combine a simplified counter with an external computer for on-line operation. The original SL 40 had a 1024 word × 12 bit computer; in its most recent version the memory has been expanded to 4096 × 12 bit words. Memory cycle time is 2 μs and add time is 6 μs; a read-only-memory of 256 × 4 bit words stores microinstructions and has a cycle time of 400 ns.

The SL 40, apart from its computational capability, is much like conventional counters with its 200-sample capacity, refrigeration, logarithmic amplifier, and '3 + 2' channels of analysis. Where it differs from predecessor counters is in the data presented

Fig. 24. Intertechnique Model SL 40 – available with 1 K or 4 K × 12 bit internal programmable computer (1968).

by its Teletype page printer. Quench correction curves – external standard ratio or channels ratio – are stored in the computer memory as third order polynomials. Once sample counting has been completed, d.p.m. values are calculated for single or dual isotope samples, making use of the chosen correction technique and the appropriate curves. Isotope ratios are also printed for dual isotope samples as is the standard deviation of the gross count in each channel, corrected for background contribution. It should be noted that standard deviation values are calculated according to proper mathematical expressions and are not numbers from 'look-up' tables which have sometimes been presented by other instruments and which are essentially always in error and for that reason have not been discussed. The SL 40 can also store results from repetitive counts and print average information. It can correct observed results from short-lived isotopes for radioactive decay. In the unit with the larger memory it is possible to store counting data from up to 80 samples, each counted up to four times, and after the last has been counted to plot on the Teletype the average of the measured or computed data from one or two isotopes as a percentage of the cumulative total. Because the SL 40 is programmable its computational capability is difficult to define and is in part limited by the commercial interest in pre-

Fig. 25. Wallac Oy 200-sample counter with pneumatically actuated sample handling mechanism (1971).

paring programs.

The Wallac DECEM-NTL[314] shown at the same time as the Philips and Intertechnique counters was a 200-sample instrument with quite acceptable performance. Modular in construction, it was available with one, two, or three pulse height analysis channels. For the most part its features were conventional, the most outstanding difference from other counters being a system for coding sample vial caps so that, rather than print out sample numbers from 1 to 200, the Teletype could print identification numbers from 1 to 10000 according to the code. The DECEM-NTL[314] was later supplemented with the DECEM-NTL[514] which offered improved counting performance, especially for carbon-14.

Most recently (1971) Wallac has announced the Model 81000 (Fig. 25), a controlled temperature 200-sample unit with the '3 + 2' channel configuration. The Model 81000 is unique among liquid scintillation counters in that most conventional switches are replaced by a binary coded plug board on which settings are made with the aid of diode pins. Linear amplification is employed together with 33 logarithmically arranged discriminators.

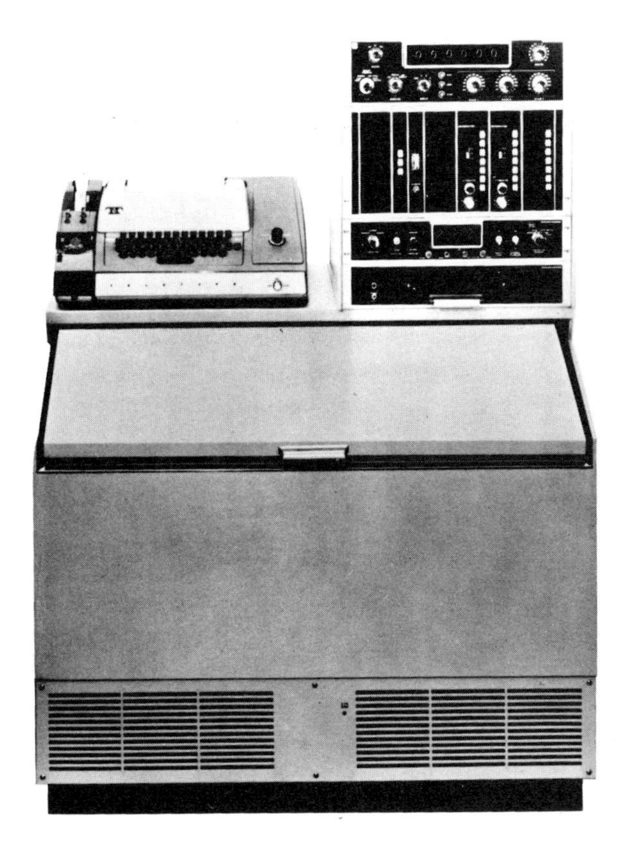

Fig. 26. Nuclear Chicago Mark II 300-sample counter (1969).

As each discriminator fires a corresponding neon lamp in a row of 33 is momentarily illuminated, the cumulative effect, especially for active samples, being to indicate the energy spectrum of an isotope under examination with the end point being represented by the last illuminated lamp and the peak by the lamps of maximum intensity. Channel settings are made with diode pins corresponding to the discriminators, using the lamps as a guide as to where the spectrum lies. The Wallac 81000 is also capable of providing d.p.m. for a single isotope only, using a third order expression for the quench correction curves. Third order coefficients are again entered with diode pins.

The Nuclear Chicago Mark II (Fig. 26), a 300-sample refrigerated counter was announced in 1969 as a replacement for the Mark I. Perhaps the most modular system yet built, the Mark II can be had with one, two, or three channels, a computational unit, a standardization module, a system for gain stabilization, and various printing devices including a lister, a Teletype, an IBM Selectric, and a solenoid actuated Olivetti Programma 101 desk calculator. By far the most interesting feature of the Mark II is its multi-user capability. Optional panels can be obtained which allow six or twelve investigators to maintain independent instrument control settings. Each may have his own selection of standardization means, preadjusted counting windows, preset time, preset count, and low level reject. A marker placed on the sample transport is coded from 1 to 12. When it arrives at the loading station it calls into operation the corresponding panel. Though each investigator can have his own selection of preadjusted windows, the same seven window settings apply to all six or twelve auxiliary panels.

Beckman has introduced (April 1971) a similar accessory for the LS-250 as well as for the simpler LS-230. The most obvious distinction between the Beckman system and that of Nuclear Chicago appears to be that the Beckman approach (intended for a maximum of six users) allows each to have variable as well as preadjusted windows whereas the Nuclear Chicago system limits multiple users only to fixed windows. The Beckman system makes no provision for individual low level rejection, but this in reality appears to be a feature that few users employ in any case. Both systems permit only one background subtraction setting for each channel. Since the different users are unlikely to have the same backgrounds for their different window settings, counting vials, and sample mixtures, background subtraction is essentially limited to a single operator.

The Nuclear Chicago Mark II was not the first counter to incorporate the Programma 101. This had been done before by Tracerlab in the Corumatic II and later, Beckman had developed a system – 'Omega' – now abandoned, for off-line processing of punched tape from the Teletype using a tape reader and a solenoid actuated Programma 101. The Programma is also used by Nuclear Enterprises as an optional data processor for the NE 8312 β/γ counter and it has been exhibited by Packard and is shown in their literature. There is no question that the Programma is a most versatile unit and far more reliable than the mechanical calculators which preceded it. However, its capacity for storage of constants and its slightly more than 100 program steps are not sufficient to permit the solution of quench curves as third order polynomials for dual isotope samples. Therefore such curves are usually treated as quadratics thereby limiting the accuracy of calculated results except over a narrow quench range.

The NE 8312 from Nuclear Enterprises (Fig. 27) handles 400 samples in 10 trays, each holding 40 vials. Separate β and γ detectors, operated at ambient temperature are side by side and samples may be directed to either or both. The amplification system used for both β- and γ-counting is described as being 'quasi-logarithmic'. Two sets of three counting windows together with independent preset count and low activity reject controls are provided; a separate preset time control is provided for each set of three window controls. Individual sample trays may be counted according to either set. Data are accumulated in three scalers and, in the basic unit, printed out on a Teletype. The γ detector is especially interesting in that its mechanism is apparently identical to that used for liquid scintillation counting, the detection element being a 3 in × 3 in sodium iodide crystal faced by two horizontally opposed photomultipliers. The crystal has a hole midway along its length and perpendicular to its axis through which the elevator may pass and in which the sample resides during counting. Output pulses from the two phototubes are summed into a single amplifier but, of course, for γ-counting there is no coincidence requirement.

In 1969 Intertechnique expanded its line of automatic liquid scintillation counters with the introduction of the SL 30. The SL 30 itself has no very unusual features – it is a 200-sample refrigerated counter with log amplifier, external standardization, '3 + 2' channels of analysis, and Teletype. What distinguishes it from previous instruments is the degree to which its data processing capability can be built up. The SL 30 performs no computation itself; it prints raw counting results and, if its Teletype is equipped with a punch, punches tape which can then be taken to a computer for further treatment. There are, however, three derivatives of the SL 30 – SL 31, SL 36 and Multi-Mat – all announced

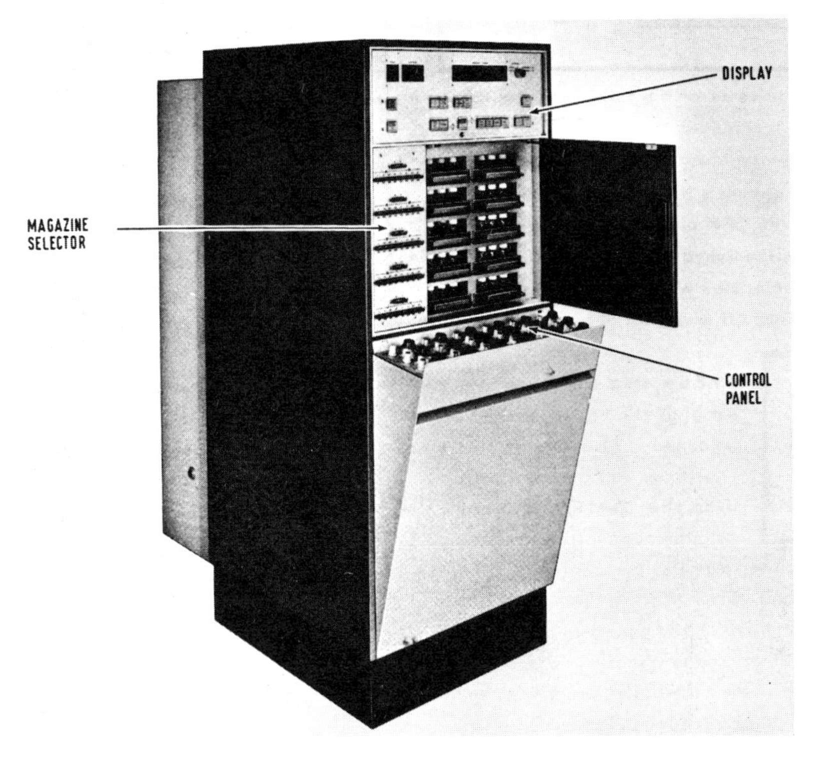

Fig. 27. Nuclear Enterprises 400-sample β/γ counter (1970). Separate counting heads are provided for each type of radiation.

in 1970–71, which have data processing potential.

The SL 31 performs simple calculations by means of a Friden Model 1151 printing calculator which is on-line to the SL 30 electronics. The 1151 can be programmed to the extent of 30 steps; it has a storage register for a single constant. Though obviously less powerful than the Programma 101, it is also much less costly and can be actuated directly through its electronics rather than by means of solenoids; system reliability is enhanced. While computational capability is rather limited, the SL 31 can provide c.p.m., d.p.m. for any one channel (assuming constant efficiency counting), cross contribution subtraction for a dual isotope experiment (again with the same assumption), external standard ratio, channels ratio, etc. Its principal disadvantage is that results appear on an adding machine tape rather than on the superior Teletype page format.

The Intertechnique SL 36 can be regarded as an SL 30 in which a hard-wired calculating unit has been interposed between the electronic chassis and the Teletype. This calculating unit is capable of providing c.p.m., d.p.m. for single or dual isotope samples counted at constant efficiency, background subtraction, external standard and channels ratios, and standard deviation.

The most elaborate use of the SL 30, and probably the liquid scintillation counting system capable of performing the most advanced computation is Intertechnique's Multi-Mat. From one to four SL 30's operate on-line to a single Multi-8 central processor. The Multi-8, a self-contained microprogrammed minicomputer has a core memory expandable from an initial 8K to 32K X 8 bit words with a cycle time of 1.1 μs. System organization is such that each counter may have up to four different data processing routines in use for the same sample load and the four routines may be applied in any order to up to eight groups of samples. Routines change when the appropriate sample number, previously entered into the computer memory via the Teletype, arrives at the loading station. Computational parameters – data points for quench correction curves, backgrounds to be subtracted, etc. – as well as instructions to select data processing routines are entered into the computer memory by typing answers in response to a series of questions posed by the computer. An illogical response is not accepted and the computer repeats the question.

The Multi-Mat is capable of constructing its own quench correction curves for single and dual isotope counting based upon either the external standard ratio or channels ratio methods. The computer is instructed that the first *n* samples are calibration standards for one isotope and that the next *m* samples are those for the second isotope; subsequent samples are corrected automatically on the basis of the calibration curves generated with these first samples. New calibration curves can be made for each group of samples or one or more curves may be permanently stored in the memory and recalled when required. Decay corrections are also made automatically using reference samples as an internal clock.

A new computer language – 'LEM-S' – is also available. It permits simplified programming by turning the Multi-8 into the equivalent of an extremely powerful desk-top calculator having the equivalent of about 600 program steps and more than 200 storage registers. A standard repertory of subroutines is provided and these may be called out either with a single instruction or by listing the appropriate mathematical expression. Printout format is also extremely flexible and the Teletype may be used not only to provide computed results but also to plot histograms, quench correction curves and curves derived from analysis of experimental data.

At the 1971 FASEB meeting both Beckman and Nuclear Chicago announced the availability of systems having many of the attributes of the Multi-Mat. At the time of writing, however, it is not known if any such systems have been delivered and as this author has also seen no commercial literature, details of these proposed assemblies are unknown.

Also at the FASEB meeting, Nuclear Chicago exhibited the Isocap/300, a 300-sample ambient counter with two counting channels. Even more recently (Brighton, September 1971) a refrigerated version employing the Mark II refrigeration unit has been shown. The most distinguishing feature of these counters is the operator's ability, again with program plugs placed on the sample transport, to call out a choice of fixed counting windows; there is no possibility of having variable windows. The basic unit is supplied with a lister but other printout devices are optional. A computation module is available as is a minicomputer for on-line operation.

Also new in 1971 is the Model BF 5000, a joint development of the Laboratorium Prof. Dr. Berthold and Frieseke and Hoepfner. Known as the Betaszint (Fig. 28), the system handles 200 samples in a conventional serpentine arrangement. Refrigeration is provided and the '3 + 2' channel configuration is employed. 'Quasi-logarithmic' amplification is achieved by combining the outputs of several elements with linear amplification. Data are put out through a Diehl Combitron printing calculator which can be programmed through its keyboard or with a paper tape reader provided for that purpose. With the Betaszint the contribution of chemiluminescence to the observed count may be estimated by feeding the output of one photomultiplier, but not the other, to a delay line. With a

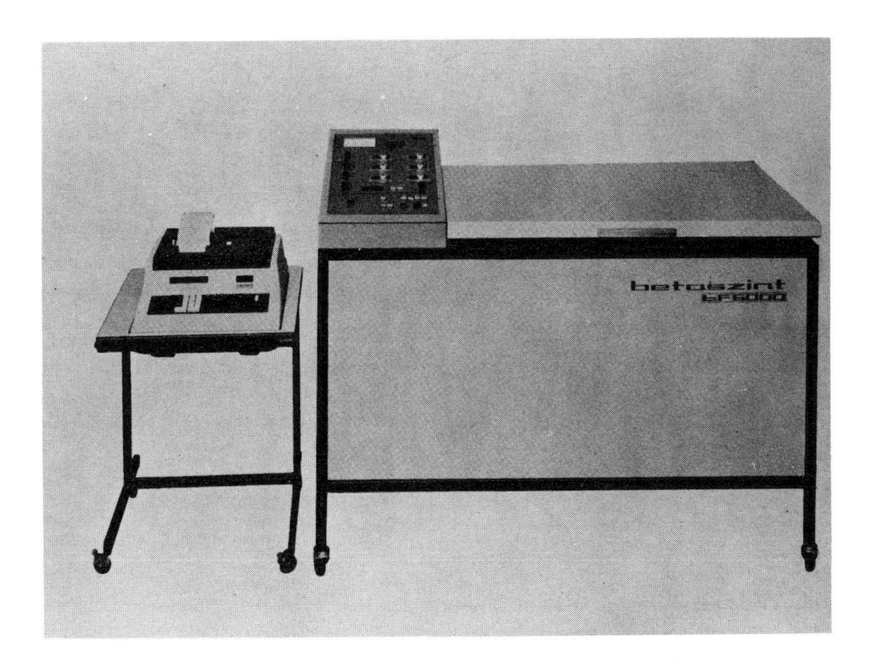

Fig. 28. Berthold/Frieseke 200-sample Betaszint (1971).

delay which is at least twice as long as the coincidence time, legitimate decay events are not counted but accidentals (in a poor sample largely chemiluminescence), whose observed rate only depends upon the rate of individual phototube pulsing and the value of the coincidence time, are counted.

REFERENCES

1 R. D. Hiebert and R. J. Watts, *Nucleonics* **11**, 38 (1953).
2 R. J. Herberg, in *Organic Scintillators and Liquid Scintillation Counting* (Eds. D. L. Horrocks and C. T. Peng), Academic Press, New York, 1971, p. 783.
3 C. H. Wang, in *The Current Status of Liquid Scintillation Counting* (Ed. E. D. Bransome), Grune and Stratton, New York, 1970, p. 305.

DISCUSSION

J. F. Stoutjesdijk: You have spoken about apparatus with logarithmic attenuators. How does apparatus with logarithmic photomultipliers manage with pulse summation, especially in ^{3}H-counting?

E. Rapkin: In the best logarithmic systems pulse summation precedes the logarithmic element, the pulses summed being linear outputs from the photomultipliers. The summed pulses then pass through the logarithmic attenuator amplifier where large pulses are compressed but small ones are hardly affected. Thus, the low energy end of the spectrum, i.e. the tritium region is 'pseudo linear' and results are almost identical to those obtained with a purely linear system. In fact, I would expect no inherent performance difference between logarithmic and linear systems, the differences which are occasionally noted being attributable to the individual photomultipliers rather than to any considerations of system logic. The principal advantage of the logarithmic system is in its simplicity.

B. E. Gordon: What is the potential problem raised by the effect of the external standard on the glass in the system?

E. Rapkin: The external standard source does induce phosphorescence in the photomultipliers and in the material of the counting vial. However, the amplitudes of the pulses created by such phosphorescence are low. By setting the lower level of the external standard counting window above the phosphorescence, it can be successfully excluded from the external standard counting results. It should be realized, however, that the frequent exposure of the photomultipliers to a γ-source probably raises the background in the tritium region by a few counts. The added background probably decays away slowly if the source is not employed.

B. W. Fox: (a) I would like to see some form of external standard ratio reject in unfavourable circumstances to avoid hidden error arising in automated systems. (b) What is your opinion of the future of external standard ratio?

E. Rapkin: (a) Such circuitry is feasible and could easily be incorporated in many existing units if there were sufficient demand. In at least one Intertechnique instrument (Model SL 40), while we do not reject samples whose external standard ratio falls outside of predetermined limits, we do print a flag to suggest to the operator that he be extra careful in interpreting the recorded result. (b) External standardization, particularly the external standard ratio, will continue to enjoy its present prominence in the foreseeable future; there seems to be no other method, so far proposed, which appears capable of replacing external standardization. It may be, however, with the improved sample quality which surely will come about as combustion techniques become more widely used, that the importance of all methods for quench correction will be diminished. Finally, it is hoped that more of the users of external standardization will quickly develop a better appreciation of the limitations of the method and the attendant errors.

B. E. Gordon: Standardization of emulsions by either internal or external standard methods is subject to error, the former by differential partitioning of sample and standard and the latter because the heterogeneous nature of the medium reacts differently to external radiation compared with a dissolved, partitioned sample.

E. Rapkin: I quite agree with this comment which should best be appended to Dr. Fox's chapter even though the remark was made at the end of my talk. I would like to add that in searching the literature during the course of writing a review on the subject of emulsion counting (as yet unpublished) I found no reference which clearly demonstrated that an external standard can be used with emulsions to provide an accurate indication of counting efficiency. On the other hand, I found numerous references which stated that the external standard was wholly unsatisfactory for this purpose.

Acknowledgement – I am grateful to the following manufacturers for permission to include photographs of their instruments: Beckman (Fig. 17), ICN Tracerlab (Fig. 19), Pye Unicam (Fig. 23), Nuclear Enterprises (Fig. 27) and Berthold/Frieseke (Fig. 28).

Chapter 6

Questions regarding the Occurrence of Unwanted Luminescence in Liquid Scintillation Samples

B. Scales

Imperial Chemical Industries Ltd., Pharmaceuticals Division, Alderley Park, Cheshire, England

INTRODUCTION

The continuing increase in the use of soft β-emitters, and the relatively high cost of the necessary counting equipment often makes it imperative that relatively unsophisticated equipment is purchased; the total number of counting heads available being much more important than the 'storage capacity' of a machine. We have available ten automatic liquid scintillation counters. The order in which we purchased these instruments and some of their more useful features are shown in Table 1. The rapid increase in the number of counting heads must gladden the hearts of manufacturers of liquid scintillation counting equipment, but it can bring unforeseen problems to the users of that equipment.

All manufacturers attempt to satisfy the most stringent requirements of their customers and, besides stability, two of the most important requirements are:

1. a high counting efficiency, and
2. a low background count.

As is the case with most other workers with soft β-emitters, the ability to create troublesome chemiluminescence was acquired soon after our entry into the field of liquid scintillation counting, but with experience we learned that in order to have acceptably low background counts, certain methods of sample preparation were best treated with extreme caution, or better still, completely ignored.

Occasionally, some workers with isotopes can accept a relatively high and fluctuating background count, but when the activity of the isotopes is limited as in tissue residue studies, or when metabolism studies or sub-micro analytical procedures are being developed using organic materials of relatively low specific activity, a low and stable background is desirable.

It is over the last two years, when different makes of equipment have become available and new techniques have appeared, that further problems have become apparent. Those problems which fit into the context of this paper are all connected with the presence of undesirable luminescence, i.e. an increase in background count.

Some of our problems have been solved; this paper attempts to show how they arose, and how they were overcome.

Table 1. Availability of automatic liquid scintillation counters.

Model	Purchased	Sample capacity	Temperature °C	Facilities
Packard 314EX	/60	100	2–4	Lister
Packard 314EX	9/61	100	2–4	Lister
Packard 3314	6/62	200	2–4	Lister, b.g., A.E.S.
Packard 3124	10/64	200	2–4	Lister, b.g., A.E.S.
Packard 3320	7/65	200	2–4	Lister, b.g., A.E.S.
Philips LSA	12/69	420	18	Teletype, Computer
Packard 3320	3/70	200	2–4	Teletype, b.g., A.E.S.
Packard 2002	1/71	100	ambient	Lister, b.g.
Packard 2002	1/71	100	ambient	Lister, b.g.
Intertechnique SL30	3/71	200	8	Teletype, b.g., A.E.S.

Lister = Lister printer
b.g. = background subtract
A.E.S. = Automatic external standardization facilities

Vial caps of different manufacture

The first problem was concerned with the phosphorescence of a certain make of vial cap. It was noticed that after a series of 50 to 60 samples had been counted overnight for 10 min each in a Packard 3320 the blank samples in the series gave background counts of the order of 150 c.p.m., i.e. from 4 to 5 times the normal count of 30 c.p.m. The sealed blanks gave normal background counts when assayed at the most diagnostic isotope settings i.e. tritium in toluene and carbon-14 in toluene, and tube noise was satisfactory.

After some time it was realized that the batch of vials being used was not from our usual source. Our new Philips Liquid Scintillation Analyzer was temporarily out of order (due to our own negligence) and these high background samples had been prepared in the last of the batch of Philips vials sent with the original purchase order. After some time we discovered that the caps of these vials had a phosphorescence which although discarded by the Philips electronics, was strong enough to pass through the coincidence system of our refrigerated Packards, when operating under normal counting conditions (see Table 2). This phosphorescence decayed very slowly at reduced temperatures (0 to 4°C).

The phosphorescence in the Philips caps was activated slightly by strip lighting and very much more so by sunlight. Irradiation by ultraviolet light at 256 nm (Chromatolite) caused gross activation of phosphorescence in both Philips and Packard caps.

Our problem had been that:

1. Carrying the vials along corridors fitted with strip lighting was sufficient to activate the phosphorescence of one make of cap and not another, and
2. This was discarded by one machine and not by another.

We tried to prevent the problem from recurring by standardizing on the type of vial and cap which appeared to suit all machines; we chose Packard vials, but I am now informed that the problem with the Philips caps was an isolated one which was rapidly overcome.

Effects of temperature on different vial caps

A second problem with the luminescence of vial caps came soon after the purchase of two room temperature machines. The vial caps which had been proved to be acceptable

Table 2. Effect of ultraviolet-induced luminescence of caps on background counts (c.p.m.).

Instrument	Cap	Time in instrument (h)				
		0	2	6	24	48
Philips L.S.A.	A	31.5	31.7	30.2	31.0	30.9
	B	29.5	30.6	30.6	30.7	30.1
Packard 3320	A	156.5	154.7	132.9	129.2	131.9
	B	28.1	29.0	27.7	27.5	29.3

Vials were low potassium-40 supplied by Packard.
Cap A supplied by Philips.
Cap B supplied by Packard.
Vials contained 20 ml toluene containing 0.6% butyl P.B.D.
All samples were counted for 10 min periods, under optimum gain conditions for carbon-14.

in all our refrigerated (controlled temperature) machines gave rise to high and unstable background counts in the two ambient temperature instruments.

The possibility of static was ruled out, and we eventually found that the sunlight, of which we had more than usual during February and March, was the chief cause of the problem. The sunlight, shining although somewhat weakly across the laboratory, was sufficient to give rise to initially 2000 c.p.m. above background. This decayed with at least two component half-lives (see Fig. 1). One half-life was of the order of seconds so that the immediate counting rate on entering the counting chamber was of the order of 30000 to 100000 c.p.m. This was not new to us; it is often seen in a refrigerated machine placed in bright sunlight. The other half-life was much longer (of the order of 2 to 3 h) and it was only observed in the ambient temperature machines where it lifted the background count by 20 to 30 c.p.m.

In this problem, therefore, there were two machines with identical electronics, yet at room temperature a long-lived component of the light-induced phosphorescence was sufficiently intense to pass the coincidence system and double the background count in the tritium channel. These problems can be solved by:

1. Increasing the threshold of the tritium and carbon-14 windows. This resulted in a lowering of the tritium efficiency from 45% to 25%, and of the carbon-14 efficiency from 88% to 84% before a normal background was achieved.
2. Reducing all sources of ultraviolet light to a minimum by using tungsten lighting, shields over fluorescent lighting and by the use of window blinds. This we feel is absolutely essential especially in the counting room.
3. Most successfully, when re-using vials – this is done on average about ten times before they are discarded – they are re-capped with cheap disposable plastic caps from the Metal Box Company. Cat. No. 1A. 22R3. These soft plastic caps do not have a cork-tinfoil insert, but rely on an annular lip of plastic for the seal. Because of the lack of a tinfoil insert, there is in our experience a slight decrease in counting efficiency of from 1 to 1.5% for both tritium and carbon-14. More importantly we have been unable to induce any phosphorescence in these caps which can be detected in any of our machines, even after intense irradiation with ultraviolet light at 265 nm or 365 nm.

Triton X-100

A further problem which would not have been readily apparent if we had not pur-

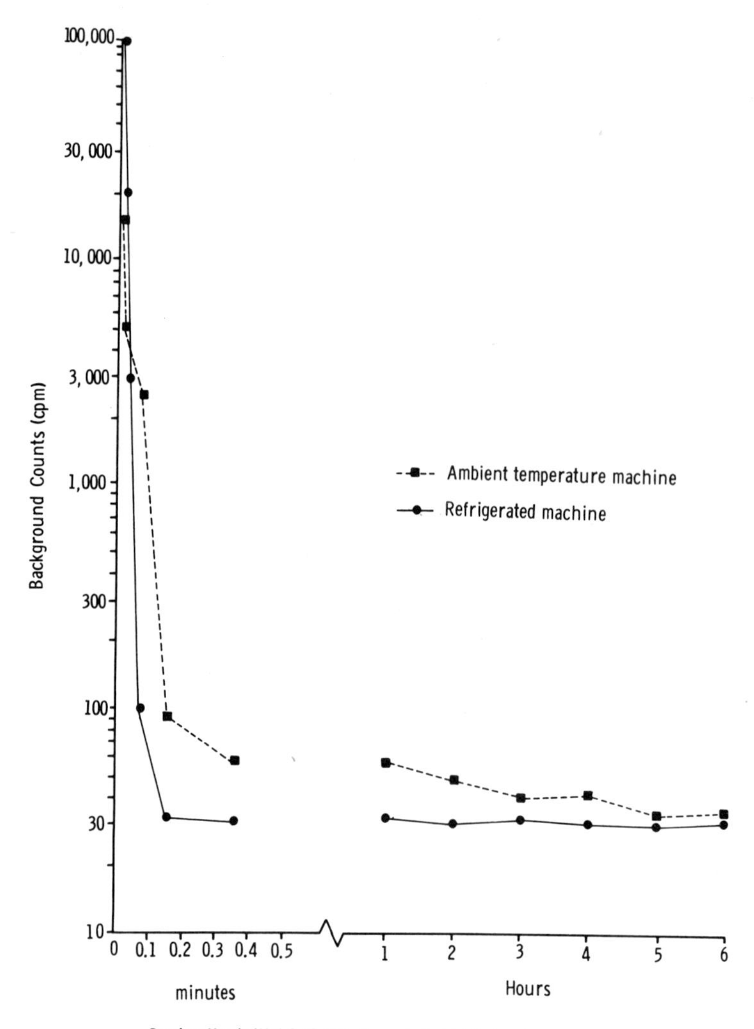

Fig. 1. Chemiluminescence of dioxane systems in refrigerated and ambient temperature machines.

chased ambient temperature machines concerned the use of Triton X-100. Although Patterson and Greene[1] state that a high background can be obtained with some batches of Triton X-100, this had never been our experience when using refrigerated equipment. However, under optimum ^{14}C-counting conditions the conventional Triton X-100–water systems were found to give backgrounds of 125 to 250 c.p.m. in the ambient temperature instruments. Background counts of 250 c.p.m. dropped with a half-life of about five days

Table 3. Effect of different batches of Triton X-100 on background counts (c.p.m.).

Instrument Packard	Temp.	Batch	Time in instrument (days) 1	3	5	8	10	15	22
2002	Ambient	New	253.5	201.1	135.7	110.3	85.5	62.1	42.2
		Old	33.3	32.8	32.9	32.3	33.0	31.9	32.7
3320	2–4°C	New	34.8	34.3	33.6	33.9	33.2	33.4	32.1
		Old	32.0	29.1	28.7	29.1	31.1	30.8	31.8

Samples were 10.0 ml of a mixture of toluene–Triton X-100 (2:1) containing 0.6% butyl P.B.D.
Batch-New contains a high proportion of luminescent impurity.
Batch-Old contains acceptably low amounts of the impurity.
All samples were counted for 100 min each under conditions for optimum carbon-14 counting.

(see Table 3). Careful examination of the background counts obtained from the same samples in the refrigerated machine showed that they were about 10 to 15% higher than expected. Also a decay of luminescence was detectable, but it occurred with a half-life of two to three weeks, and it was not therefore noticeable when using the refrigerated machine that a faulty batch was in use. The normal, acceptable background counts obtained on both refrigerated and room temperature machines with good quality Triton X-100 are shown in Table 3 for comparison.

The luminescence from this impurity was such that its contribution to the background of the tritium channel could only be removed by increasing the threshold to such a level that the ^{3}H-counting efficiency was reduced by 2/3. The ^{14}C-counting efficiency was also reduced by about 1/6 before its background became normal.

This fluorescent impurity could not be removed by absorption onto silica as suggested by Patterson and Greene.[1] The fluorescent impurity appears to be a contaminant occurring during manufacture since it does not form on keeping Triton X-100 in darkness or in sunlight and, when present, its proportion appears to remain constant. It would, therefore, appear that selection of batches of Triton X-100 prior to purchase is the best way of ensuring the quality.

Dioxane

Until recently, in common with many other workers we used dioxane-naphthalene – PPO-POPOP or -butyl PBD scintillators for aqueous samples containing tritium, carbon-14, calcium-45 or iodine-131. For this purpose the dioxane was purchased in 25 gallon drums from BASF. This product has a very low water content (less that 0.1%) and usually needs no special purification, even for ^{3}H-counting, especially as the dioxane-based scintillators were used at the rate of 20 l/week. Occasionally, if extra high purity material is required, in order to decrease the amount of chemiluminescence produced by somewhat alkaline solutions, we obtain a very good clean-up by a crude zone-melting procedure, carried out by alternately freezing and thawing winchesters of dioxane. This is done quite simply by wheeling trolleys loaded with the winchesters into the cold room (0 to 4°C) for freezing overnight; any liquid impurities are decanted the following morning and the solid contents then thawed out during the day. The process is then repeated until no liquid impurities are apparent. This results in good quality dioxane which is very satisfactory for liquid scintillation counting with little effort and hazard. However, the ready and often uncritical acceptance of Triton X-100 by many workers resulted in a decreased usage of dioxane and as a result of the increased peroxide formation we started getting high backgrounds due to chemiluminescence, even at neutral pH (6.5 to 7.5). Once again this luminescence was only apparent in ambient temperature machines in which, of course, the luminescence decayed fairly rapidly. Thus at neutral pH it was not noticeable after 30 min. At pH 8.5 or greater, even in refrigerated machines it can take at least 2 to 3 days for acceptable background counts to be obtained and often much longer. The addition of antioxidants such as B.H.T. or recommended sulphydryl reagents and chelating agents such as diethyl dithiocarbamate,[2] although reducing this chemiluminescence also significantly reduced counting efficiency in the tritium channel.

Injection tubes for GLC

The problems discussed above were not isolated problems; they did in fact occur within a period of two months. The final problem I will mention was encountered at the same time as the problems of luminescence with vial caps, Triton X-100 and peroxide formation in dioxane, so that at one time we had at least three unwanted phenomena taking place in any one vial.

This last problem occurred when we were investigating various analytical techniques intended for use eventually with non-radio-labelled drugs undergoing toxicological and clinical evaluation. The technique in question involved a series of extractions and back-extractions from biological fluids, TLC of the extracts and after elution from the plates, the formation of a derivative for GLC. The application of the samples to the Pye GLC machine was by use of a 'solid injector'.

It is imperative in such a complicated method which has to be carried out on the nanogram level that each stage works efficiently and that any slight variation can be understood and overcome. To this end, the use of radio-labelled materials is invaluable even though the count rate can be quite low with specific activities in the 2 to 20 μC/mg range.

Whilst checking for possible losses of a drug during evaporation of 30 μl of solvent from the Pye solid injection tubes, and also checking that drug placed in the tubes was volatilized in the pre-heater of the GLC column, it was observed that the small glass 'solid injection' tubes, when placed in a vial containing scintillator, could give rise to anything from 0 to 15000 c.p.m. in the tritium channel and 0 to 5000 c.p.m. in the carbon-14 channel. This count occurred in all machines irrespective of temperature and scintillation medium. The tubes had no phosphorescence of their own and in dioxane–water required the presence of naphthalene, or better, naphthalene-butyl PBD for maximum luminescence. It is inconceivable that the tubes have a high potassium-40 content, since the energy spectrum is not that of potassium-40 and the count is too high for natural potassium-40.

Some batches of injection tubes when cleaned in dilute hydrochloric acid and methanol had a normal tritium and carbon-14 background, but those cleaned for us in Haemasol and water, to give tubes which were satisfactory for GLC use, occasionally but not always gave this high background of from 1000 to 15000 c.p.m. We were therefore initially very suspicious of the use of Haemasol. Attempts to induce this prolonged luminescence by baking Haemasol and methanol onto GLC injection tubes and liquid scintillation vials have not been very successful. All we have achieved is a count of 400 c.p.m. which decays over a 30 min period. We have subsequently tried to remove this low luminescence from the injection tubes by acid washing and so far have been quite unsuccessful. The problem of high luminescence from some of these glass injection tubes remains a mystery.

CONCLUSION

During this talk I have repeatedly stated that these problems showed up when ambient temperature machines were used. I am not suggesting that such machines are a bad buy as do many of our users, but I would rather suggest that they did bring to light sloppy techniques, and that much more care is required in liquid scintillation counting than many workers are prepared to admit.

REFERENCES

1 M. S. Patterson and R. C. Greene, *Anal. Chem.* **37**, 854 (1964).
2 *Chemical Notes:* A summary of current practices in sample preparation for liquid scintillation counting – March 1970. Issued by the Packard Instrument Company.

DISCUSSION

Anon: Comment: We have had considerable problems with materials in plastic bottles leaching out into solvents and producing excessively high background counts. Recently, different batches of solid scintillators, when dissolved in solvent were found to produce a very high, long-lived luminescence. This was eventually traced to the use of scoops made out of plastic bottles which were allowed to remain in contact with the organic scintillators.

F. E. L. ten Haaf: Comment: I should like to make a brief comment on the fluorescent vial caps as I happen to know the inside story about them. The vials and the caps are made for us by another manufacturer and delivered to us in batches of a few hundred thousand at a time.
We designed them rather carefully and the material for the caps has been given due consideration. Several series were delivered and found to be good, but suddenly, without consulting us, the manufacturer changed the material for the caps. Before this was discovered a few thousand vials had been delivered to customers and I am fully aware that, although we replaced them with good ones, this has caused you and other users considerable embarrassment. I can assure you that we will try very hard not to make a habit of it.

Chapter 7

The Estimation of ATP, ADP and AMP in Human Plasma using Luciferin/Luciferase and a Scintillation Counter

P. I. Parkinson* and E. Medley

The Middlesex Hospital, London, England

INTRODUCTION

The measurement of adenosine triphosphate (ATP) using luciferin and the enzyme luciferase results in an emission of photons and therefore lends itself to scintillation counting.

$$\text{ATP} + \text{Luciferin} \xrightarrow[O_2]{\text{Luciferase}} \text{Adenyl luciferin} + \text{photons}$$

We are measuring ATP in amounts of 10^{-12} to 10^{-9} mole in aqueous solution using a method adapted from that described by Stanley and Williams.[1]

The technique of treating whole blood to obtain deproteinized plasma is crucial as destruction of the cellular elements of blood and degradation of ATP occur readily and distort results.

TECHNIQUE

Deproteinized plasma is obtained in the following way: 7 ml of blood is drawn into a syringe containing 1 ml of normal saline and 21 mg of EDTA (pH 7.4) at a rate of 1 ml/5 s. Air is drawn into the syringe and the syringe is inverted eight times. EDTA binds calcium ions and thus inhibits both coagulation and ATPase activity (see Table 1). The contents of the syringe are placed in a siliconized glass tube and spun at 5000 G for 30 min. Table 2 illustrates the effect of spinning six sequential samples from the same patient in siliconized glass or plastic tubes. The increase in counts in plastic tubes we attribute to release of ATP from platelets. 3 ml of the supernatant is measured into a plastic tube to which 5 ml of 0.5 m perchloric acid is added. The effect on the ATP counts of leaving pooled plasma to stand at room temperature is shown in Table 3. The tube is shaken and spun at 5000 G for 10 min. This supernatant is decanted and neutralized to pH 7.0 with potassium hydroxide (see Table 4). The precipitate of potassium perchlorate is removed by cooling and spinning at 5000 G for 10 min and the supernatant is again decanted. The volume is made up to 8.5 ml with distilled water. To this is added 2 ml of 0.55 m triethanolamine buffer and 1 ml of a solution containing 4 mmole magnesium sulphate, 1 mmole potassium chloride and 0.01 mmole phosphoenol pyruvate(PEP)/10 ml.

---

* Charles Wolfson Fellow in Clinical Cardiology.

Table 1. Effect of varying the mixing of EDTA with blood.

	Counts
Syringe not inverted	22 824
Syringe inverted twice	544 720
Syringe inverted four times	717 747
Syringe inverted eight times	832 539

Table 2. Effect of spinning samples in siliconized glass or plastic tubes.

	Counts
Siliconized glass tubes	883 585
	847 070
	892 162
Plastic tubes	2 332 645
	1 885 000
	1 615 185

Table 3. The effect on ATP counts when plasma is left standing at room temperature (pooled plasma).

Time standing	Counts
13 min	50 120
25 min	51 747
29 min	38 808
53 min	37 989
62 min	31 943
75 min	27 840

Table 4. The inhibiting effect of potassium perchlorate on the luciferase reaction.

Dialysis fluid + 10^{-10} mole ATP + $HClO_4$ + KOH + spin	Dialysis fluid + 10^{-10} mole
Counts	Counts
11 074	75 119
9 842	85 646
10 806	1100 466
10 697	58 850

Three 1 ml aliquots of this solution are taken for triplicate estimations of ATP, 0.02 mg of pyruvate kinase (PK) is then added to convert adenosine diphosphate (ADP) to ATP and the samples are left to incubate at room temperature for 15 min before counting again in triplicate. 0.2 mg of myokinase (MK) is then added and the samples are left to incubate for 1 h to convert adenosine monophosphate (AMP) to ATP:

$$\text{AMP} + \text{ATP} \xleftrightarrow{\text{MK}} 2\ \text{ADP}$$

$$\text{ADP} \xrightarrow[\text{PK}]{\text{PEP} \rightarrow \text{Pyruvate}} \text{ATP}$$

and a further three aliquots are taken for counting. Subtraction of the figures obtained gives an estimate of ADP and AMP.

The remaining 2.5 ml of deproteinized plasma solution in one tube is mixed with the remnants from other samples from which four 1 ml aliquots are estimated for their ATP content. To another four aliquots of the same plasma solution a known amount of ATP is added and they too are estimated for their ATP content. Subtraction of the resulting figures provides an internal standard.

Firefly extract is prepared by grinding the tails of 10 to 14 fireflies (Sigma FFT) into a paste with arsenate buffer which is then made up to 5 ml with more arsenate buffer. This is spun at 5000 G for 10 min. The straw-coloured supernatant is decanted into a plastic container and left at 8°C for 2 h.

The machine, a Packard Tricarb model 3320 liquid scintillation spectrometer, is given the following settings: preset count 900000 or time 20 s; pulse height discriminator window 60 to 65 (see Fig. 1); gain 100%. The coincidence switch is turned off. The counting chamber is at 8°C. The counting window indicates time.

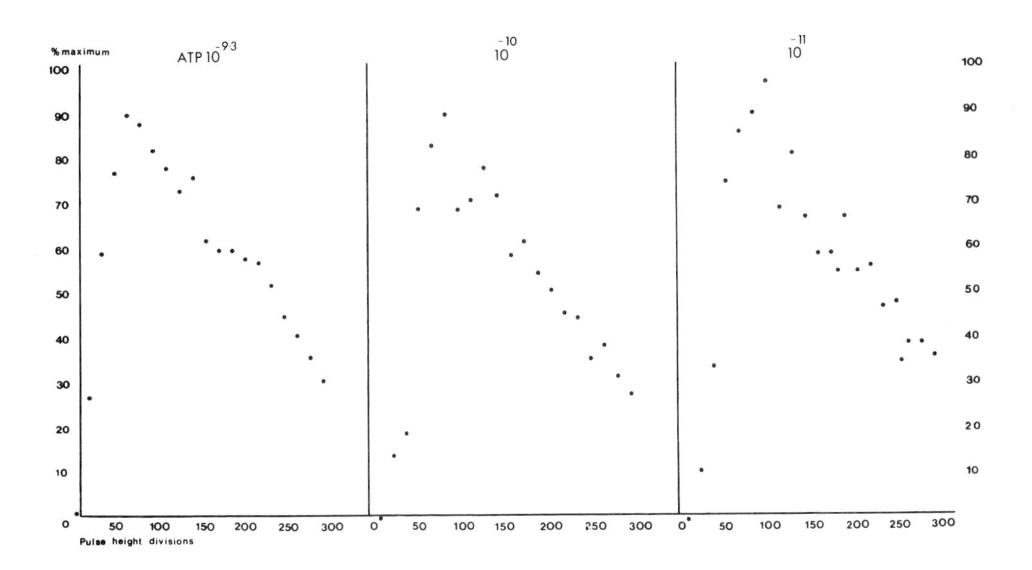

Fig. 1. Pulse height spectra for out of coincidence counting of the luciferin/luciferase system with varying amounts of ATP.

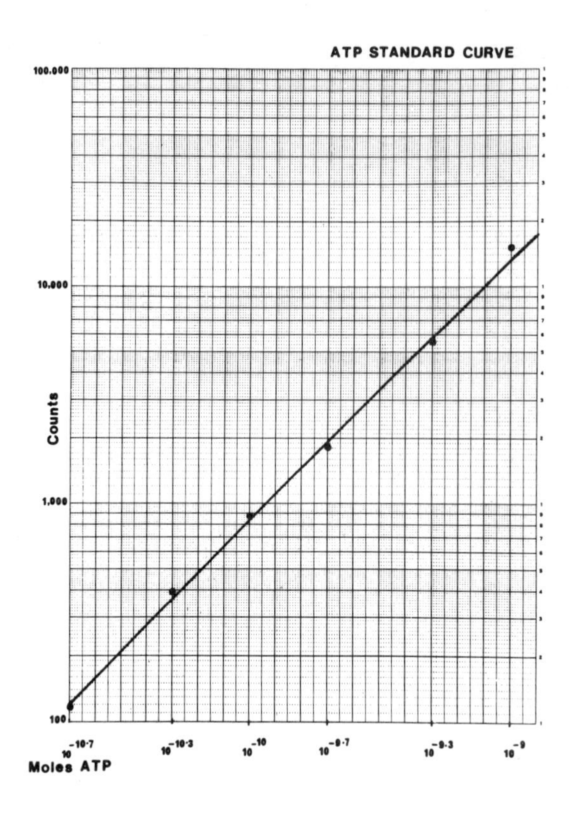

Fig. 2. ATP standard curve.

Each vial contains 1 ml of arsenate buffer, 1 ml of phosphate buffer and 1 ml of saline containing 90 mmole/l of sodium chloride. For ATP estimations 'plasma solution' replaces the saline and the volume of phosphate buffer is adjusted to accommodate the addition of standard solutions of ATP.

The vials to be counted include a calibration curve with a range from 10^{-11} to 5×10^{-9} mole (in triplicate using dilutions of ATP in phosphate buffer, see Fig. 2), the unknown plasmas, an internal standard and four blanks with no plasma nor ATP added. These are left to equilibrate at room temperature for 15 min.

At the commencement of counting an empty vial is placed in the counting chamber. As 10 s is indicated, 0.05 ml of firefly extract is added to the next vial to be counted. The cap is screwed on and the vial is shaken three times and placed in position.

The calibration curve is plotted on semilog paper. The mean value of the counts obtained from the plasma samples are multiplied by an inhibition factor. This is obtained by comparing the counts of the internal standard with the counts of an equivalent amount of ATP in the calibration curve. The corrected figure gives an estimate of the amount of ATP in 1 ml of the plasma solution. The amount in moles can be read from the calibration curve. By applying the formula:

$$X \times 10^{-n} \times V \times \frac{Z+1}{Z} \times 1/3 \times 10^{6}$$

where X = antilog of ATP/ml of plasma solution
V = total volume of plasma solution (usually 11.5 ml)
$n = 10^{-9}, 10^{-10}, 10^{-11}$

$$Z = \frac{100 - \text{packed cell volume}}{100} \times 7$$

a result in μm/ml of true plasma is obtained, and the result may be expressed in μg by multiplying by 551.2.

RESULTS

Using this method the range of normal values for ATP in human plasma is 0.2 to 1.2 μg/ml. Normal values for ADP in human plasma are 0.05 to 0.7 μg/ml and there is negligible AMP in resting human plasma.

The method has a standard error of ±13% at 0.2 μg (one S.D.) and ±7% at 2.5 μg (one S.D.). As the values for ADP and AMP are obtained by subtraction of repeated estimations of ATP the sensitivity of the method for ADP is a function of the sum of ATP and ADP and the sensitivity of the method for the estimation of AMP is a function of the sum of ATP, ADP and AMP.

When nucleotide is added to whole blood as it is being drawn into the sampling syringe the method gives a 60% recovery of ATP, 80% recovery of ADP and a 60 to 80% recovery of AMP.

REFERENCE

1 P. E. Stanley and S. G. Williams, *Anal. Biochem.* **29**, 381 (1969).

DISCUSSION

P. Stanley: Have you found that dust and/or micro-organisms in samples and the solutions used for bioluminescence assays present a problem?
I believe that a number of problems have yet to be solved before the estimation of AMP and ADP can be reliably made on samples from a range of systems. Presently internal standards offer the best way out of the problem.
Pulse pile-up is a problem that many people are not aware of in bioluminescence assays. Since large numbers of photons are produced the electronic system in the spectrometer becomes saturated and the pulse height spectrum becomes distended and distorted.

P. I. Parkinson: In reply to your first point, we have noticed no such problems. We find that there is minimal degradation of ATP even at low concentrations at room temperature, and what there is, is virtually undetectable for the time of the assay. We were aware of this problem, and although we do not formally sterilize our glassware, it is dried at 200°C and stored in a clean and dust-free cupboard.
Concerning your second point, we have found that the estimation of ADP is quite reliable, but the myokinase reaction for the conversion of AMP tends to vary from assay to assay.

Chapter 8

Semi-automatic Microtransferator and Cell for the Bioluminescence Assay of ATP and Reduced NAD with Scintillation Counters

E. Schram and H. Roosens

Vrije Universiteit Brussel, Belgium

INTRODUCTION

In previous articles[1,2] we have described the adaptation of the firefly–luciferase reaction to the bioluminescence assay of ATP with scintillation counters. It was concluded from our experiments that one is not bound to the measurement of the peak intensity of the luminescence curve (Fig. 1(a)), but that, under the proper conditions, the luminescence decreases exponentially over a rather long period of time while remaining proportional to the ATP concentration. This enables one to use the *integral mode* of counting (Fig. 1(b)), which is the only one compatible with the use of scintillation counters, and at the same time makes *flow-monitoring* possible. A lin-log plot of the luminescence versus time gives straight lines which run perfectly parallel over a wide concentration range and whose slopes are proportional to the concentration of the enzyme preparation. A similar technique was used for the bioluminescence assay of reduced NAD(P) and FMN by means of bacterial luciferase.

The main advantages of scintillation counters when used for the present purpose are their high sensitivity, the large concentration range which can be assayed, the digital display and/or output of the results, and their suitability for kinetic experiments or flow-monitoring (when used in the 'repeat' mode).

In view of the above experiments a special cell has been developed, suitable for flow as well as discrete measurements. A fixed cell, remaining in the counter, was preferred to vials in order, chiefly, to avoid phosphorescence phenomena. At the same time a device was built for the transfer of small volumes of liquid.

CELL

The cell represented in Fig. 2 has been built to fit in the sample drawer of an Intertechnique SL 20 counter. This kind of counter offers the advantage that the counting chamber is easily accessible and that the connecting lines can be kept short. The body of the cell is made of quartz tubing in order to avoid long-lived phosphorescence after exposure to light (on installation or for maintenance). It occupies an eccentric position, closer to the photomultiplier that is used for the measurements; the dimensions of the holder, however, are similar to those of a regular counting vial so that it can be inserted without

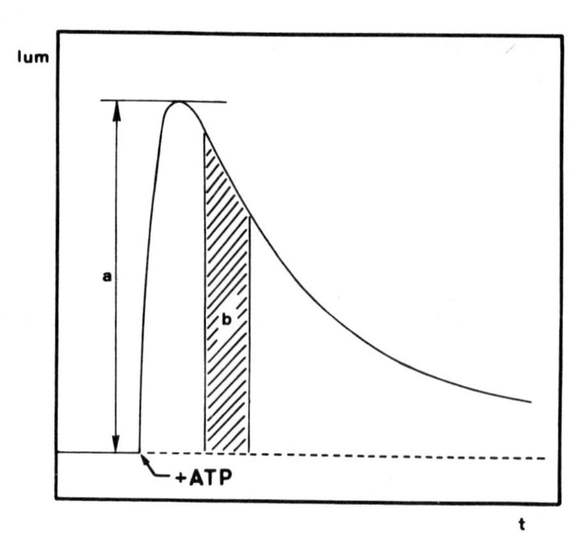

Fig. 1. Luminescence curve. a = peak intensity; b = integration of light intensity.

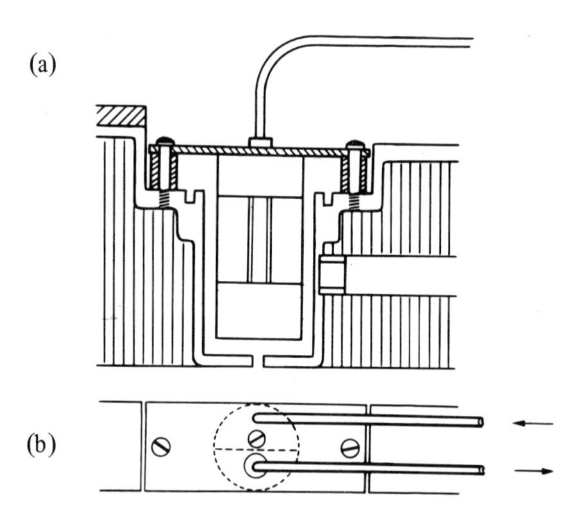

Fig. 2. Cell for discrete or flow measurement of bioluminescence; a) side view, b) top view.

any modification of the counter. The tendency has been towards a reduction of the cell volume, which may be as small as 50 μl, thus reducing the enzyme consumption. On filling, the liquid level rises somewhat above the transparent segment, avoiding the need for a quantitative transfer. The effective volume and the geometry remaining constant, the reproducibility of the results is much increased above that attained with individual vials.

TRANSFER DEVICE

Many techniques have been used for the transfer of samples, and are becoming increasingly popular with the present expansion of automatic sample analysers. The sample solution is usually circulated by means of automatic pipettes (syringes), peristaltic pumps, compressed air or vacuum, or a combination of these systems. The flow of liquid is either unidirectional or bidirectional, intermittent or continuous. Most of these devices, unfortunately, are not adequate for the transfer of very small volumes and often require sample solution for rinsing. Our own efforts have tended towards the development of a reliable system which would not have these disadvantages.

Transfer of the sample solution into and out of the cell is achieved respectively by suction and pressure by means of a motor-driven microsyringe. When the cell is being emptied an excess of air helps to expell the last droplets of liquid from the outlet tubing, leaving less than a few tenths of a per cent of the sample solution in the cell and connecting lines. Hence no rinsing is necessary when samples of similar concentration are counted in succession. One rinse with distilled water by means of the transfer syringe is fully satisfactory in other cases. No trouble or contamination was encountered with the present device after several months of continuous operation.

Table 1. Reproducibility of results (measurements obtained on loading and unloading the same sample several times in succession).

Time (min)	Observed counts (10^3)	Theoretical[a] counts (10^3)	Deviation (10^3)
4	260.1	259.8	+ 0.3
6	246.0	245.9	+ 0.1
8	234.6	233.7	+ 0.9
10	222.0	221.7	+ 0.3
12	209.1	210.2	− 1.1
14	199.9	199.4	+ 0.5
16	188.6	189.1	− 0.5
18	179.5	179.4	+ 0.1

[a] As deduced from regression line.
Mean number of counts: 217.4×10^3.
Standard deviation: theoretical ($\sqrt{\text{mean}}$): 467 or 0.22%
observed: 584 or 0.27%

RESULTS

Reproducibility

The reproducibility of the measurements was checked by loading and unloading the same sample several times in succession and observing the deviation with respect to the luminescence decay curve (calculated according to the method of least squares). These are shown in Table 1.

In other tests the luminescence decrease was calculated for different samples when recounted after the same time interval. The ratios found did not differ by more than 1% from the mean.

High voltage and discriminator settings

A thorough study of the optimal counting conditions was performed, using the

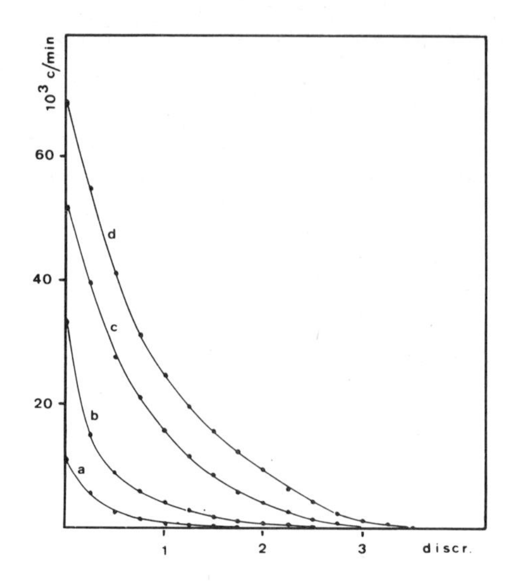

Fig. 3. Energy distribution of background (SL 20 Intertechnique counter). a = 1000 V; b = 1100 V; c = 1200 V; d = 1300 V.

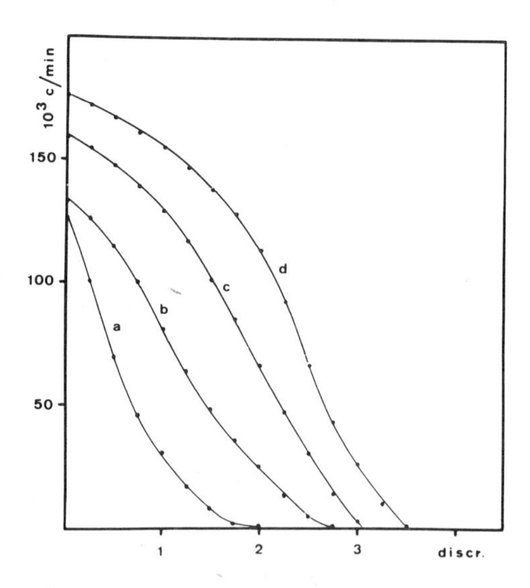

Fig. 4. Energy distribution of bioluminescence (SL 20 Intertechnique counter). a = 1000 V; b = 1100 V; c = 1200; d = 1300 V.

SL 20 Intertechnique scintillation counter. Overlapping of the photomultiplier noise and luminescence spectra is an important factor in this respect (measurements are performed with the coincidence circuit disabled). The statistical error was therefore determined for several luminescence to noise ratios as a function of the lower discriminator and high voltage settings. Increasing the high voltage above 1000 V did not improve the counting statistics. Moreover memory effects occurring at higher count rates could be eliminated at this voltage, and the dynamic range of the assay thus increased (see Figs. 3, 4 and 5).

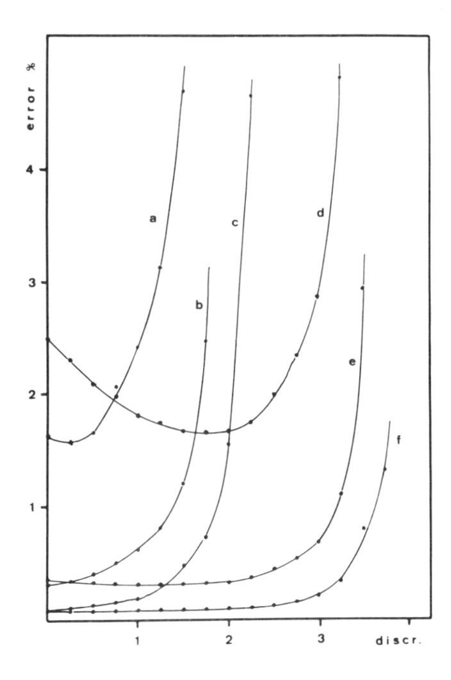

Fig. 5. Standard deviation at different ATP concentration levels and HV settings. a, b, c = 1000 V; d, e, f = 1300 V. a, d concentration so as to give a luminescence equal to the background at 1000 V; b, e = 10 × a; c, f = 100 × a.

Sensitivity

The sensitivity of bioluminescence as an analytical tool depends not only on the proper use of the counting equipment, but also on the reagents. For our determinations endogenous ATP was separated from commercial firefly luciferase by gel filtration on BioGel P-10, with an appreciable reduction of background. Luciferin is eliminated as well (it has a similar molecular weight) but is readily replaced by synthetic luciferin at even higher concentration in order to reach maximum efficiency. The eluate from the filtration column does not need to be concentrated since diluted enzyme solutions correspond to a more adequate luminescence half-life.

When taking all necessary precautions, amounts of ATP of the order of 10^{-15} mole can be measured. In many cases, however, such efforts will prove unnecessary since with regular commercial enzyme preparations, 10^{-13} mole of ATP can be assayed with a statistical accuracy of better than 2%.

APPLICATIONS

The present method for ATP determination has been used for the study of amino acid activation in *Bacillus stearothermophilus,* the periodicity of photophosphorylation in *Acetabularia mediterranea,* the development of amphibian eggs (*Xenopus laevis*); simultaneous determination of AMP and ADP after their enzymatic conversion to ATP has been achieved in blood and brain.

REFERENCES

1 E. Schram, in *Liquid Scintillation Counting* (Ed. E. D. Bransome), Grune and Stratton, New York, 1970, p. 129.

2 E. Schram *et al.*, in *Organic Scintillators and Liquid Scintillation Counting* (Ed. D. L. Horrocks and C. T. Peng), Academic Press, New York, 1971, p. 125.

DISCUSSION

F. E. L. ten Haaf: At what temperature are the measurements performed?

E. Schram: Luciferase shows an optimum activity around 23°C. All measurements are therefore performed at room temperature. Care should be taken to equilibrate all solutions at this temperature before mixing.

F. E. L. ten Haaf: When working with the very low pulse amplitudes resulting from the single photons from the sample, one is liable to have noise from the photomultiplier dynodes in addition to the thermal noise from the photocathode. Is this background sufficiently constant to enable correction for it?

E. Schram: No trouble was encountered as far as the stability of the background is concerned. As the photomultiplier noise is strongly temperature-dependent one should, of course, let the counter reach its equilibrium temperature after switching on. Dynode pulses are partly eliminated by discrimination (see text for background and luminescence spectra).

J. P. Verhassel: For these measurements do you use the output of just one photomultiplier or do you combine the output from both tubes?

E. Schram: Both methods are suitable but all counters cannot be used in both ways. With the Intertechnique SL 20 counter which we are using, only the tube with the best performance characteristics is used. In this case the counting cell is put closer to that tube in order to improve geometry.

B. Scales: You suggest that cyclic AMP can be measured making use of your system, since the sensitivity of the method is very high. Can you please briefly outline the enzyme system which you prefer for the conversion of cyclic AMP to adequate amounts of ATP?

E. Schram: Cyclic AMP has to be converted to 5′-AMP which can be assayed quantitatively with luciferase. However, the commercial enzyme used for this conversion still contains phosphatase activity and should therefore be further purified. Several authors have recently used their own enzyme preparations.

E. D. Saggerson: Is the firefly–luciferase reaction special for adenine nucleotides or will others, such as GTP, interfere?

E. Schram: The reaction is specific for ATP; although contamination of the enzyme with nucleoside diphosphokinase will lead to interference from other nucleoside triphosphates, these are generally present in small amounts compared with adenine nucleotides. Moreover, as the interconversion is not instantaneous the luminescence yield is only a small fraction of that obtained for an equivalent amount of ATP.

SECTION III

METHODS OF SAMPLE PREPARATION OF INORGANIC MATERIALS INCLUDING CERENKOV COUNTING

Chapter 9

Methods of Sample Preparation of Inorganic Materials including Cerenkov Counting

A. Dyer

Department of Chemistry, University of Salford, Salford, England

INTRODUCTION

This review will concern itself with the radiochemical determination of inorganic isotopes by liquid scintillation counting. It will include cases where the isotopes are recovered as inorganic materials from biological materials. The recovery of calcium-45 as calcium chloride, from milk, and that of iron isotopes, as ferric chloride, from blood are two examples of this general technique.

Some fifty isotopes have been estimated by liquid scintillation counting or closely related procedures. The use is not confined solely to methods whereby a scintillator solute-solvent system is activated by a β^--particle. The references cited herein include the determination of γ, β^+, α and electron capture isotopes as well as cases where a liquid scintillation spectrometer is used to record photons from Cerenkov radiation and plastic scintillators.

The advantages of using liquid scintillation methods over the more traditional methods of counting inorganic isotopes can be summarized as follows:

1. speed and reproducibility of sample preparation;
2. the counting geometry tends to maximize counting efficiency when dealing with (i) low activity samples, (ii) low energy emitters;
3. convenient method for small volumes of solution;
4. automatic nature of the modern equipment gives a gain in convenience and ease of data processing when large numbers of samples are to be monitored;
5. possibility of adaptation for flow cell use;
6. suitable for fundamental studies of nuclear processes and absolute counting.

FUNDAMENTAL STUDIES

As most of the earlier literature involves the use of liquid scintillation procedures for fundamental investigations a brief survey of this will be considered first.

Many workers have found it a useful method for absolute counting and related studies. Particularly well investigated are α-emitting isotopes as studied by Horrocks and his co-workers.[56] They have examined astatine-217, rhodium-220, thorium-232, uranium-233, uranium-236, plutonium-236 plutonium-238, plutonium-239 and curium-242. Similar studies have been made by Basson[4] on astatine-211, and by Joon and Deurloo,[73,74] Flynn *et al.*[34] on plutonium-239. Stoicovici and Uray[128] also studied uranium isotopes.

Horrocks[55,57] has also paid attention to the β^--emitters, etc.; sulphur-35, iron-55, nickel-63, yttrium-90, ruthenium-106, cadmium-109/silver-109, tin-113/indium-113m, caesium-137/barium-137m, samarium-151, plutonium-241, californium-252. Analogous work on sodium-22, sodium-24, phosphorus-32, cobalt-60, iodine-131 and gold-198 has been done by Steyn,[126,127] and on zirconium-95/niobium-95 by Ludwick.[87] Josefowicz[75] has also examined phosphorus-32.

Ryves[113] has used samples deposited upon polymer film, which are then immersed in a PPO/POPOP in toluene scintillator, to obtain absolute counting efficiencies close to 100% for the isotopes sodium-24, phosphorus-32, cobalt-60, thallium-204.

Joon and Deurloo[74] commented on the stability of carrier-free solutions and recommended the use of tri-n-octyl phosphine oxide as a stabilizer when counting zirconium-95, ruthenium-106, cerium-144 and plutonium-239. Erdtmann[29] showed how similar problems may be overcome by additions of carrier or powdered silica.

Half-life studies of isotopes held in liquid scintillator solution are known for the isotopes nickel-63,[58] rubidium-87[35] and samarium-147.[142] Horrocks[60,61] has examined pulse height energy relationships for electrons and photoelectrons in liquid scintillation media for strontium-85, cadmium-109/silver-109, tin-113/indium-113m, indium-114, barium-133, caesium-137/barium-137m, mercury-203, astatine-217/thorium-229, thorium-232, uranium-233, americium-241 and curium-242. Similar studies by Flynn[34] (plutonium-239) and by Seliger[121] (phosphorus-32, caesium-137/barium-137m, polonium-210) may also be noted.

All of these investigations necessitate the incorporation of the inorganic isotope into mutual solution with a scintillation medium and often form the basis for routine sample preparation. This aspect of these fundamental studies is covered in the subsequent text.

PRACTICAL ASPECTS

Isotopes can be incorporated into phosphors as solids, liquids or gases.

Solid samples

Generally the solid is dispersed, usually as a fine powder, in the scintillator medium. The methods of dispersion can be summarized.

Use of a gelling agent. This has been used by Yarbrough *et al.*[144] to estimate calcium-45 and strontium-89 in unicellular algae. The algae are dehydrated, decolourized and then dispersed, in a PPO/DMPOPOP in toluene scintillator, by Cab-O-Sil. Turpin and Bethune[136] used the same suspension agent for calcium-45 containing solids obtained from urine and plasma, whereas Ott *et al.*[100] evolved a similar method for strontium-89 in bone ash. (Note also Scales and McIntosh).[117]

The isotopes iron-55 and iron-59 can be counted in a like manner. Eakins and Brown[26] suggest a suspension of a ferriphosphate complex, whereas Graber *et al.*[44] use a ferric benzene phosphinate. (See also Katz *et al.*)[76]

An earlier gelling agent was Thixcin (a castor-oil derivative) which White and Helf[51,141] used with a PPO in toluene scintillator for sodium-22, chlorine-36, nickel-63, strontium-90/yttrium-90, barium-133 and lead-210 as simple salts.

Fleishman *et al.*[33,122] report the use of a methyl methacrylate gel, formed by warming together the labelled solid, polymer and a *p*-terphenyl/POPOP in toluene phosphor. They claim 100% efficiency for potassium-40, strontium-90 and caesium-137. Recently Benakis[6] proposed a similar gel of polyolefinic character, used with both PPO/POPOP and butyl PBD/POPOP, to estimate inorganic samples. Materials estimated were calcium-45, strontium-90 as sulphates, iron-55 and iron-59 as ferriphosphate or ferric benzene phosphinate complexes, caesium-137 as perchlorate and iodine-131 as sodium iodide. (N.B. Ihle[68] has used a PVC-acetate copolymer for α-emitter estimations).

Use of glass fibre filter paper. Johnson and Smith[71] deposited barium sulphate containing sulphur-35 on glass fibre filter paper, homogenized it in a naphthalene/PPO/POPOP in dioxane scintillator, and claimed a convenient and efficient method. Lerch and Cosandey[83] counted phosphorus-32 and calcium-45 as calcium phosphate in a similar manner. Hutchinson[67] preferred calcium chloride deposition (for calcium-45) when dealing with urine, plasma, stool digests, and tissue extracts. He used a PPO/POPOP in toluene scintillator. Cramer and Ross[18] ashed biological tissue and bone containing calcium-45 and compared cellulose to glass fibre papers (as well as other methods). They found cellulose paper superior as it is not prone to shed lint, although glass fibre paper was the most efficient. Fairman and Sedlet[31] have used an anion exchange paper dispersion to determine technetium-99.

Simple mixing. Dyer *et al.*[24] have counted iron ore, containing sodium-22, by placing it in a scintillator and have found this simple method as successful as chemical extraction of the sodium-22 as sodium chloride followed by incorporation into a Hayes and Gould solution, or, directly G-M counting the ore. Nisson and Benson[98] used this simple approach, with a Bray scintillator, for plant roots containing potassium-42.

Liquid samples

Use of a labelled compound soluble in a scintillator solvent. The ideal counting conditions would be to produce an inorganic material of sufficient solubility in a scintillator solvent. The nearest approach to this is the use of PBD in toluene to take up chlorine-36 in the form of silicon tetrachloride. This was the method of Ronzani and Tamers[111] as a dating technique for sodium chloride in ground waters.

Use of aqueous solutions of inorganic salts. This again is a simple method, a blending solvent being used to form a homogeneous sample in a 'cocktail'. There are several techniques described for the counting of calcium-45 in this way. Samples of biological origin (serum, urine, bone, blood and milk) are variously treated to yield a precipitate of calcium oxalate which is then dissolved in aqueous mineral acid and blended with ethanol in a toluene-based phosphor. Methods of this type are due to Carr and Parsons,[14] Sarnat and Jeffay,[115] Vemmer and Guette,[139] and Kumar.[80]

Similar methods are available for iron-55 and iron-59 from Dern and Hart,[21] Jenner and Obrink[70] and Perry and Warner.[105] The iron is as chloride or perchlorate and ascorbic acid may be added as a stabilizer.

Dyer *et al.*[23] have given counting conditions for sodium-22, silver-110m, promethium-147 and thallium-204 whereby aqueous salt solutions are taken into a Hayes and Gould type cocktail. (Note also Klumpar and Majerova[78] for calcium-45, strontium-90/yttrium-90 and thallium-204).

Plutonium isotopes, from biological sources, have been estimated in this way by Toribara *et al.*[133,134] as chlorides and by Lindenbaum and Lund[86] as nitrates.

Table 1. Use of complexing agents.

Isotopes	Cocktail	Experimental details	Comments	Reference No.
Iron-59	PPO/POPOP (Hayes and Gould)	Iron–*ortho*-phenanthroline complex, 50% efficiency	Iron metabolism studies	82
Iron-55	DMPOPOP/PPO/ naphthalene/toluene	Iron–di(2-ethyl hexyl phosphoric acid) (HDEHP) complex	Blood	17
Nickel-63	NE 240	$[NiPy_4]$ $(CNS)_2$ complex	Effluent from AGR reactors	48
Plutonium-249, Americium-241	PPO/POPOP/ naphthalene/dioxane	Tri-n-octyl phosphine oxide (TOPO) complex	Comparison with other methods	68
Plutonium-249	*p*-terphenyl/POPOP/ toluene	Acid extraction from ashed samples by HDEHP	Biological samples	77
Plutonium-241, nickel-63, samarium-151, sulphur-35, ruthenium-106/rhodium-106	*p*-terphenyl/POPOP/ xylene	Sm, Pu as HDEHP in HCl, nickel-63 as di-octyl phosphate complex, sulphur as Na_2SO_4, Ru/Rh *p*-toluidine complex	Absolute disintegration studies	55
Plutonium-239, curium-242, thorium-232, uranium-233, astatine-217, californium-252, yttrium-90	DMPOPOP/PPO/ toluene	HDEHP extraction in xylene. Quenching studied	α-resolution studies, etc.	56, 57
Plutonium-239, ruthenium-106, cerium-144, zirconium-95, uranium	DMPOPOP/PPO/ naphthalene/dioxane	TOPO - carrier free, comments on stability		74
Cerium-144, cobalt-60, sodium-22	POPOP/PPO/naphthalene/ dioxane/xylol/ethanol	HDEHP complexes	Counting loss studies	29
Promethium-147	DMPOPOP/PPO	HDEHP complex	Urine assay	88
Promethium-147	POPOP/PPO/naphthalene/ dioxane	HDEHP in 0.1 M HNO_3	Fission products	9

Table 1. Use of complexing agents (contd.).

Americium-241, curium-244, europium-152	PPO/POPOP	HDEHP complex from aqueous oxalate and sulphate solution at pH 3.6 and constant ionic strength	Ion-exchange effluent	3
Thorium-232, nickel-63, cadmium-109, barium-133, americium-241, samarium-113, indium-113m, mercury-203	*p*-terphenyl/POPOP/xylene DMPOPOP/PPO/toluene	HDEHP complexes	Half-life and other fundamental studies	56,61
Thorium-232	*p*-terphenyl/POPOP/toluene	DOP complex		41

Table 2. Use of metal organic salts.

Isotopes	Cocktail	Experimental details	Comments	Reference No.
Caesium-137/barium-137m	DMPOPOP/PPO/toluene	Salt of 2-ethyl-hexanoic acid (octoate)		60
Rubidium-87	*p*-terphenyl/POPOP/toluene	Octoate	Half-life studies	35
Calcium-45	PPO/POPOP/toluene + ethanol	Octoate from oxalate in HCl	Biological fluids	89
Strontium-90/yttrium-90	PPO/POPOP/toluene	Octoate	Assays	138
Mercury, cadmium, calcium, potassium, rubidium		Metal loaded scintillators using octoates		20, 112
Samarium-147	*p*-terphenyl/POPOP/toluene	Octoate	Half-life studies	142
Plutonium-239		Octoate		34
Nickel-63	PPO/POPOP/toluene + ethanol	n-caproate		40

Table 3. Complexing or extraction with emulsifier added.

Isotopes	Cocktail	Experimental details	Comments	Reference No.
Phosphorus-32	PPO/POPOP/toluene	Phosphomolybdic acid, extracted n-butanol/ethyl ether/ethanol. Add Hyamine 10-X	Food analysis	27
Calcium-45		Add Triton X-100	Urine and serum	96
Calcium-45, iron-55, rubidium-87, samarium-147, strontium-90, strontium-89, chlorine-36	PPO/bis MSB/ xylene	Ca, Fe as fluoride, Rb, Cl, Sr, Y as chloride, Pm co-precipitated as oxalate with praesodymium, then as EDTA complex. All with Triton N-101 added.	Low-level counting	92
Iron-55, Iron-59	PPO/POPOP/toluene	Ferrous citrate + Hyamine 10-X	Plasma samples	91
Chromium-51, manganese-54, iron-55, zinc-65, yttrium-88, iodine-125	PBD or PPO/POPOP/ toluene/ethanol	Cr as chloride or chromate, Mn as nitrate, Fe, Zn, Y as chloride, I as potassium iodide, Triton X-100 added	Electron capture nuclides	130
Cobalt-60	BBOT/toluene	Co^{II} thiocyanato complex with Triton X-100	Low-level counting of environmental waters	15
Chromium-51		Cr as EDTA complex. Triton X-100 added	Simultaneous measurement with carbon-14.	123

Table 4. Solvent extraction procedures.

Isotopes	Cocktail	Experimental details	Comments	Reference No.
Calcium-45	PPO/toluene	Calcium perchlorate extracted by tributyl phosphate (TBP)	Biological sources	66
Calcium-45	PPO/α-NPO/toluene *p*-terphenyl/α-NPO/	Calcium chloride extracted by dibutyl phosphate (DBP). Quenching studies	Agricultural materials, plants and soils, etc.	47
Strontium-90/yttrium-90		DBP extraction		36
Plutonium-241	*p*-terphenyl/POPOP/ xylene	DBP extraction from 1 M HCl		54
Yttrium-90, plutonium		TBP extraction	Agricultural materials	99
Americium-241, uranium-233	DMPOPOP/PPO/ naphthalene/dioxane or BBOT/toluene	Tri-isoctylamine + 30% DMF extraction		120
Thorium, uranium	*p*-terphenyl/α-NPO/ toluene	TBP extraction		2, 30

Low levels of uranium have been counted by Levin[84] in aqueous solutions blended in a dimethoxyethane/naphthalene/PPO/POPOP cocktail. Other authors mention similar determinations of manganese-56,[22] bismuth-210[32] and polonium-210.[32]

Absolute standardization studies, as mentioned earlier, often use related procedures. Seliger[121] incorporated cobalt-60, caesium-137/barium-137m, gold-198 and polonium-210 as chlorides and Steyn[126,127] mentioned the use of sodium-22, sodium-24 as acetate, phosphorus-32 as phosphoric acid and iodine-131 as sodium iodide. Ludwick[87] used a Davidson and Feigelson[20] cocktail for niobium-95/zirconium-95 estimation.

Use of complexing or solubilizing agents. These methods depend upon the use of an organic material to produce a complex, or ion-association complex, of a metal isotope which then has an appropriate scintillator solvent solubility. Obviously a colourless end-product is advisable, and hence phosphate complexing agents have been used widely. Nevertheless, at least two viable methods using coloured complexes are reported. Harvey and Sutton[48] used the pale blue tetrapyridino nickel-63^{II} thiocyanate and Leffingwell *et al.*,[82] the red iron-59 *ortho*-phenanthroline complex. A summary is given in Table 1.

Use of an organic salt of the metal isotope. Here a metal salt of an organic acid (e.g. an octoate) often has sufficient solubility in the alkylbenzenes to facilitate liquid scintillation counting. (See Table 2 for details).

Some authors have found it an advantage to add a further emulsifying and solubilizing agent to the counting mixtures or that a subsequent solvent extraction step is advisable. These instances are given in Tables 3 and 4.

Sulphur-35 determinations. Radin and Fried[108] converted ^{35}S-containing materials to sulphuric acid prior to counting as suspensions in Primene 81-R in a terphenyl/POPOP in toluene scintillator, whereas Lloyd and Rees-Evans[85] suggested conversion to lithium sulphate. Jeffay *et al.*[69] used a modified Piries reagent to convert biological material to magnesium sulphate; this is then extracted into a glycerol/ethanol/DMF mixture which is miscible with PPO/POPOP in toluene.

Gaseous samples

The application of the liquid scintillation technique for the estimation of radioactive gases depends upon two techniques:

1. 'trapping' the gas, i.e. stoichiometric conversion into a liquid or solid compatible with a scintillation mixture;
2. dependence upon the gas having a reproducible solubility in a scintillator solvent.

Trapping techniques have been used by Urone *et al.*,[137] whereby $^{35}SO_2$ was taken into $Na_2Hg_2Cl_4$ or H_2O_2/H_2SO_4 solutions and counted in a DMPOPOP/PPO/naphthalene in dioxane mixture. (N.B. Rapkin[22] has suggested that the use of ethanolamine as a trapping agent may be better). Mahadevappa and Eager[90] converted $H_2{}^{35}S$ to $Na_2S_2O_3$ and $Na_2S_4O_6$ which was then trapped by hyamine hydroxide in a Hayes and Gould scintillator. Gordon *et al.*[43] used a similar method but concluded that it was better to rely on the dissolution of the gas in toluene contained in a serum-capped vial.

As the rare gases have appreciable solubilities in toluene, Horrocks and Studier[59] have been able to count krypton-85, xenon-131m and radon-222 in this way. (See also Curtis *et al.*,[19] Schwendiman and Mishima,[119] Moghissi,[92] as well as Noguchi).[97] Note: These methods use a pre-evacuated counting vial, or ampoule, into which the gas and scintillant are injected (or pre-frozen). PPO and POPOP are the preferred phosphors.

Other methods

Use of a scintillating ion-exchange resin. Heimbuch[50] has developed a resin bead copolymerized with a phosphor (9,10-diphenylanthracene or *p*-terphenyl). This can be used for both anion and cation exchange, even in concentrated acid and alkali conditions. Such isotopes as sulphur-35, chlorine-36, nickel-63, strontium-90/yttrium-90, iodine-131, neptunium-237 and plutonium-239 were counted in this way.

Use of a suspended scintillator. Anthracene, of the blue-violet fluorescence grade, may be wetted by a Triton type detergent and suspended in an aqueous labelled sample. The method was developed by Steinberg,[124,125] who used it for phosphorus-32, calcium-45 and iodine-131. Myers and Bush[95] used the technique for beryllium-7, $H_3{}^{32}PO_4$, $H_2{}^{35}SO_4$, strontium-90/yttrium-90, caesium-137 and polonium-210. They examined the effect on counting efficiency due to the presence of methyl and ethyl alcohols, acetic acid, and acetone.

Use of scintillator granules or shavings. Tsirlin *et al.*[135] counted aqueous salt solutions (sodium-22, potassium-40) by placing granules of anthracene in them. Sax *et al.*[116] determined atmospheric krypton-85 in a gas-tight vial full of plastic scintillator shavings.

Flow cells. Continuous monitoring can be achieved by passing solutions of isotopes through either (a) a tube constructed of plastic scintillator, or (b) a vial containing rods or granules of scintillator. Schram[118] recently has commented on this method, which has been used by Tkachuk[132] (potassium-40, strontium-90/yttrium-90), Sanson and Taylor[114] (phosphorus-32, calcium-45) and Pickering *et al.*[107] (calcium-45 in bone dissolved in 1 M nitric acid).

PROBLEMS AND GENERAL COMMENTS ON THE COUNTING OF INORGANIC MATERIALS

Solutes. It can be seen from previous details that PPO and POPOP are often compatible with inorganic sample determinations, but that *p*-terphenyl can often be a cheaper alternative. This should be remembered in designing scintillator systems for inorganic use provided that determinations are not to be carried out in a machine whose sample chamber is maintained at a low temperature which will cause *p*-terphenyl to come out of solution.

Solvents. Toluene, xylene and dioxane seem adequate in most cases. Recently, Gomez *et al.*[42] have reported the use of benzo- and acetonitriles as scintillator solvents. They found them particularly useful for the dissolution of ferric, lithium, mercuric, antimony, and copper chlorides, and zinc acetate.

Vials. The use of siliconized vials may be useful to prevent adsorption of metallic isotopes onto the glass, as suggested by Petroff *et al.*[106]

Temperature control. Horrocks[63] recently advised that a reasonable control of temperature was required in the counting of uranium-233 as DBP complex. This may well be a universally applicable comment.

Quenching and its correction. Obviously colour quenching can be a problem and chemical quenching can arise in the presence of acids, alkali, inorganic salts or complexing and blending agents. Forster[37] made a detailed study of the quenching arising in the counting of β^--emissions from promethium-147. Blair and Murrenhoff[7] also considered this specific problem, and Peng[104] has made some more general comments. Many of the authors quoted herein consider quenching problems which, of course, are more noticeable with the low β^--energy of calcium-45 and the electron capture nuclides (e.g. iron-55). Generally it has

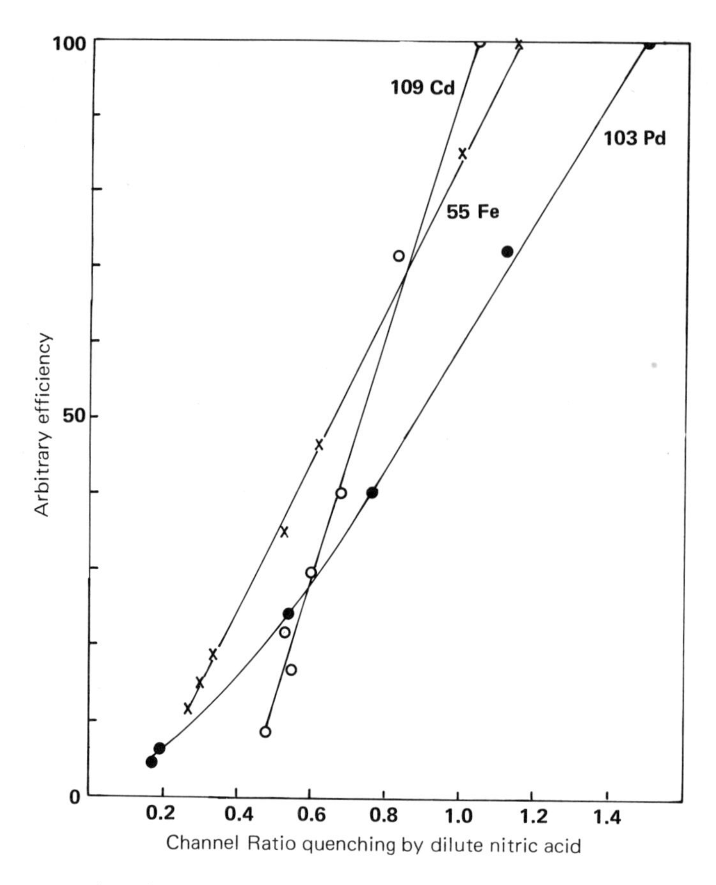

Fig. 1. Some examples of quench correction.

been concluded that the standard methods of quench correction, such as that of channels ratio, are applicable. Some examples[25] are given in Fig. 1.

Simultaneous counting of two or more isotopes. Yerick and Ross[145] as well as Rhodes[109] examined the simultaneous counting of the isotopes iodine-125, iodine-128, iodine-131. Zaduban[146] used an amplitude attenuation method to deal with iodine-131 in the presence of tritium and carbon-14. Muller,[94] in his studies of bone metabolism, designed a triple-tracer method for phosphorus-32, calcium-45, strontium-89. Double isotope studies have been carried out by Ludwick[88] (zirconium-95, niobium-95), by Hoyer[64] (iron-55, iron-59), and Katz *et al.*[76] (iron-55, iron-59).

Kobayashi and Maudsley[79] recommend the use of an Enberg plot to determine counting conditions for simultaneous estimations of two isotopes.

The iron isotopes. The counting of iron isotopes has received much attention in the recent literature and the general methods are listed in Table 1 or in the text. Practitioners should be aware of the comments on the reproducibility of iron-55 measurements in the presence of iron-59, which have been made by Hoyer.[64] Horrocks[62] has suggested the possibility of maximizing iron-55 counting efficiency to 90% by including the K, L, M X-rays and Auger electrons, in addition to the 5.9 keV K-X manganese electrons normally counted.

Chemiluminescence. This can arise and we have noted its occurrence in counting zinc-65 and barium-133 in a Hayes and Gould system.

Comparative studies. Some authors have made detailed investigations of the liquid scintillation technique in comparison with other counting techniques for the counting of inorganic isotopes. These have been generally favourable and are summarized in Table 5.

CERENKOV COUNTING MAKING USE OF LIQUID SCINTILLATION EQUIPMENT

Introduction. In 1953 Belcher[5] reported the detection of a faint luminescence from aqueous solutions of high energy β^--emitters. He concluded that this was mainly due to Cerenkov radiation which arises as a blue-white light emitted by the interaction of β^--particles of energy >300 keV with liquids such as water. Cerenkov radiation is thought to be the source of the sudden light flashes we all have experienced in our eyes. The light is in the near ultraviolet and is detectable by modern automatic liquid scintillation spectrometers designed to cope with very low intensity light pulses. The problems of detecting these weak emissions have been likened by Parker and Elrick[101-103] to those encountered in tritium counting and they have used tritium instrument settings for the determination of phosphorus-32, iodine-131 and other isotopes. (See also Hoffman).[53]

Problems. The Cerenkov light is highly directional and of heterogeneous wavelength. These problems have been approached by Haberer and Kölle[45,46] who suggested the use of the water-soluble sodium salt of 1-naphthylamine 4-sulphonic acid as a wavelength 'shifter'. Its function is to absorb some of the directional ultraviolet radiation and re-emit it isotropically in the visible region. Läuchli[81] reported the similar use of 7-amino 1,3-naphthalene disulphonic acid. Also, Elrick and Parker[28] have studied in detail the use of the double sodium, potassium salt of 2-naphthylamine disulphonic acid. Other improvements in detection efficiency can be achieved by the use of plastic vials.[28,45,65,81]

Problems arising due to variation in refractive index[46,28] and density[28] have been considered and Clausen[16] reported on the effects of the presence of organic solvents, inorganic salts, acids, and alkalis, on the estimation of phosphorus-32. Similar investigations were made by Haviland and Bieber[49] (organic solvents) and by Johnson[72] (isopropanol, ethanol, chloroform, and formic acid).

The method has proved to be almost entirely free from chemical quenching, but colour quenching can arise, although De Volpi *et al.*[22] reported the counting of a pink manganese-56 solution without difficulty. The correction of colour quenching in Cerenkov counting lends itself to treatment by standard means. Francois[38] used internal standardization to measure potassium-42 in urine, and colour quench correction[129] curves have been used by various investigators[140] for phosphorus-32 and by Moir[93] for potassium-42 from biological materials. External standardization can be used as by Brownell and Läuchli[10] for chlorine-36 and rubidium-86 in plant materials. Elrick and Parker[28] recommended radium-226 as the external γ-source. They critically reviewed the quench correction methods available and offered an alternative spectrophotometric method for samples possessing similar optical absorption spectra. These authors have provided excellent reviews on the theory and use of Cerenkov counting.[102,103]

The method has been used, with success, to deal with large volumes of solution (Francois[38] for potassium-42 and Yamada[143] for strontium-90/yttrium-90) but Rippon[110] considers it a not very suitable method for fission products in effluents. Other uses include

Table 5. Comparative studies.

Authors	Reference No.	Isotopes considered	Techniques compared
Aubouin	1	Phosphorus-32, sulphur-35, potassium-42, calcium-45, iron-59, strontium-90/yttrium-90, thallium-204	G.M. Plastic scintillator. Labyrinth plastic scintillator
Burt and Gibson	11	Chlorine-36, strontium-90/yttrium-90. caesium-137	Plastic beads, plastic flow cell, anthracene filled cell
Carr and Nolan	12	Calcium-45, strontium-85	G.M. Gas flow counters, scintillator beads, scintiglas, gel, glass fibre filter paper
Carr and Parsons	13	Calcium-45	Plastic scintillators. General review of liquid scintillation counting methods for calcium-45

the determination, by Braunsberg and Guyver,[8] of phosphorus-32 and sodium-24 in tissue slices, of phosphorus-32 in chromatographic media by Haviland and Bieber,[49] and in tissue extracts by Threlfall *et al.*[131] (See also Garrahan and Glynn).[39] Hoch *et al.*[52] offered a method for iodine analysis in biological samples by neutron activation analysis followed by scintillation counting of Cerenkov radiation.

In conclusion, it can be said that Cerenkov counting offers a useful and convenient adjunct to the method of liquid scintillation and one might expect its use to grow considerably for β^--emitting isotopes whose E_{max} is > 0.5 MeV, i.e. high enough to produce detectable Cerenkov radiation.

REFERENCES

1 G. Aubouin, NP-16483, CEA Grenoble, January 1965, DT1E.
2 R. C. Axtmann and L. Cathey, *Intern. J. Appl. Radiation Isotopes* **4**, 261 (1959).
3 A. Aziz, S. J. Lyle and S. J. Naqvi, *J. Inorg. Nucl. Chem.* **30**, 1013 (1968).
4 J. K. Basson, *Anal. Chem.* **28**, 1472 (1956).
5 E. H. Belcher, *Proc. Roy. Soc. (London)* **A216**, 90 (1953).
6 A. Benakis, in *Liquid Scintillation Counting, Volume 1,* (Ed. A. Dyer), Heyden & Son Ltd., London, 1971, p. 97.
7 A. F. Blair and A. P. Murrenhoff, *Ber. Kernforschanlage Juelich* 377 (1966).
8 H. Braunsberg and A. Guyver, *Anal. Biochem.* **10**, 86 (1965).
9 R. D. Britt, *Anal. Chem.* **33**, 602 (1961).
10 J. R. Brownell and A. Läuchli, *Intern. J. Appl. Radiation Isotopes* **20**, 797 (1969).
11 A. K. Burt and J. A. B. Gibson, AERE, R-4638 (1964).
12 T. E. F. Carr and J. Nolan, *Proc. Summer School,* Beckman (1967).
13 T. E. F. Carr and B. J. Parsons, *Radioisotope Sample Meas. Tech. Med. Biol. Proc. Symp. Vienna,* 1965, p. 457.
14 T. E. F. Carr and B. J. Parsons, *Intern. J. Appl. Radiation Isotopes* **13**, 57 (1962).
15 H. C. Claessen, *Anal. Chim. Acta* **52**, 229 (1970).
16 T. Clausen, *Anal. Biochem.* **22**, 70 (1968).
17 F. J. Cosolito, N. Cohen and H. C. Petrow, *Anal. Chem.* **40**, 213 (1968).
18 C. F. Cramer and B. H. Ross, *Intern. J. Appl. Radiation Isotopes* **21**, 237 (1970).
19 M. L. Curtis, S. L. Ness and L. L. Bentz, *Anal. Chem.* **38**, 636 (1966).
20 J. D. Davidson and P. Feigelson, *Intern. J. Appl. Radiation Isotopes* **2**, 1 (1957).
21 R. J. Dern and W. L. Hart, *J. Lab. Clin. Med.* **57**, 322 (1961).
22 E. Rapkin, *Intern. J. Appl. Radiation Isotopes* **15**, 69 (1964); (manganese-56, see also A. De Volpi and K. G. A. Porges, *Intern. J. Appl. Radiation Isotopes* **16**, 496 (1965)).
23 A. Dyer, J. M. Fawcett and D. U. Potts, *Intern. J. Appl. Radiation Isotopes* **15**, 377 (1964).
24 A. Dyer, D. F. Ball and J. Fitton, unpublished work.
25 A. Dyer and M. A. Williams, unpublished work.
26 J. D. Eakins and D. A. Brown, *Intern. J. Appl. Radiation Isotopes* **17**, 391 (1966).
27 M. Ellis, S. N. Wampler and R. H. Yager, *Anal. Chim. Acta* **34**, 169 (1966).
28 R. H. Elrick and R. P. Parker, *Intern. J. Appl. Radiation Isotopes* **17**, 361 (1966).
29 G. Erdtmann, *Radiochim. Acta* **2**, 215 (1964).
30 G. Erdtmann and G. Herrmann, *Z. Elektrochem.* **64**, 1092 (1960).
31 W. D. Fairman and J. Sedlet, *Proc. 9th Ann. Conf. Bioassay Anal. Chem., San Diego,*

USAEC, Rept. T1D 7696, 1963, p. 10.
32 W. D. Fairman and J. Sedlet, *Anal. Chem.* **40**, 2004 (1968).
33 D. G. Fleishman and L. G. Shakhidzhanian, *At. Energ.* **6**, 669 (1959).
34 K. F. Flynn, L. E. Glendenin, E. P. Steinberg and P. M. Wright, *Nucl. Instr. Methods* **27**, 13 (1964).
35 K. F. Flynn and L. E. Glendenin, *Phys. Rev.* **116**, 744 (1959).
36 H. Foreman and M. B. Roberts, see data in ref. 22.
37 G. V. Forster, *Thesis,* Purdue Univ., Indiana, 1963 (*Nucl. Sci. Abstr.* **18**, No. 3704).
38 B. Francois, *Intern. J. Appl. Radiation Isotopes* **18**, 525 (1967).
39 P. J. Garrahan and I. M. Glynn, *J. Physiol. (London)* **186**, 55 (1966).
40 C. E. Gleit and J. Jumot, *Intern. J. Appl. Radiation Isotopes* **12**, 66 (1961).
41 L. E. Glendenin, *Ann. N. Y. Acad. Sci.* **91**, 166 (1961).
42 E. Gomez, J. Freer and J. Castrillon, *Intern. J. Appl. Radiation Isotopes* **22**, 243 (1971).
43 B. E. Gordon, H. R. Lukens and W. Ten-Hove, *Intern. J. Appl. Radiation Isotopes* **12**, 145 (1961).
44 S. Graber, L. C. McKee and R. M. Heyssel, *J. Lab. Clin. Med.* **69**, 170 (1967).
45 K. Haberer and W. Kölle, *Atompraxis* **11**, 664 (1965).
46 K. Haberer, *Packard Tech. Bull.* No. 16 (1966); *Atomwirtschaft* **10**, 36 (1965).
47 J. E. Hardcastle, R. J. Hannapel and W. H. Fuller, *Intern. J. Appl. Radiation Isotopes* **18**, 193 (1967).
48 B. R. Harvey and G. A. Sutton, *Intern. J. Appl. Radiation Isotopes* **21**, 519 (1970).
49 R. T. Haviland and L. C. Bieber, *Anal. Biochem.* **33**, 323 (1970).
50 A. H. Heimbuch, H. Y. Gee, A. De Haan and L. Leventhal, IAEA SM 61/65 (Conf. 650507-24), 1965 (*Nucl. Sci. Abstr.* **19**, No. 32098).
51 S. Helf, C. G. White and R. N. Shelley, *Anal. Chem.* **32**, 238 (1960).
52 F. L. Hoch, R. A. Kuras and R. D. Jones, *Anal. Biochem.* **40**, 86 (1971).
53 W. Hoffmann, *Radiochim. Acta* **4**, 117 (1965).
54 D. L. Horrocks and M. H. Studier, *Anal. Chem.* **30**, 1747 (1958).
55 D. L. Horrocks and M. H. Studier, *Anal. Chem.* **33**, 615 (1961).
56 D. L. Horrocks, *Rev. Sci. Instr.* **35**, 334 (1964).
57 D. L. Horrocks, *Rev. Sci. Instr.* **34**, 1035 (1963).
58 D. L. Horrocks and A. Lee Harkness, *Phys. Rev.* **125**, 1619 (1962).
59 D. L. Horrocks and M. H. Studier, *Anal. Chem.* **36**, 2077 (1964).
60 D. L. Horrocks, *Nucl. Instr. Methods* **30**, 157 (1964).
61 D. L. Horrocks, *Nucl. Instr. Methods* **27**, 253 (1964).
62 D. L. Horrocks, *Intern. J. Appl. Radiation Isotopes* **22**, 258 (1971).
63 D. L. Horrocks, in *The Current Status of Liquid Scintillation Counting,* (Ed. E. D. Bransome), Grune and Stratton, New York, 1970, p. 32.
64 F. Hoyer, *Nucl. Instr. Methods* **68**, 351 (1969).
65 W. Huelsen and V. Prenzel, *Anal. Biochem.* **26**, 483 (1968).
66 E. R. Humphreys, *Intern. J. Appl. Radiation Isotopes* **16**, 345 (1965).
67 F. Hutchinson, *Intern. J. Appl. Radiation Isotopes* **18**, 136 (1967).
68 H. R. Ihle, M. Karayannis and A. P. Murrenhoff, *Radioisotope Sample Meas. Tech. Med. Biol. Proc. Symp. Vienna,* 1965, p. 485.
69 H. Jeffay, F. O. Olubajo and W. R. Jewell, *Anal. Chem.* **32**, 306 (1960).
70 H. Jenner and K. J. Obrink, *Scand. J. Clin. Lab. Invest.* **14**, 466 (1962).

71 D. R. Johnson and J. W. Smith, *Anal. Chem.* **35**, 1991 (1963).
72 M. K. Johnson, *Anal. Biochem.* **29**, 348 (1969).
73 K. Joon and P. A. Deurloo, *Kjeller Rept.* KR-100, 1965; (*Chem. Abs.* **65**, 1720h).
74 K. Joon and P. A. Deurloo, *Intern. J. Appl. Radiation Isotopes* **16**, 334 (1965).
75 E. T. Josefowicz, *Nukleonika* **5**, 713 (1960).
76 J. H. Katz, M. Zoukis, W. L. Hart and R. J. Dern, *J. Lab. Clin. Med.* **63**, 885 (1964).
77 R. F. Keoug and G. J. Powers, *Anal. Chem.* **42**, 419 (1970).
78 J. Klumpar and A. Majerov, *Jaderna Energie* **8**, 356 (1962).
79 Y. Kobayashi and D. V. Maudsley, in *The Current Status of Liquid Scintillation Counting,* (Ed. E. D. Bransome), Grune and Stratton, New York, 1970, p. 76.
80 M. A. Kumar, *Intern. J. Appl. Radiation Isotopes* **17**, 556 (1966).
81 A. Läuchli, *Intern. J. Appl. Radiation Isotopes* **20**, 265 (1969).
82 T. P. Leffingwell, R. W. Riess and G. S. Melville, *Intern. J. Appl. Radiation Isotopes* **13**, 75 (1962).
83 P. Lerch and M. Cosandey, *Atomlight* **52**, 1 (1966).
84 L. Levin, *Anal. Chem.* **34**, 1402 (1962).
85 R. A. Lloyd and D. B. Rees-Evans, *Intern. J. Appl. Radiation Isotopes* **16**, 393 (1965).
86 A. Lindenbaum and C. J. Lund, *Radiation Res.* **37**, 131 (1969).
87 J. D. Ludwick, *Anal. Chem.* **32**, 607 (1960).
88 J. D. Ludwick, *Anal. Chem.* **36**, 1104 (1964).
89 L. Lutwak, *Anal. Chem.* **31**, 340 (1959).
90 D. S. Mahadevappa and R. L. Eager, *Current Sci. (India)* **34**, 15 (1965).
91 M. Miller, J. G. Kereiakes and B. I. Friedman, *Intern. J. Appl. Radiation Isotopes* **20**, 133 (1969).
92 A. A. Moghissi, in *The Current Status of Liquid Scintillation Counting,* (Ed. E. D. Bransome), Grune and Stratton, New York, 1970, p. 86.
93 A. B. Moir, *Intern. J. Appl. Radiation Isotopes* **22**, 213 (1971).
94 W. A. Muller, *Radioisotope Sample Meas. Tech. Med. Biol. Proc. Symp. Vienna,* 1965, p. 475.
95 L. S. Myers and A. H. Bush, *Anal. Chem.* **34**, 342 (1964).
96 A. Nadarajah, B. Leese and G. F. Joplin, *Intern. J. Appl. Radiation Isotopes* **20**, 733 (1969).
97 M. Noguchi, *Radioisotopes (Tokyo)* **13**, 362 (1964).
98 P. Nisson and A. A. Benson, *Intern. J. Appl. Radiation Isotopes* **15**, 505 (1964).
99 D. G. Ott, *Oklahoma Conf. Radioisotopes in Agriculture,* AEC TD1-7578, 1959, p. 49.
100 D. G. Ott, C. R. Richmond, T. T. Trujillo and H. Foreman, *Nucleonics* **17**, 106 (1959).
101 R. P. Parker and R. H. Elrick, *Brit. J. Radiol.* **39**, 236 (1966).
102 R. P. Parker and R. H. Elrick, in *The Current Status of Liquid Scintillation Counting,* (Ed. E. D. Bransome), Grune and Stratton, New York, 1970, p. 110.
103 R. P. Parker, *Digitechniques Tech. Rev.* Intertechnique, Paris, 1969.
104 C. T. Peng, *Anal. Chem.* **32**, 1292 (1960).
105 S. W. Perry and G. J. Warner, *Intern. J. Appl. Radiation Isotopes* **14**, 397 (1963).
106 C. P. Petroff, P. P. Nair and D. A. Turner, *Intern. J. Appl. Radiation Isotopes* **15**, 491 (1964).

107 D. E. Pickering, H. L. Reed and R. L. Morris, *Anal. Chem.* **32**, 1214 (1960).
108 N. S. Radin and R. Fried, *Anal. Chem.* **30**, 1926 (1958).
109 B. A. Rhodes, *Anal. Chem.* **37**, 995 (1965).
110 S. E. H. Rippon, *Nucl. Instr. Methods* **21**, 185 (1963).
111 C. Ronzani and M. A. Tamers, *Radiochim. Acta* **6**, 206 (1966).
112 A. R. Ronzio, *Intern. J. Appl. Radiation Isotopes* **4**, 196 (1959).
113 T. B. Ryves, *J. Sci. Instr.* **37**, 201 (1960).
114 B. F. Sansom and P. J. Taylor, *Nature* **211**, 626A (1966).
115 M. Sarnat and H. Jeffay, *Anal. Chem.* **34**, 643 (1962).
116 N. I. Sax, J. D. Denny and R. R. Reeves, *Anal. Chem.* **40**, 1915 (1968).
117 B. Scales and D. A. D. McIntosh, *J. Pharmacol. Exp. Therap.* **160**, 249 (1968).
118 E. Schram, in *The Current Status of Liquid Scintillation Counting,* (Ed. E. D. Bransome), Grune and Stratton, New York, 1970, p. 95.
119 L. C. Schwendiman and J. Mishima, *Waste Management Res. Abs.* No. 6, IAEA/HSW/5, 1971, p. 20.
120 C. A. Scott and M. L. Good, *J. Inorg. Nucl. Chem.* **29**, 255 (1967).
121 H. H. Seliger, *Intern. J. Appl. Radiation Isotopes* **8**, 29 (1960).
122 L. G. Shakidzhanian, D. G. Fleishman, V. V. Glazunov, V. G. Leont'ev and V. P. Nesterov, *Dokl. Akad. Nauk. SSSR* **125**, 1 (1969).
123 G. Sheppard and C. G. Marlow, *Intern. J. Appl. Radiation Isotopes* **22**, 125 (1971).
124 D. Steinberg, *Advances in Tracer Methodology, Volume 1,* Plenum Press, New York, 1963, p. 93.
125 D. Steinberg, *Anal. Biochem.* **1**, 23 (1960).
126 J. Steyn, *Proc. Phys. Soc. (London)* **A69**, 865 (1956).
127 J. Steyn and F. J. Haasbroek, *Proc. 2nd U.N. Intern. Conf. Peaceful Uses At. Energy, Geneva,* **21**, 95 (1958).
128 S. Stoicovici and I. Uray, *Nucl. Instr. Methods* **51**, 25 (1969).
129 R. D. Stubbs and A. Jackson, *Intern. J. Appl. Radiation Isotopes* **18**, 857 (1967).
130 A. Tada and T. Radoszewski, in *Liquid Scintillation Counting, Volume 1,* (Ed. A. Dyer), Heyden and Son Ltd., London, 1971, p. 49.
131 G. Threlfall, D. M. Taylor and A. T. Buck, *Amer. J. Pathol.* **50**, 1 (1967).
132 R. Tkachuk, *Can. J. Chem.* **40**, 2348 (1962).
133 T. Y. Toribara, D. A. Morken and C. Predmore, U.R. 607, Conf. W7401, (*Nucl. Sci. Abstr.* **16**, No. 11533).
134 T. Y. Toribara, D. A. Morken and C. Predmore, *Talanta* **10**, 205 (1963).
135 Y. A. Tsirlin, A. Ya. Gel'fman, N. B. Bykovets and V. M. Solomonov, *Zavodsk. Lab.* **31**, 704 (1965); (*Chem. Abs.* **63**, 5209f).
136 R. A. Turpin and J. E. Bethune, *Anal. Chem.* **39**, 362 (1967).
137 P. Urone, J. B. Evans and C. M. Noyes, *Anal. Chem.* **37**, 1104 (1965).
138 G. S. Uyesugi and A. E. Greensberg, *Intern. J. Appl. Radiation Isotopes* **16**, 581 (1965).
139 H. Vemmer and V. O. Gutte, *Z. Tierphysiol. Tierernaehr. Futtermittelk.* **18**, 346 (1963).
140 H. Vemmer and V. O. Gutte, *Atompraxis* **10**, 475 (1964).
141 C. G. White and S. Helf, *Nucleonics* **14**, 46 (1956).
142 P. M. Wright, E. P. Steinberg and L. E. Glendenin, *Phys. Rev.* **123**, 205 (1961).
143 S. Yamada, *J. Phys. Soc. Japan* **17**, 865 (1962).

144 J. D. Yarbrough, A. F. Findeis and J. C. O'Kelley, *Intern. J. Appl. Radiation Isotopes* **17**, 453 (1966).
145 R. E. Yerick and H. H. Ross, *Oakridge Radioisotope Conf. Res. Appl. Phys. Sci. Eng., Gatlinberg, Tenn.*, 1963, p. 20.
146 M. Zaduban, *J. Radioanal. Chem.* **5**, 97 (1970).

DISCUSSION

F. E. L. ten Haaf: Have you any experience with water-soluble scintillators? I think there has been a recent publication on this subject, but I have not yet found out the details.

A. Dyer: No, we have not tried this. The publication you refer to is presumably that of Haberer and Kolle, *Atompraxis* **11**, 664 (1965) who refer to an improvement in the Cerenkov counting efficiency of certain β-emitters brought about by the addition of a water-soluble fluorescent compound to the aqueous solution.

B. E. Gordon: The paper by Leffingwell (T. P. Leffingwell *et al., Intern. J. Appl. Radiation Isotopes* **13**, 75 (1962)) reported an increased counting efficiency when colourless Fe^{+++} *ortho*-phenanthroline was converted to red Fe^{++} *ortho*-phenanthroline. This is impossible to explain by increased pulse height. This increase arose because the counter was set for carbon-14 counting and the red colour simply lowered the pulse height so that the peak shifted into the carbon-14 channel. A far better answer would be to leave the iron in the colourless ferric form as the *ortho*-phenanthroline complex, and open the upper discriminator to include the higher energy portion of the spectrum.
The only other answer I can think of is that uncomplexed *ortho*-phenanthroline is a far better chemical quencher than complexed Fe^{++} *ortho*-phenanthroline is a colour quencher. This is a less likely explanation.

B. W. Fox: What blender was used when chemiluminescence was observed?

A. Dyer: The chemiluminescence with zinc-65 and barium-133 occurred in a Hayes and Gould cocktail i.e. ethyl alcohol was the blending solvent. The zinc and barium were present as chlorides.

B. W. Fox: Comment: We have been interested in trying to find a solid peroxide scavenger to keep with solvents. We find that granulated zinc kills peroxides in dioxane and alkoxy-alcohols within 24 h.

Chapter 10

Liquid Scintillation Counting of Low Levels of Carbon-14 for Radiocarbon Dating

R. Burleigh

The British Museum,
London, England

INTRODUCTION

Before describing the preparation and measurement of samples I shall attempt to indicate the purpose of this work by outlining briefly the principles of radiocarbon dating.

The possibility of radiocarbon dating by measurement of naturally occurring carbon-14 was first suggested about 1949 by W. F. Libby and his co-workers at the Institute for Nuclear Studies, University of Chicago.[1,2] Earlier work by Korff and others some years previously had shown that secondary neutrons are produced in appreciable numbers at a few miles altitude in the Earth's atmosphere by the action of primary cosmic radiation. From consideration of a number of possible nuclear reactions which might occur involving these neutrons, Libby concluded that the relatively long-lived radioactive isotope carbon-14 would be continuously produced in small amounts as a result of neutron capture by ordinary stable atmospheric nitrogen-14. He predicted that the carbon-14 atoms formed in this way would be rapidly oxidized to carbon dioxide which would become mixed and uniformly distributed by atmospheric circulation and exchange processes in a short time in comparison with the average lifetime of a carbon-14 atom of about 8300 years, so that an equilibrium would be established between the atmosphere and ocean water and the biosphere such that the rate of production of carbon-14 would equal the loss from the system by radioactive decay.

The mechanism by which the cosmic ray-produced carbon-14 is oxidized to carbon dioxide before entry to the exchange reservoir is still imperfectly understood although various processes have been suggested.[3] However Libby showed first, using an isotopic enrichment process, that carbon-14 is present in the biosphere in detectable amounts and secondly, by measuring modern living material from different locations on the Earth's surface, that essentially uniform global mixing does occur. Natural carbon-14 was found to have a low specific activity of the order of 7×10^{-12} Ci/g of carbon so that only a very small proportion of atmospheric carbon dioxide is labelled with carbon-14.

Carbon-14 enters the biosphere during photosynthetic uptake of atmospheric carbon dioxide by green plants. Because such plants are at the base of the food chain, all living organisms are theoretically in radiocative equilibrium with atmospheric carbon-14. When an organism ceases to exchange with the carbon-14 reservoir at death, then disregarding here the possibility of subsequent physical contamination by material of different age, the only way in which the incorporated carbon-14 activity can be altered is exponentially with time by radioactive decay. From a knowledge of the length of the half-life (taken as 5570±30 years) and the so-called contemporary assay, and by measuring the activity remaining in once living samples such as wood, charcoal, bone, etc., the time which has elapsed since exchange ceased can be readily determined with reasonable accuracy.

There are, of course, uncertainties in the interpretation of whether the age obtained for the sample really dates the particular context from which the sample came but this is largely dependent on careful selection of material for measurement and is outside the scope of the present discussion.

After some twenty years the general validity of the carbon-14 dating method has not been seriously challenged, although there have been a number of severe practical and theoretical difficulties such as those relating to the establishment of a suitable modern reference standard due to the diluting effect of large scale combustion of fossil fuel over the last 150 years, accurate determination of the half-life of carbon-14, isotopic fractionation effects in nature, and more recently the realization that there have been fluctuations in the rate of carbon-14 production, so that for some periods in the past, carbon-14 years are not equivalent to solar years.[4] Although the relative sequence of radiocarbon dates is still correct, many important archaeological chronologies cannot be absolutely dated until there is an adequate means of correcting for these past fluctuations and several laboratories are currently devoting much effort to the elucidation of this.

The most important practical requirement for radiocarbon dating is still measurement of the very low activity levels involved with sufficient accuracy – ideally to better than 1%, which is equivalent to ±80 years – and which demands great stability of operation of the counting equipment over long periods. The sample for measurement must obviously consist of carbon in the most concentrated form which is practicable. Until fairly recently, because of the low energy of the carbon-14 β-particles, most radiocarbon dating measurements have been made by gas counting, usually with relatively large, low background proportional counters surrounded by an anti-coincidence counter arrangement and enclosed in massive steel or aged lead shielding weighing several tons. The central counter is typically filled with up to about 10 l of methane, acetylene or carbon dioxide at 2–3 atmospheres pressure, prepared from the original excavated sample material. Most of the eighty or so radiocarbon laboratories now operational still use the gas counting method, but laboratories being newly set up are now using liquid scintillation counting systems from the start and some older established laboratories, like ourselves, have adopted liquid scintillation counting to replace ailing conventional gas counter systems. Apart from the high figures of merit easily attainable, the attraction is largely that the automatic liquid scintillation counter enables frequent check measurements of the background and modern reference standard and known age samples to be made, so that the performance of the system can be closely and continuously monitored over the period during which the unknown samples are measured. This is not possible with a gas counter system in which sample changing is much less straightforward and background variations may be more critical. In other words, for low level work of the nature involved in radiocarbon dating, the safeguards afforded by liquid scintillation counting appear to be much better. There are a number of other important advantages, for example the scintillation system can be kept in continuous use whereas gas counters often require long periods for pumping out and refilling, during which time the electronic equipment is idle. The chemistry of sample preparation for scintillation counting is generally more complicated but once satisfactorily prepared, the sample can be remeasured directly as many times as one may require and subsequently easily stored. Gas samples usually have to be repurified between each measurement and the time, effort and storage space involved is generally greater than for the more elaborate but single operation of preparing a sample for scintillation counting. Removal of radon, often present in the gas counting samples, is accomplished by storing the gas for about one month before counting, during which time the level of radon normally found in the samples decays to below detectable limits. Radon appears to be excluded during the synthesis of benzene for liquid scintillation counting so that no storage period is necessary before counting. Finally there is no need for the construction and support of the massive and costly shielding used to screen a gas counter system.

Earlier successful use has been made of liquid scintillation techniques for radiocarbon dating by a number of workers among whom Audric and Long[5,6] used liquid acetylene; Barendsen[7] used liquid carbon dioxide, Delaney and McAulay[8] synthesized methyl alcohol and Tamers[9] synthesized benzene using a diborane-activated catalyst. Generally because of various preparation hazards, low yields or lengthy catalysis times, none of these methods offered an attractive alternative to gas counting at the time. It was largely the work of Noakes *et al.*,[10] who developed a much more convenient and effective vanadium-activated silica–alumina catalyst for trimerization of acetylene to benzene together with the improvements which had by then taken place in commercially available liquid scintillation counters, which led to a general renewal of interest in liquid scintillation counting among radiocarbon dating laboratories.

TECHNIQUE

The synthesis in use at the British Museum Research Laboratory is based on the large scale high-vacuum line preparation of acetylene from carbon dioxide via lithium carbide first developed by Barker[11] at the British Museum for gas proportional counting and later improved by him to give almost 100% yields of acetylene (Barker, unpublished work). A highly efficient silica–alumina catalyst is then used to convert acetylene to benzene for counting. All radiocarbon dating laboratories using liquid scintillation counting of benzene now follow the final version or minor variants of Barker's acetylene chemistry. Various detailed improvements have been made at the catalysis stage resulting in better yields of high purity benzene (judged mainly on quenching criteria although gas chromatographic analysis has also shown that only very small traces of impurities are present), greatly reduced risk of isotopic fractionation and reduction of the time needed for the catalysis.

Sample materials such as wood and charcoal and other plant remains are first pretreated with dilute acid and in some cases also with dilute alkali, for removal of possible contaminants accumulated during burial. All our samples are archaeological materials and usually will have been penetrated by carbonates or more recent organic substances (humic acids) transported by movement of soil water. To obtain a meaningful date these intrusive substances must clearly be removed as completely as possible. Bone and antler samples are demineralized with cold dilute hydrochloric acid at reduced pressure to remove the inorganic carbonate and phosphate components, leaving the protein (collagen) which has been found to be the most reliable fraction of these two materials (Krueger).[12] More detailed consideration of the complexities of sample pretreatment (Olson)[13] would not be appropriate here.

After washing and oven drying, up to about 12 g of pretreated sample is ignited in a silica crucible supported inside a 3 l stainless steel 'bomb' containing a suitable excess of pure oxygen, by passing an electric current through an element suspended in the crucible and covered by the sample. This combustion method is much more convenient and effective than the alternative large scale tube furnace (Barker *et al.*).[14] As the bomb is filled to a pressure of about 10 atm with pure, aged oxygen, combustion is virtually instantaneous and generally little or no ash remains. The resulting gas mixture is pumped out with a large, 450 l/min, rotary pump, over a period of about 45 min, through a series of traps cooled with dry ice–acetone slush and liquid nitrogen for separation of water and carbon dioxide respectively from excess oxygen which is pumped away. Usually 12 g of pretreated sample materials will yield at least 9 l of carbon dioxide and this does not need to be purified before conversion to acetylene. For the age range normally expected 9 l of carbon dioxide will ultimately provide a large enough benzene sample to obtain an acceptably small associated probable error in one set of measurements. As the overall yield is predictable it is easiest to control the approximate final sample size at the start of the synthesis and any excess carbon dioxide over 9 l is generally not used. For samples which are too small to yield 9 l of carbon dioxide, the final benzene volume is made up to the standard 5.5 ml by adding 'dead' Analar benzene before mixing with scintillator solution. The volume of carbon dioxide obtained from the sample combustion is measured to determine the amount of lithium metal needed for formation of lithium carbide. Experience has shown that the highest yields of acetylene are obtained when 100% stoichiometric excess of lithium metal is used. After cutting a suitable quantity from an ingot kept under liquid paraffin, the surface of the metal is scraped to remove any carbonates and then washed with petroleum ether to remove oil residues. The lithium is then heated under vacuum to 650°C in a stainless steel furnace of 3 l capacity, with a Meker burner. Lithium melts at 179°C and at about this temperature occluded hydrogen is released and is pumped away. After about 10 min heating when the pressure in the system is about 10^{-2} mm Hg, the furnace is closed off from the pumps and the carbon dioxide sample is slowly admitted over about 45 min, keeping the pressure at about 10 cm Hg and the temperature constant at 650°C. The temperature can be judged by the cherry red colour of the molten lithium when viewed under dim light through the glass inlet tube and with a little experience the reaction conditions can be controlled without difficulty. When all the carbon dioxide has been reacted, after a further period of pumping at 650°C, the furnace is allowed to cool to room temperature, usually overnight. Distilled water is then run in slowly to regenerate acetylene. The resulting mixture of acetylene, hydrogen from the excess lithium and water vapour is pumped through a series of cooled traps to strip out the acetylene, in a similar way to that outlined above for extraction of carbon dioxide from the combustion bomb. Under favourable conditions the yield of acetylene is consistently 98% and seldom falls below 96%. The acetylene is passed through two columns of glass beads coated with 50% w/v sodium hydroxide solution and syrupy *ortho*-phosphoric acid respectively, and through a number of dry ice–acetone traps for final drying and is then absorbed onto a vanadium-activated silica–alumina catalyst. Storage bulbs at several points on the vacuum line allow the synthesis to be interrupted at each main stage and there is usually a different sample at each of these points in the process at any one time.

The silica–alumina catalyst is a commercial product though it is not generally available and we have been fortunate to secure some years' supply of this material. The catalyst is in the form of semi-opaque spherical beads ranging in diameter from about 2 to 5 mm. These are activated by soaking for a few hours in a solution containing 10% w/v of vanadium pentoxide in 100 vol hydrogen peroxide (both GPR grade) diluted approximately fourfold with distilled water. The beads are then drained, washed and oven dried. The vanadium pentoxide solution is quite difficult to prepare satisfactorily but can be re-used several times. Before use the catalyst is finally dried under vacuum at 300°C for about 1½ h. The tube in which the catalyst is finally dried before absorption of acetylene has been designed and constructed by N. D. Meeks for efficient heat exchange. Absorption of acetylene onto the catalyst tends to be exothermic but higher yields of benzene are obtained if the catalyst is kept at ambient temperatures during the absorption. The tube consists of a small double concentric cylinder some 15 × 4 cm with the walls, which are about 1½ average bead diameters apart, sealed at the bottom and joined at the top to an inlet tube having a detachable high-vacuum stopcock. Construction of the tube is difficult but it is extremely efficient and quite robust. The catalyst is held in a thin layer in the annular space between the walls of the tube and the whole assembly is cooled with running tap water. 5 g of activated catalyst are used per litre of acetylene. Initially the acetylene absorption is very rapid but as the pressure falls below about half an atmosphere, the rate of uptake decreases until finally after about 1½ h there is a residual pressure of 6 to 8 cm which is a low percentage loss as the dead volume of the manifold is small. If the acetylene pressure is raised to 1 atm or above benzene 'rains' from the catalyst onto the cooled tube walls. However, our attempts based on this and similar observations with other catalysts, to construct an apparatus in which acetylene absorption and benzene recovery are simultaneous have so far proved to have no advantages in time saved or higher benzene yields.

Benzene is recovered by heating the tube to 150°C in a tube furnace, the desorbed benzene being collected in a liquid nitrogen cold finger. Approximately 85 to 90% of the available benzene is desorbed in the first 10 min but to achieve the minimum acceptable yield of 95% heating must be continued for at least 1 h.

The catalysis mechanism is probably that gaseous acetylene molecules become attached to the catalyst and polarized, and other acetylene molecules link with these until, when three are attached, the benzene structure is formed by angular rotation (Noakes *et al.*).[1] Conversion of acetylene to benzene is most probably almost quantitative but, probably due to the large surface area of the catalyst (450 m^2/g) complete benzene recovery is not possible. A semi-micro combustion apparatus was constructed for the quantitative re-conversion of benzene to carbon dioxide so that mass spectrometric measurements of the carbon-13/carbon-12 ratios could be made for comparison with the carbon-13/carbon-12 abundance ratios of the bomb carbon dioxide from the initial sample combustion. Re-conversion of benzene to carbon dioxide was necessary for this comparison due to the irregularity of the cracking pattern of benzene in the mass spectrometer. Measurements made on a considerable number of samples over a period with an AEI MS 20 double collection mass spectrometer have shown that no significant fractionation occurs during the synthesis of benzene although the overall yield from carbon dioxide to benzene may be as low as 92% theoretical. This also has the practical advantage that, for the routine correction of dates for isotopic fractionation in nature, a mass spectrometer sample can simply be taken from the bomb carbon dioxide and it is not necessary to reconvert an aliquot of each benzene sample synthesized.

The activated catalyst is yellow-brown in colour but becomes black after use probably as a result of slight reduction of vanadium (P. R. Mitchell, personal communication). It is possible to re-use the catalyst but this is not done to avoid any cross-contamination of samples arising from memory effects.

At the end of the benzene recovery period dry air is admitted to the collection manifold and the benzene is allowed to warm up to room temperature. The collection tube is then detached from the vacuum line and the benzene is poured into a tared counting vial for accurate weighing. The preferred sample size is 5.54 ml, corresponding to 4.86 g of benzene. Samples smaller than this are made up with 'dead' benzene to 5.5 ml. 9.5 ml of scintillator solution consisting of PPO in scintillation grade toluene is added, to give a final concentration of PPO equivalent to 4.0 g/l and total sample volume of 15 ml. No wavelength shifter is added and the samples and scintillator solutions are handled in ordinary air. Flushing with an inert gas does not appear to be necessary and would probably cause unacceptable evaporation losses with little improvement in quenching. Standard low potassium glass counting vials with plastic caps lined with cork-backed foil are used. A group of these having comparable backgrounds has been laboriously selected and these same vials are re-used continuously with occasional replacement of the caps. There would be some advantage in using lower background quartz vials but none of those available have closures which will satisfactorily prevent evaporation losses. Plastic vials which might be better still are not sufficiently resistant to the toluene–benzene samples over the long counting periods necessary although the possible use of some newly available plastic vials is being investigated.

The samples are counted in a Model 3315 Packard Tricarb liquid scintillation spectrometer having selected quartz-faced photomultiplier tubes (EMI 6255B), AES and teletype output. With the high voltage, gain and window settings used, the efficiency to an unquenched carbon-14 standard of known activity is about 66% and the efficiency to tritium is negligible. This is important because the distilled water used to generate acetylene from lithium carbide is derived from London tap water which is not tritium-free so that there is a level of tritium in the samples of approximately the same order as the expected carbon-14 activity. The operating temperature is 0°C and there are no measurable evaporation losses from the samples over a period of several weeks at least.

Samples are counted, usually for one week, in groups consisting of a sealed, unquenched carbon-14 standard, modern reference standard, background, sealed known age sample and a group of three or four unknowns. Apart from the 'hot' carbon-14 standard which is stopped by a preset count of 900000 after about 13.75 min, at least fifteen individual counts of 100 min each are accumulated for all samples in the group.

External standard channels ratio quench measurement is made at the end of each week's run to avoid additional background variations due to imprecise repositioning of the AES source if this is repeatedly moved while sample counting is in progress. There is only a few per cent of quenching and although the samples do not vary greatly and there would be a slightly lower background with the AES source removed, lack of correction for quenching could introduce age errors of perhaps 40 years which is of the same order as the precision of measurement attainable. Old or small samples do not provide enough carbon-14 counts to permit reliable quench correction by internal channels ratio.

The particular background accumulated each week is used for that week's counting only because, although there is little variation from the nominal background of 7.7 counts/min, changes of 0.1 c.p.m. are important when unusually old or small samples are being measured. In an eight-sample group counted for one week, about 12000 background counts and up to about 50000 net counts for each sample are accumulated, depending on the size and age of the samples. The upper detectable age limit is about 45000 years before present (B.P.) although this could be extended by at least one half-life by increasing the amount of benzene but keeping the total sample volume constant. Performance of our Tricarb, which is used exclusively for radiocarbon dating, has been extremely stable and practically free from faults over a period of five years and although the gain settings are checked every week the balance point seldom requires any adjustment at all.

A matrix of data in the form of eight-track punched paper tape comprising sample identifications, accumulated counts, quench data, benzene weights and isotopic fractionation values is processed each week by University of London Atlas Computing Service using an Algol program written for us by A. J. Barker of King's College, London which is stored on magnetic tape. The print out gives a complete statistical analysis of the count data, a number of parameters needed for monitoring system performance and the calculated sample ages and associated age errors. At present no data are permanently stored in the computer. Usually the results are available within a day and the cost of computation is very modest.

Most of our work is related to the research programmes of the British Museum's own Antiquities Departments and is concerned largely with the dating of prehistoric cultures of the British Isles, parts of Europe and some other areas, and investigation of carbon-14 reservoir fluctuations back to 3000 B.C. using closely dated material from Egyptian and other well documented contexts. Special projects in human palaeontology, dating the extinction of post-Pleistocene mammals, comparisons with other dating methods such as thermoluminescence and some fundamental work on carbon-14 dating problems, are also in long-term progress. About 150 to 200 samples are dated in a year and the results are published periodically in the journal *Radiocarbon* and some also appear in articles in various leading archaeological journals.

ACKNOWLEDGEMENTS

No work of this kind is possible without a team effort. We are indebted to Mr. H. Barker under whose direction the dating work is carried out, for much helpful criticism and guidance. Mr. N. D. Meeks has skillfully constructed much of the glass high-vacuum system and painstakingly synthesized hundreds of benzene samples. More recently Miss M. A. Pulle has also synthesized samples and both of them have read and commented on this paper. Mr. A. J. Barker has been to much trouble to write a comprehensive Algol computer program for us. We are also indebted to our colleagues in other departments of the British Museum and elsewhere for advice on many matters relating to the dating work.

REFERENCES

1 W. F. Libby, *Radiocarbon Dating (2nd edition)*, University of Chicago Press, Chicago and London, 1952.
2 W. F. Libby, *Phil. Trans. Roy. Soc. London* **A269**, 1 (1970).
3 T. F. Dorn, A. W. Fairhall, W. R. Schell and Y. Takashima, *Radiocarbon* **4**, 10 (1962).
4 I. U. Olsson (Ed.), *Radiocarbon Variations and Absolute Chronology, Proc. 12th Nobel Symposium, Uppsala, 1969,* John Wiley, New York and Almqvist and Wiksell, Stockholm, 1970.
5 B. N. Audric and J. V. P. Long, in *Radioisotope Conference, Volume 2,* Butterworths, London, 1954, p. 134.
6 B. N. Audric, in *Liquid Scintillation Counting,* Pergamon, London, 1958, p. 288.

7 G. W. Barendsen, *Rev. Sci. Instr.* **28**, 430 (1957).
8 C. F. G. Delaney and I. R. McAulay, *Sci. Proc. Roy. Dublin Soc.* **A1**, 1 (1959).
9 M. A. Tamers, *Science* **132**, 668 (1960).
10 J. E. Noakes, S. Kim and J. J. Stipp, in *Proc. 6th Intern. Conf. on Radiocarbon and Tritium Dating, Washington, 1965*, p. 68.
11 H. Barker, *Nature* **172**, 631 (1953).
12 H. W. Krueger, in *Proc. 6th Intern. Conf. on Radiocarbon and Tritium Dating, Washington, 1965*, p. 332.
13 E. A. Olson, *Thesis*, Columbia University, 1963.
14 H. Barker, R. Burleigh and N. Meeks, *Nature* **221**, 49 (1969).
15 J. Noakes, S. Kim and L. Akers, *Geochim. Cosmochim. Acta* **31**, 1094 (1967).

DISCUSSION

P. Stanley: Since the counting period is so long is the evaporation of your scintillator solution significant?

R. Burleigh: Our counter is operated at 0°C and we have shown by careful periodic weighing of samples that evaporation losses are very small – a few milligrams over a period of several months if the vial caps are tightened up well. There is no measurable loss of scintillator solution at all over the normal counting period of 1 to 2 weeks.

P. Stanley: Do you employ 5 ml vials attached to an optically dense plastic base to cut down optical cross talk between the photomultiplier tubes or do you employ conventional sized vials?

R. Burleigh: Standard 20 ml low potassium glass vials selected for comparable backgrounds and containing a total volume of 15 ml of scintillator solution are used. The walls of the horizontally opposed photomultiplier tubes are optically masked from each other.

P. Stanley: Do you use a reduced high voltage supply to the photomultiplier tubes and operate at balance-point settings?

R. Burleigh: The high voltage supply to the photomultiplier tubes has been reduced by several hundred volts below the factory optimized setting for tritium. This has been done principally to minimize the background (which comes down to about 7.7 c.p.m. with selected quartz face photomultiplier tubes) but also eliminates possible interference from any tritium introduced during the benzene synthesis. Under the operating conditions used there is negligible efficiency to tritium and about 65% carbon-14 efficiency at balance-point settings giving a figure of merit of about 600 – a great improvement on the performance of most gas proportional counters used for radiocarbon dating.

Chapter 11

Liquid Scintillation Counting of Calcium-45 in Biological Samples containing Environmental Strontium-90

J. Nolan

MRC Radiobiology Unit,
Harwell, England

INTRODUCTION

The method described, basically a modification of that due to Carr and Parsons,[1] was developed to measure low specific activities of calcium-45 in human excreta containing strontium-90. Of the published methods for the liquid scintillation counting of calcium-45,[1-9] only those involving the suspension of a calcium salt in a scintillator[3,4,9] had high enough calcium capacities for our purposes. Suspended sample methods are, however, prone to variations in counting efficiency that are not measurable by standard methods for quench correction when the calcium content exceeds 300 to 400 mg. The errors involved are small for calcium-45 alone but lead to serious inaccuracies when strontium-90 is present. Methods in which true solution of a calcium salt is obtained allow reliable quench correction[10] but have only been used for samples containing up to about 500 mg of calcium.

In the present method calcium is separated from biological samples as the oxalate, converted to the chloride and dissolved in ethanol. A toluene-based scintillator solution is then added. Up to 1.4 g of calcium as chloride can be dissolved in 20 ml of scintillator system and the calcium-45 counted with an efficiency of 60 to 84%. Quench correction by the external standard method is shown to give reliable results both with calcium-45 alone and in the presence of strontium-90, a constituent of contemporary biological samples. Internal and external standard measurements of the calcium-45 counting efficiency of the samples are in good agreement. The detection limit of the method, i.e. when the calcium-45 rate = 10% of the background, is approximately 3.3 p Ci/g calcium.

EXPERIMENTAL

Reagents

Scintillator solution: butyl PBD (CIBA), 20 g/l in analar toluene.
Ethanol: absolute alcohol (BPC).
Buffered ammonium oxalate: pH 4.2. Add 4% w/v oxalic acid solution to 4% w/v ammonium oxalate solution (approximately 80 ml/l) until pH = 4.2.
Ammoniated water: approximately 1 to 250 dilution of '0.880' ammonia in distilled water.
Indicator: 1% w/v solution of bromocresol green in ethanol.

Sudan IV quencher solution: 0.005% w/v solution of Sudan IV (BDH) in scintillator solution.

Synthetic urine and faecal ash solutions: add the following components to approximately 4 l of distilled water, heat to 80 to 90°C and add concentrated HCl with stirring until a solution is obtained. Make up to a final volume of 5 l with water.

1. Faecal Ash		2. Urine Ash	
NaCl	15.2 g	NaCl	587 g
KCl	44.8 g	KCl	305 g
Saturated $MgCl_2$	204 ml	Saturated $MgCl_2$	94 ml
Fe_2O_3	1.22 g	$CaHPO_4$	43 g
H_3PO_4 (88%)	92.0 g	H_3PO_4	163 g
H_2SO_4 (98%)	20.5 g		
$CaCO_3$	80.0 g		

Procedure

After an initial concentration of calcium, if necessary the entire process may be carried out in a single centrifuge tube of suitable volume. ¾ in B24 stoppered test tubes (Quickfit and Quartz MF 25/2/6) are suitable when samples contain up to 100 mg of calcium, and 250 ml centrifuge bottles (MSE heat-resistant glass) may be used for the larger amounts.

1. Dry ash biological samples, dissolve in dilute hydrochloric acid, centrifuge to remove insoluble debris and transfer a suitable aliquot to the appropriate size of tube or bottle. Add calcium chloride solution if necessary to make the total calcium at least 20 mg in the smaller tubes or 200 mg in the 250 ml bottles.
2. Add a small excess of ammonium oxalate solution (a hot 10 to 15% solution can be used if the sample volume is large) and adjust the pH to approximately 4.2 with 0.880 ammonia solution (green to bromocresol green). Add buffered ammonium oxalate solution to roughly double the volume and stand for 30 min or until cooled to room temperature. Centrifuge at 1800 r.p.m. for 5 min, reject the supernatant and drain for 2 to 3 min.
3. Re-suspend the precipitate in buffered ammonium oxalate solution, centrifuge and drain as in (2).
4. Dissolve the precipitate in a minimum of 60% hydrochloric acid (heat if necessary) and repeat stages (2) and (3) above. (Buffered ammonium oxalate solution only should be added).
5. Wash the precipitate twice with ammoniated water, drain for 2 to 3 min.
6. Place the tubes in an oven at 160°C until thoroughly dry (2 to 3 h) and transfer to a muffle furnace at 525°C for at least 4 h but preferably overnight.
7. Dissolve the residue in a minimum of 50% hydrochloric acid and dry at 160°C.
8. Add ethanol (12.0 ml), stopper and shake at 40 to 50°C until solution is complete. (If rubber bungs are used they should be covered with a polythene sheet). Add the scintillator solution (8.0 ml), mix and transfer to counting vial. Black or reddish residues of carbon or iron are best removed by centrifugation after standing for 12 h. When strontium-90 is present allow 14 days for equilibration of strontium-90 with yttrium-90 before counting.

Preparation of quenched standards

Three sets of quenched standards were used each containing from 1.0 to 4.0 ml of Sudan IV quencher solution substituted for scintillator. One set was for background measurements and the others contained known activities of either calcium-45 or strontium-90.

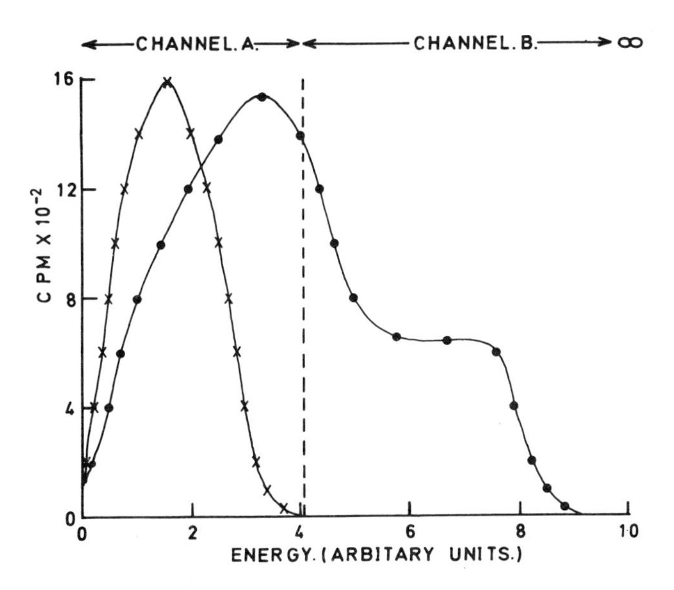

Fig. 1. Spectra of calcium-45 and strontium-90 (yttrium-90). X = calcium-45, ● = strontium-90 (yttrium-90).

Counting conditions and analysis of results

1. Instrument settings for external standard will depend on the counter used. On an early model 'Beckman Liquid Scintillation System' the gain was adjusted to give an external standard ratio of about 1.5 for the least quenched standard. Because of the statistical variation of the external standard ratio on this instrument each sample was counted three times and a mean ratio taken.
2. Adjust channel A to include all counts due to calcium-45 using the least quenched standard and channel B to contain events of a higher energy (see Fig. 1).
3. Count samples for a suitable time together with the quenched series appropriate to the expected calcium-45 and strontium-90 count rate of the samples (see Figs. 2 and 3).
4. From the data from the quenched series establish the following relationships:
   (a) Efficiency of calcium-45 in channel A with external standard ratio.
   (b) Efficiency of strontium-90 in channel A with external standard ratio.
   (c) Efficiency of strontium-90 in channel B with external standard ratio.
   (d) Background in channel A with external standard ratio.
   (e) Background in channel B with external standard ratio.

In the work reported here the paper tape output from the counter was fed to an IBM 360 computer which fitted the above curves to three-term polynomials and calculated the d.p.m. of calcium-45 and strontium-90 in each sample together with the corresponding

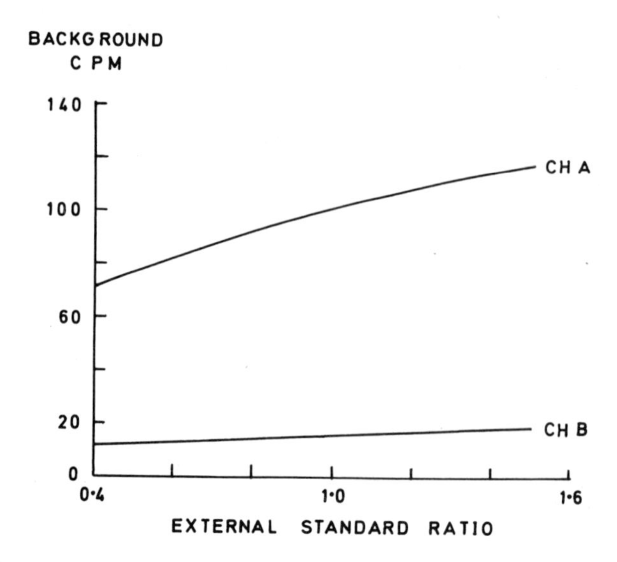

Fig. 2. Effect of quenching on background in channels A and B.

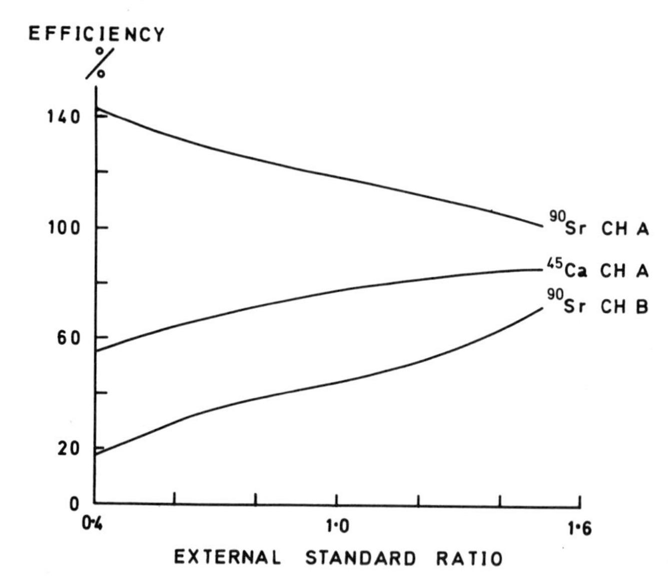

Fig. 3. Effect of quenching on calcium-45 and strontium-90 (yttrium-90) counting efficiencies.

standard errors and the standard error of the mean for duplicate samples. The statistical calculation assumed all counts to have a Poisson distribution and included errors in curve fitting.

RESULTS

Chemical recovery. No significant loss of calcium could be detected by flame photometric measurements but the recovery of strontium was sometimes reduced to about 85%. Since the method was designed only to correct the calcium-45 counts for the contribution due to strontium-90 the reason for this was not thoroughly investigated. The solubility of strontium oxalate is significant over 30°C, therefore more careful temperature control during precipitation might ensure a quantitative recovery of strontium.

Effect of quenching on background and counting efficiencies. Figures 2 and 3 show the effect of quenching by Sudan IV on the background, calcium-45 and strontium-90 counting efficiencies in channels A and B. The decrease in background with increased quenching, especially in channel A is considerable and leads to significant corrections at low calcium-45 count rates.

Strontium-90 (E_{max} = 0.516 MeV) is counted together with its daughter yttrium-90 (E_{max} = 2.27 MeV) at equilibrium giving an efficiency of over 100% in terms of d.p.m. strontium-90. Quenching increases the contribution of strontium-90 (yttrium-90) in channel A. Contemporary biological samples e.g. food or faeces contain up to 16 pCi strontium-90/g calcium (36 d.p.m. strontium-90/g calcium).

Recovery of calcium-45 from urine and faecal samples. Table 1 shows the recovery of calcium-45 from a range of urine and faecal ashes containing different levels of strontium-90. The recovery of calcium-45 is seen to be good and essentially unaffected by the presence of strontium-90 at levels much higher than those experienced in experimental samples. Synthetic urine and faecal ashes were prepared using the figures given in *Documenta Geigy* for the average daily output in man of Na^+, K^+, Mg^{2+}, Ca^{2+}, Fe^{3+}, P and S.

Synthetic samples were used because of the difficulty of obtaining human samples in large quantities; they proved rather more difficult to process than experimental samples.

Table 1. Recovery of calcium-45 from urine and faeces containing strontium-90.

Sample		Ca content mg	d.p.m. ^{45}Ca		d.p.m. ^{90}Sr	Recovery %
			Calculated	Added		
	1	19	704 (49)	696	29	101
Synthetic	2	19	717 (61)	696	290	103
Faecal Ash	3	640	670 (21)	696	29	96
	4	640	670 (20)	696	290	96
Human	5	1288	−5.9 (5.2)	–	26	–
Faecal Ash	6	1288	131 (6.7)	134	26	98
	1	25	698 (63)	696	29	100
	2	25	722 (73)	696	290	104
Synthetic	3	633	688 (30)	696	29	99
Urine Ash	4	633	700 (35)	696	290	101
	5	1265	670 (15)	696	29	96
	6	1265	671 (20)	696	290	96

Figures in parenthesis are standard errors of the mean for duplicate samples.

Stability of samples. Table 2 shows the observed c.p.m. and calculated d.p.m. of a pair of samples counted at various times after preparation. No significant variation of the calcula-

Table 2. Observed c.p.m. and calculated original d.p.m. of calcium-45 in two samples counted at intervals after preparation.

Days after preparation	c.p.m. calcium-45 observed		Original d.p.m. calcium-45 calculated	
	Sample 1	Sample 2	Sample 1	Sample 2
47	9434	9516	12977	13036
61	8966	8801	13070	12962
74	8469	8366	13130	13011
101	7475	7421	13022	12906
122	6845	6794	13063	12965
135	6553	6489	13080	13004
176	5441	5393	13017	12932
203	4096	4086	13087	13013
253	4062	4009	13020	12850

Table 3. d.p.m. calcium-45 measured in duplicate samples of urine and faecal ashes.

Sample	Calcium content mg	d.p.m. calcium-45 calculated	Standard error of the mean
Urine 1	890	681 707	58[a]
Urine 2	900	642 627	25
Urine 3	1165	613 629	25
Faeces 1	1297	185 188	22
Faeces 2	1130	158 161	19
Faeces 3	1160	186 211	55[a]

[a] For these samples the difference between duplicates was just significant ($p = 0.05$).

ted original d.p.m. with time is observed. The uncorrected c.p.m. give a radioactive half-life for calcium-45 of 164.9±1.0 days which is in excellent agreement with the published figures.

Reproducibility. Table 3 shows some typical experimental results from duplicate samples of human urine and faecal ashes. The standard errors of the mean shown were computed assuming that all errors were due to counting statistics and curve fitting errors. The results marked *a* are those in which the difference between duplicates was significant at the $p = 0.05$ level. The standard error of the mean is, in these cases, expanded by the appropriate heterogeneity factor. During the counting of about 300 samples the $p = 0.05$ level

Table 4. Comparison of calcium-45 counting efficiency in samples prepared from urine and faeces measured by the external and internal standard methods.

Sample	Calcium-45 counting efficiency	
	External standard	Internal standard
Urine Ash 1	71.0	70.6
2	71.1	71.5
3	68.8	69.1
4	69.1	69.2
Faecal Ash 1	61.9	63.4
Ash 2	64.9	63.4
3	65.4	66.4
4	66.6	66.4

was exceeded in about one in seven measurements indicating that additional sources of error were present. The absolute difference between duplicates was, however, mostly quite small and only four measurements needed repeating.

Calcium-45 measurements by internal standard. Table 4 shows the calcium-45 counting efficiency of some experimental samples as measured by the external and internal standard methods. The agreement between the two methods is seen to be good.

Possible interferences. *Chemical:* no samples of biological origin so far encountered have caused any serious difficulties with the separation of calcium or excessive quenching in the counting sample. *Radioactive:* any isotope significantly precipitable as the oxalate could of course cause errors and should therefore be removed prior to precipitation. No significant interferences have been encountered in contemporary biological samples.

DISCUSSION

The modifications to the Carr and Parsons method described above have considerably increased the experimental possibilities of the use of calcium-45 as a tracer particularly in the field of bone metabolism. The ability to allow for the effects of quenching on background and strontium-90 interference considerably increases the precision of measurements made on samples containing large amounts of calcium since even with careful preparation considerable variations in quenching occur with such samples. Measurements of calcium-45 in biological samples containing as little as 40 pCi calcium-45/g calcium can readily be made.

ACKNOWLEDGEMENTS

I would like to thank Mr. D. G. Papworth for writing the computer program used in this work and Dr. G. E. Harrison for his help and advice in the writing of this manuscript.

REFERENCES

1 T. E. F. Carr and B. J. Parsons, *Intern. J. Appl. Radiation Isotopes* **13**, 206 (1962).
2 D. Steinberger, *Anal. Biochem.* **1**, 23 (1960).
3 S. Helf, C. G. White and R. N. Shelly, *Anal. Chem.* **32**, 238 (1960).
4 F. J. Sandalls, *The Use of Gel Suspensions in Scintillation Counting of Low Energy Beta Emitters in Bioassay,* AERE R–5474, HMSO, 1967, p. 163.

5 L. Lutwak, *Anal. Chem.* **31**, 340 (1959).
6 M. Sernat and H. Jeffay, *Anal. Chem.* **34**, 643 (1962).
7 J. E. Hardcastle, R. L. Hannapel and W. H. Fuller, *Intern. J. Appl. Radiation Isotopes* **18**, 193 (1967).
8 E. R. Humphries, *Intern. J. Appl. Radiation Isotopes* **15**, 69 (1964).
9 F. J. Sandalls, *The Determination of Calcium-45 in Urine by Liquid Scintillation Counting,* AERE M–1845, HMSO, 1967.
10 A. W. Rogers and J. F. Moron, *Anal. Biochem.* **16**, 206 (1966).

DISCUSSION

J. F. Stoutjesdijk: Is it not possible to dissolve the ash immediately in hydrochloric acid? At ITAL, Wageningen, we ash several grams of plant material in a glass scintillation vial at about 400°C, dissolve the ash in hydrochloric acid and add a dioxane scintillation mixture. The hydrochloric acid produces considerable but constant quenching and corrections are not necessary.

J. Nolan: Yes, it is possible to do what you suggest for small biological samples, containing up to perhaps 20 mg calcium. It would, however, be impractical for the low specific activity samples with which I am concerned. It would be very difficult to evaporate 8 l of human urine and ash it in a counting vial! Also the colour of the dissolved residue would severely reduce counting efficiencies and thereby seriously affect our limits of detection.

B. Scales: Could you please give some indication of the long-term stability of your calcium chloride-containing samples. Is there any deterioration of the counting efficiency over a period of months?

J. Nolan: We could detect no change in counting efficiency or background in samples over a period of at least a year. The vials were closed with polythene inserts and capped with standard metal foil lined tops; they were stored at approximately 10°C in the dark.
I think the explanation for the negligible amount of evaporation of the samples is that the polythene and cap material are less permeable to the 12:8 ethanol:toluene mixture than they are to more or less pure toluene.

Chapter 12

The Simultaneous Determination of Plutonium Alpha Activity and Plutonium-241 in Biological Materials by Gel Scintillation Counting

J. D. Eakins and A. E. Lally

Health Physics & Medical Division, UKAEA Research Group, Atomic Energy Research Establishment, Harwell, England

INTRODUCTION

Plutonium-241 is produced from plutonium-239 in nuclear reactors by successive neutron captures. It has a physical half-life of 13.2 years and decays predominately by β-emission to americium-241, an α-emitter with a half-life of 458 years. A very small proportion, about 2.3×10^{-3} %, decays by α-emission to uranium-237.

The maximum permissible body burden (MPBB) recommended by the International Commission on Radiological Protection[1] is 0.9 μCi, compared with 0.04 μCi for plutonium-239. However, the determination of plutonium-241 in biological materials is becoming increasingly important as the burn-up level of uranium in power reactors increases. As Dalton *et al.*[2] have pointed out, the advent of the advanced gas-cooled reactor with proposed fuel irradiations of the order of 12000 MW days/metric ton will result in each MPBB of α-active plutonium being associated with 5.2 MPBB's of plutonium-241. In addition, the use of plutonium-239 as a nuclear fuel will greatly increase the relative hazard of plutonium-241.

Because of the low energy of its β-emission (E_{max} = 21 keV), plutonium-241 is usually detected either by internal proportional or liquid scintillation counting. Gale and Peaple[3] have reported the use of both internal proportional and plastic scintillation counters for the measurement of plutonium-241 surface contamination on smears. Dalton *et al.*[2] have described a small volume low background internal proportional counter for the determination of low levels of plutonium-241 electrodeposited on stainless steel. However, difficulties have been experienced with this method due to variations of source thickness causing changes in the counting efficiency.

The use of liquid scintillation for the determination of plutonium-241 was first reported by Horrocks and Studier.[4] These workers modified a Packard Tricarb liquid scintillation spectrometer to improve the light collection efficiency and incorporated plutonium-241 into a xylene-based liquid scintillator as the dibutyl phosphate complex. Ludwick[5] used this procedure for the determination of plutonium-241 in urine, after measuring the plutonium α-activity of an electrodeposited source by nuclear track counting. The plutonium was removed from the stainless steel backing by an acid treatment and after several chemical steps it was extracted into dibutyl phosphate for incorporation into a toluene-based scintillator. A minimum detectable level of 2.2 pCi of plutonium-241 was reported, with a counting time of 24 h.

Several workers have reported the use of liquid scintillation counting for the determination of plutonium α-activity in urine. Toribara *et al.*[6] used a toluene-based scintillator and after the addition of an aqueous solution of plutonium, added absolute alcohol to obtain a single-phase system. Because of the presence of trace quantities of iron in the plutonium, phosphoric acid was also added to produce a colourless solution. Lindenbaum and Lund[7] described the analysis of small samples of animal tissue in which the oxidation of the samples was carried out directly in the counting vial prior to the addition of liquid scintillator. Keough and Powers[8] dry-ashed tissue samples and added a nitric acid solution of the ash to a counting vial containing di(2-ethylhexyl) phosphoric acid dissolved in a toluene-based scintillator. The plutonium was extracted into the organic phase and liquid scintillation counting performed with both phases present in the vial. In all these methods the counting efficiency for plutonium α-activity was almost 100 %.

The method described in this chapter utilizes a simple gel scintillation counting technique simultaneously to determine both plutonium α-activity and plutonium-241 in a two-channel Packard Tricarb Model 3214 liquid scintillation spectrometer.

PREPARATION OF SOURCE FOR COUNTING

Chemical separation of plutonium

Any method can be used to extract the plutonium from the biological material, provided the end product is an acidic aqueous solution of ionic plutonium containing little or no iron. The present authors use the methods described in AERE–AM 103[9] to separate plutonium from urine and faeces. In the final stage of both methods the plutonium is eluted from an anion exchange column with concentrated hydrochloric acid containing a small amount of hydriodic acid. This solution is evaporated to near dryness under an infrared lamp and 1 ml of nitric acid is added to oxidize any residual hydriodic acid. The solution is again evaporated almost to dryness and transferred with water washings to a 10 ml centrifuge tube.

Preparation of gel source of ferriphosphate complex

Eakins and Brown[10] described the preparation of a white iron complex of diphosphatoferric acid which was used to determine iron-55 and iron-59 in blood by gel scintillation counting. It was found that this complex will coprecipitate the actinide elements quantitatively and is thus a useful method of introducing them into a scintillation system for gel counting. 2 mg of iron carrier are added to the acid aqueous solution of plutonium and ferric hydroxide is precipitated by the addition of ammonia. The precipitate is washed with water and dissolved in 0.25 ml of orthophosphoric acid. Absolute ethanol containing ammonium chloride is added to precipitate the ferriphosphate complex, which is centrifuged and washed with methylated spirit. The complex is slurried with 4 ml of NE 220* liquid scintillator and transferred to a counting vial containing Cab-O-Sil† silica powder as a gelling agent. The centrifuge tube is washed out twice with 4 ml volumes of scintillator, the washings being transferred to the counting vial which is mechanically shaken to produce a uniform gel for scintillation counting. The procedure is shown in outline in Fig. 1.

* NE 220 liquid scintillator is a dioxan-based scintillator manufactured by Nuclear Enterprises (G.B.) Ltd., Sighthill, Edinburgh 11, Scotland.

† Cab-O-Sil silica powder is supplied by the Packard Instrument Company Inc., York House, Empire Way, Wembley, Middlesex, England.

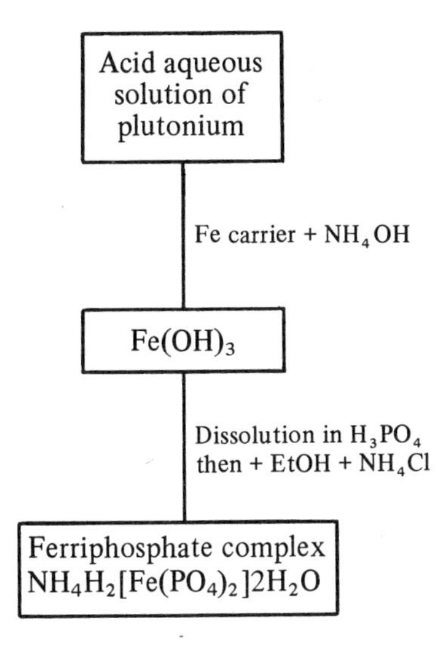

Fig. 1. The preparation of the ferriphosphate complex from a solution of plutonium.

SELECTION OF OPTIMAL COUNTING CONDITIONS

The term 'plutonium α-activity' is used in this chapter because plutonium-239 contains variable amounts of plutonium-240 and plutonium-238, depending on its source. The α energies of plutonium-239 and -240 are almost identical at 5.15 and 5.16 MeV respectively. The energy of the plutonium-238 α-particle is 5.49 MeV but this causes little shift of the gain spectrum compared with plutonium-239. Plutonium-239/240 containing only a few per cent of plutonium-238 was used for all the experiments described in this report and this is referred to as plutonium-239. Where the plutonium is of unknown isotopic composition the term plutonium α-activity is used.

Gain scans of plutonium-239 and plutonium-241

In order to obtain the optimum gain settings for α- and β-emitting plutonium a source of each was prepared as described above. Several gain scans were carried out at different high voltage settings at the full instrument window of 50–1000. The maximum count rate and peak separation were obtained with a high voltage setting of 4.3 on the instrument in use. These gain scans are shown in Fig. 2 where it can be seen that the optimum gain for plutonium-239 is about 3% and that for plutonium-241 is 60%.

Optimum window settings for plutonium-239

With the lower window or bias level at 50 and the gain at 3%, the upper level was reduced successively and the count rate at each setting recorded. In Fig. 3, curve A is a plot of count rate versus upper window setting and it can be seen that few counts are lost until the upper level is reduced to 500. The upper bias level was therefore fixed at 600 and the lower level raised as shown in Fig. 3, curve B. The count rate does not begin to drop appreciably until the lower level reaches 350 and the counting window was therefore taken initially as 300–600.

If E is the counting efficiency and B is the background, the sensitivity of the counter

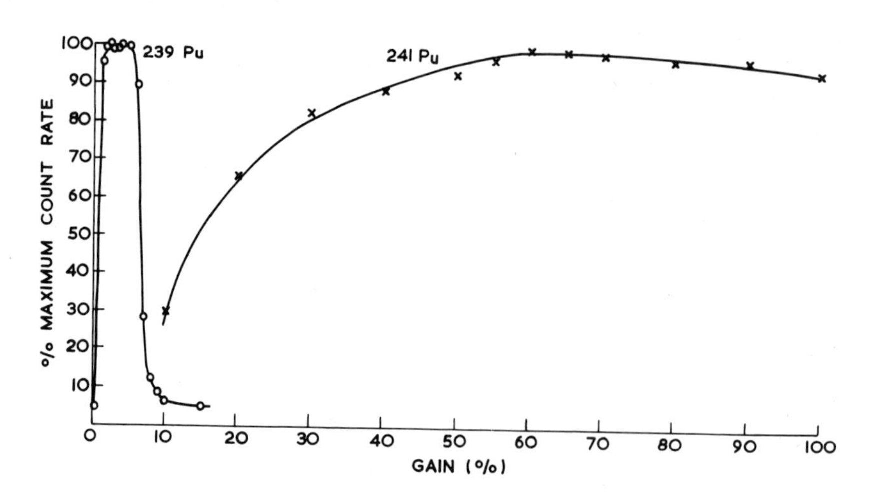

Fig. 2. Gain scans of plutonium-239 and plutonium-241.

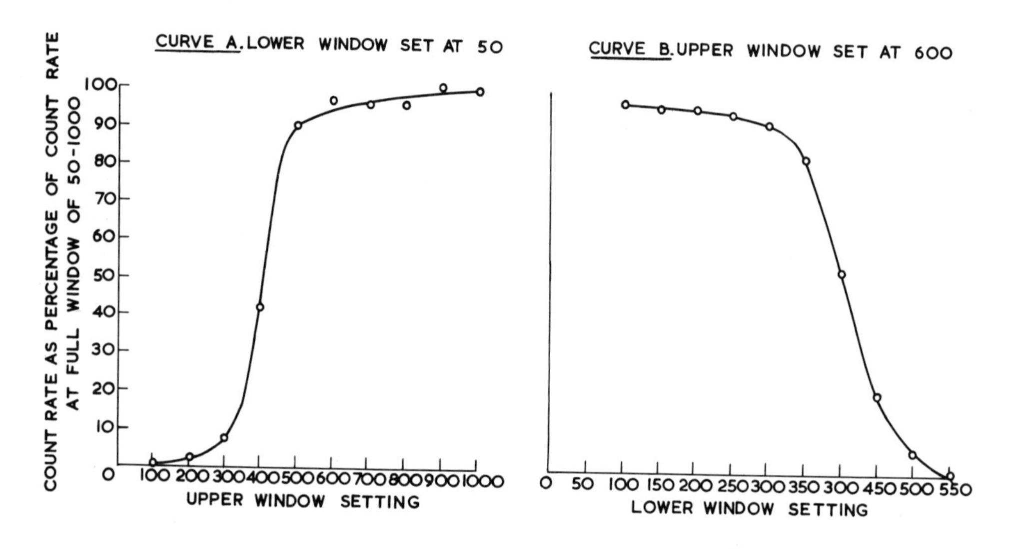

Fig. 3. Effect of window closure on count rate of plutonium-239.

is greatest when E^2/B is a maximum. The counting efficiency of the plutonium-239 source at a gain of 3% with a window width of 300–600 was 92.5%. The background count rate at these settings from a gel containing no added plutonium was 6.7 c.p.m. and hence E^2/B was 1277. Closing the window still further reduced both the counting efficiency and background as shown in Table 1. The maximum counting sensitivity was with a window of 300–500 and this was adopted as the optimum setting for plutonium-239.

Optimum window settings for plutonium-241

The background count rate of a gel containing no plutonium was 16.3 c.p.m. at the optimum gain of 60% with the full instrument window of 50–1000. The counting efficiency

Table 1. Window settings for counting plutonium-239 at maximum sensitivity.

Window	Counting efficiency (%)	Background (c.p.m.)	E^2/B
300–600	92.5	6.7	1277
300–550	89.7	5.6	1437
300–500	86.4	4.4	1696
350–500	69.0	3.2	1490
350–450	55.3	2.2	1274

for the plutonium-241 source was 21% and no reduction of the window was possible without a considerable loss in counting efficiency with little reduction in background. E^2/B was therefore at a maximum of 27.6 for plutonium-241 with a full window of 50–1000.

BACKGROUND MEASUREMENTS AND COUNTING EFFICIENCIES

Background measurements on urine samples from unexposed persons

In order to determine the mean background from urine, six 1400 ml (nominal 24 h) urine samples from people not occupationally exposed to plutonium were analysed by the method for plutonium in urine described in AERE–AM 103.[9] The sources were prepared for gel scintillation counting as described on p. 156. The gain and window settings were adjusted to count plutonium α-activity in the red channel of the scintillation spectrometer and plutonium-241 in the green channel. The sources were counted for 100 min and the results are shown in Table 2.

These backgrounds compare favourably with the figures of 4.4 and 16.3 c.p.m. previously obtained by counting ferriphosphate gels prepared directly from iron carrier solution.

Table 2. Background count rates obtained from 1400 ml urine samples.

Sample number	Red channel (c.p.m.)	Green channel (c.p.m.)
1	4.9	16.8
2	4.4	16.4
3	4.5	16.8
4	4.5	16.6
5	5.0	16.8
6	4.4	16.8
Mean and SD	4.6±0.3	16.7±0.2

Counting efficiency of plutonium-239 coprecipitated with ferriphosphate complex

To obtain a mean counting efficiency for plutonium-239 coprecipitated on the ferriphosphate complex, six samples of 2 mg of iron carrier in acid solution were spiked with

Table 3. Counting efficiency for plutonium-239.

Sample number	Counting efficiency (%) Red channel	Green channel
1	87.7	99.1
2	90.5	99.2
3	85.2	98.2
4	86.9	98.6
5	91.7	102.6
6	88.2	100.7
Mean and SD	88.4±2.2	99.9±1.5

53.3 pCi of plutonium-239 and prepared for gel scintillation counting. Both channels of the counter were set at 3% gain, with a reduced window of 300–500 in the red channel but with a full window of 50–1000 in the green channel. The red channel counts therefore gave the counting efficiency for plutonium-239 at maximum sensitivity and the green channel gave the maximum counting efficiency obtainable. The results of this experiment are shown in Table 3. The counting efficiency in the green channel is virtually 100%, indicating that the coprecipitation of plutonium with the ferriphosphate complex is quantitative.

The above sources were then recounted with the red channel on the same gain and window settings, but with the green channel set for plutonium-241 with a gain of 60% and a full window of 50–1000. The mean count rate above background observed in this channel was 4% of the count rate in the plutonium-239 channel. This could be a genuine contribution of the α-activity in the plutonium-241 region or it could possibly be due to a small amount of plutonium-241 in the plutonium-239 spike solution.

Counting efficiency of plutonium-241 coprecipitated with ferriphosphate complex

Six samples were prepared containing 253 pCi of plutonium-241 with 2 mg of iron as the ferriphosphate complex. These were counted in the liquid scintillation counter for 100 min with the red channel set for counting plutonium α-activity and the green channel

Table 4. Counting efficiency for plutonium-241.

Sample number	c.p.m.	Efficiency (%)
1	117.0	20.8
2	116.1	20.7
3	121.0	21.6
4	118.0	21.0
5	118.7	21.2
6	118.2	21.1
	Mean and SD	21.1±0.3

for plutonium-241. There were no significant counts in the red channel above background, showing that the plutonium-241 was not contributing any counts to this channel. The counts obtained in the green channel and the counting efficiency for plutonium-241 are shown in Table 4.

RADIOCHEMICAL RECOVERIES FROM URINE

Plutonium-239 spikes

Six 1400 ml urine samples from unexposed persons were spiked with 10.66 pCi of plutonium-239 and analysed according to the procedure described. The gelled sources of ferriphosphate complex were counted for 50 min and the amounts of plutonium-239 detected and the radiochemical recoveries are shown in Table 5.

Table 5. Radiochemical recovery of plutonium-239 from 1400 ml urine samples.

Sample number	Plutonium-239 found (pCi)	Radiochemical yield (%)
1	10.00	93.8
2	9.05	84.9
3	9.86	92.5
4	9.10	85.4
5	9.82	92.1
6	9.19	86.2
	Mean and SD	89.2±3.8

Plutonium-241 spikes

The above experiment was repeated with six 1400 ml urine samples spiked with 253 pCi of plutonium-241 and the sources were counted for 50 min. The results are shown in Table 6.

Table 6. Radiochemical recovery of plutonium-241 from 1400 ml urine samples.

Sample number	Plutonium-241 found (pCi)	Radiochemical yield (%)
1	211	83.6
2	214	84.7
3	208	82.4
4	224	88.4
5	202	80.7
6	210	82.9
	Mean and SD	83.8±2.4

High level mixed spikes

Six 1400 ml urine samples were spiked with 10.66 pCi of plutonium-239 and 127 pCi of plutonium-241 and analysed. It was assumed that 4% of the plutonium-239 counts in the red channel of the liquid scintillation spectrometer would appear in the green channel with the plutonium-241 counts and this was allowed for in calculating the results shown in Table 7.

Table 7. Recovery of plutonium from high level plutonium-239 and plutonium-241 spiked urine samples.

Sample number	Plutonium-239 recovery (%)	Plutonium-241 recovery (%)
1	84.9	92.8
2	81.1	81.1
3	86.6	90.4
4	84.9	84.3
5	89.6	93.9
6	81.5	81.5
Mean and SD	84.8±2.9	87.3±5.2

Low level mixed spikes

The previous experiment was repeated with six 1400 ml urine samples spiked with 2.13 pCi of plutonium-239 and 12.7 pCi of plutonium-241. The results are given in Table 8.

Table 8. Recovery of plutonium from low level plutonium-239 and plutonium-241 spiked urine samples.

Sample number	Plutonium-239 recovery (%)	Plutonium-241 recovery (%)
1	95.2	80.4
2	76.8	81.2
3	100.0	95.3
4	87.2	82.8
5	102.3	91.4
6	96.6	96.8
Mean and SD	93.0±8.3	88.0±6.2

ANALYSIS OF URINE SAMPLES FOR PLUTONIUM α-ACTIVITY AND PLUTONIUM-241 FOLLOWING AN INCIDENT

Urine samples were received from six subjects following acute exposure to plutonium. As the subjects had been treated with DTPA it was necessary to oxidize the urine to break down complexed plutonium. The samples were evaporated to dryness with nitric acid and the residue heated at 500°C for 30 min. The ash was dissolved in nitric acid and calcium

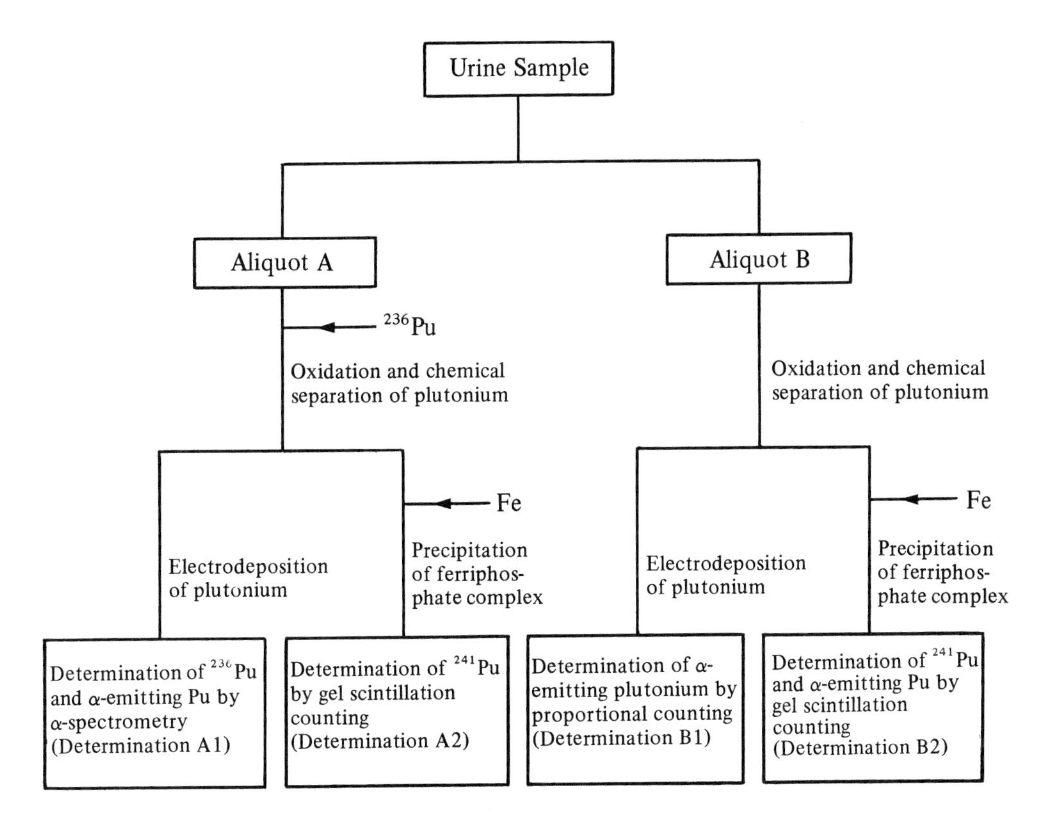

Fig. 4. Procedure used to analyse urine samples following acute exposure to plutonium and subsequent treatment with DTPA.

phosphate precipitated by the addition of ammonia. The analyses were then completed as described in AERE–AM 103[9] and in this chapter.

The urine samples were split into two equal aliquots. The first aliquot was spiked with plutonium-236 as an internal standard. After chemical separation but prior to the addition of iron to prepare the ferriphosphate complex, half of this aliquot was taken and the plutonium electrodeposited for α-spectrometry. The radiochemical recovery of plutonium α-activity in the urine was determined from the yield of plutonium-236. The mean radiochemical recovery obtained by this method was 79.5%. The other half of this aliquot was prepared as the ferriphosphate complex and the plutonium-241 determined by gel scintillation counting. The second aliquot was analysed in exactly the same way but without the addition of plutonium-236. In this case the plutonium α-activity on the electrodeposited source was determined by proportional counting. It was also determined simultaneously with the plutonium-241 by scintillation counting of the gelled source. Three results were thus obtained for the plutonium α-activity and two for the plutonium-241. The analysis is shown diagrammatically in Fig. 4 and the results are shown in Tables 9 and 10.

The errors quoted in Tables 9 and 10 are 2σ and refer only to the counting statistics. Although the results are quoted in pCi/l, it should be noted that the volumes of urine

Table 9. Plutonium α-activity in urine.

Sample number	Determination A1 by plutonium-236 and α-spectrometry (pCi/l)	Determination B1 by proportional counting (pCi/l)	Determination B2 by gel scintillation counting (pCi/l)	Mean (pCi)
1	28.6±1.1	29.7±1.8	33.7±1.4	30.7
2	38.7±1.2	36.9±2.2	42.4±1.5	39.3
3	21.9±1.8	14.4±0.9	14.1±3.9	16.8
4	25.3±1.3	24.4±1.5	27.1±2.3	25.6
5	14.2±1.2	15.4±0.9	14.1±2.2	14.6
6	28.7±1.7	26.4±1.6	27.1±4.1	27.4

Table 10. Plutonium-241 in urine.

Sample number	Determination A2 (pCi/l)	Determination B2 (pCi/l)	Mean (pCi/l)
1	692±13	648±12	670
2	844±14	912±15	878
3	339±34	406±41	372
4	590±20	543±18	567
5	364±18	321±16	343
6	598±25	518±22	558

received ranged from 0.25 to 1.0 l and that because of the subdivision of the samples, the final counting sources in each case represented the plutonium in urine volumes of from 62.5 to 250 ml. This accounts in some measure for the relatively large errors obtained. The mean recovery of 79.5% is lower than that found with spiked urine samples but this is probably due to losses occurring during the complete oxidation of the urine samples prior to analysis.

COMMENTS

The procedure described can adequately detect α-emitting plutonium at the 1 pCi level and plutonium-241 at the 10 pCi level of activity. The minimum detectable level (equivalent to a count rate equal to two standard deviations of the background of the counter) for plutonium α-activity is 0.3 pCi with a counting period of 100 min and 0.15 pCi for 300 min. The corresponding figures for plutonium-241 are 1.8 and 1.0 pCi respectively. It should be stressed that these limits can only be obtained with a stable background. The most likely cause of variable background is the liquid scintillator and it is important to process blank samples with the active specimens to establish the background for each batch. Dioxan-based scintillators are particularly prone to variations in background due to oxygen absorption and such scintillators should be thoroughly purged by bubbling with nitrogen both before and after use.

For routine urine analysis the procedure is not sufficiently sensitive for the detection of plutonium-239 at the levels suggested by Dolphin *et al.*[11] i.e. taking a reference level of 0.2 pCi/24 h. However, as the MPBB for plutonium-241 is about 22 times greater than that for plutonium-239, a corresponding reference level of 4.4 pCi could be detected with a 2σ error of ±30% using a counting period of 300 min. If necessary, prior to the formation of the ferriphosphate complex, an aliquot of the solution can be taken for electrodeposition and detection of plutonium-239 by solid state α-counting, when a minimum detectable level of 0.02 pCi is readily obtainable.[12] However the gel scintillation procedure is sufficiently sensitive for the analysis of faecal and nose blow samples, where much higher levels of activity may be expected, and also for the analysis of urine samples following an acute exposure.

Although other workers[4,5] have claimed counting efficiencies of 30% for plutonium-241 compared with the gel scintillation counting efficiency of 21%, the use of gel counting eliminates the necessity for quench corrections and simplifies the chemistry involved in the preparation of the source.

REFERENCES

1 I.C.R.P. Publication 2, *Report of Committee II on Permissible Dose for Internal Radiation,* Pergamon Press, London, 1959.

2 J. C. Dalton, B. J. McDonald and V. Barnes, *Biological Monitoring for Plutonium-241,* I.A.E.A. Symposium on Radioisotope Sample Measurement Techniques in Medicine and Biology, Vienna, May 1965.

3 H. J. Gale and L. H. J. Peaple, AERE–R 4113, 1962.

4 D. L. Horrocks and M. H. Studier, *Anal. Chem.* **30**, 1747 (1958).

5 J. D. Ludwick, *Health Phys.* **6**, 63 (1961).

6 T. Y. Toribara, D. A. Morken and C. Predmore, UR–607, 1962.

7 A. Lindenbaum and C. J. Lund, *Radiation Res.* **37**, 131 (1969).

8 R. F. Keough and G. J. Powers, *Anal. Chem.* **42**, 419 (1970).

9 J. D. Eakins, A. E. Lally, A. Morgan and F. J. Sandalls (Eds.), AERE–AM 103, 1968.

10 J. D. Eakins and D. A. Brown, *Intern. J. Appl. Radiation Isotopes* **17**, 391 (1966).

11 S. A. Beach, G. W. Dolphin, K. P. Duncan and H. J. Dunster, AHSB (RP) R 68, 1966.

12 F. J. Sandalls and A. Morgan, AERE–R 4391, 1964.

DISCUSSION

T. H. Bates: Comment: At Windscale we are employing this technique satisfactorily in the bioanalytical laboratory using a more recent Packard Tricarb Model and we obtain a counting efficiency of 37% for plutonium-241 and 95% for plutonium-α. By use of a set of accurately calibrated standard solutions of plutonium-239 and plutonium-241 blended so as to give β/α ratios ranging from 5 to 40 we found no statistical bias to suggest that there was any detectable contribution in the β-counting channel from the α-emitting isotope. Apparent counts in the β-counting channel from americium-241 and plutonium-238 standards could also be entirely accounted for by the presence of trace amounts of contaminating plutonium-241 determined by mass spectrometry.

Chapter 13

Techniques for Counting Carbon-14 and Phosphorus-32 Labelled Samples of Polluted Natural Waters

E. J. C. Curtis and I. P. Toms

Water Pollution Research Laboratory of the Department of the Environment, Stevenage, Hertfordshire, England

INTRODUCTION

There is considerable interest at present in aquatic biological systems, especially in the field of pollution where organic wastes, inorganic nutrients and algal productivity of natural waters are receiving particular attention. Problems with these are likely to throw an increasing strain on natural resources in the future as industrialization and urbanization continue to increase. Two key elements essential to all biological systems are carbon and phosphorus. In aquatic systems available carbon occurs as carbon dioxide, bicarbonate and carbonate, and in diverse organic compounds which may be present from polluting discharges, or from the breakdown of biological material. Phosphorus is intimately involved in all biological energy transformations, and in aquatic systems, where it is usually present in very low concentrations, it can be a limiting factor in biological production by photosynthetic or heterotrophic growth. The ability to measure accurately carbon and phosphorus, and their rates of uptake by microbial populations is thus essential to an understanding of any aquatic biological system.

Classical methods of analysis have serious limitations in the study of most natural aquatic systems where important substances are often present at levels less than 1 or 2 mg/l of water. In many waters the accurate estimation of soluble orthophosphate is difficult and tedious. A similar situation occurs with the chemical assay of most organic compounds likely to be present, and the test for biochemical oxygen demand (BOD) is non-specific for biodegradable organic material and is unreliable at best, especially at low concentrations. Additionally, for an understanding of aquatic ecosystems it is necessary to know both the standing crop (biomass) of the organisms, and the metabolic activity or productivity of the system. Using classical methods it is difficult to assess accurately the biomass of either autotrophic systems (algae and rooted plants) or heterotrophic ones (such as bacterial growths and slimes). Primary production – the photosynthesis of cellular carbon from carbon dioxide together with the release of oxygen – is due primarily to algae in deep water masses. This can be measured by three principal methods:

1. from the direct increase in suspended solids (a method which is unreliable when the weight of algae can be only a few mg/l);
2. from the rate at which dissolved oxygen is produced (this method is insensitive at low

numbers of algae and can be erroneous when the water is supersaturated with oxygen);
3. from the change in carbon dioxide concentration (this may be chemically calculated from pH changes, the accuracy depending on the buffering capacity of the water and the accuracy with which the dissociation constants of carbonic acid are known).

The use of radionuclides as tracers provides a powerful technique for the study of substances present in low concentrations, and both carbon-14 and phosphorus-32 have been widely used in biological studies, though the assay techniques have been subject to limitations. Substances have been assayed using either end-window Geiger-Müller counting of biological materials and evaporated solutions (a method subject to errors from self absorption, the geometry of the counter and dead time), or by liquid scintillation counting of non-aqueous solutions (which is usually limited to the use of materials miscible with organic solvents and is subject to loss of efficiency and reproducibility owing to chemical quenching). The present paper discusses techniques involving Cerenkov counting of aqueous phosphorus-32 and solubilization techniques which allow liquid scintillation counting of biological materials, in particular of phosphorus-32 in biological slime growths and carbon-14 in algae. The techniques discussed could be applied to studies involving various aquatic systems.

MATERIALS AND METHODS

Automatic liquid scintillation counter. For phosphorus-32 assay a Tracerlab Coru/matic II automatic twin channel counter was used at an EHT setting of 360 (equivalent to 1100 V). The channel settings are tabulated in Table 1.

Table 1. Channel settings for phosphorus-32 counting.

Channel	Coarse gain	Fine gain	Threshold	Window
Scintillation	16	1.5	025	Out
Cerenkov	64	1.0	025	1000

Manual liquid scintillation counter. For carbon-14 assay an EKCO M5402 liquid scintillation head with a single photomultiplier was coupled, via an interface circuit, to an EKCO N610 B scaler/timer with pulse height analyser. Two reproducible channels were selected by fixing the threshold, and then counting with and without a higher energy discriminator. As shown in Fig. 1, this enabled the whole pulse spectrum to be counted in integral channel C_I, and only the lower energy region in channel C_{II}. The criteria of Bush[1] were used to select the actual channel widths which are set out in Table 2. To obtain steady consistent backgrounds on the instrument 'smoothing and filtering' of the mains electricity supply was necessary.

End-window counter. An EKCO N530F scaler/timer connected to a 2B2 thin end-window (window thickness 1.9 mg cm^{-2}) Geiger-Müller tube.

Ultrasonic cleaning bath. This consisted of a KS 100 ultrasonic generator connected to a KS 101 cleaning bath, both items manufactured by Kerry Ultrasonics Ltd., Hitchin.

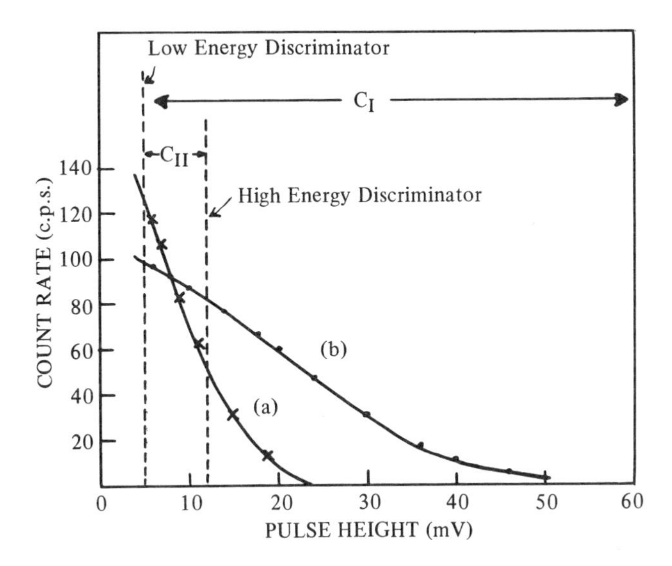

Fig. 1. Pulse spectra of (a) quenched and (b) unquenched carbon-14.

Table 2. Channel settings for counting carbon-14.

Channel	EHT	Amplifier gain	Threshold	Window
C_I	1255 V	1000	5 mV	Out
C_{II}	1255 V	1000	5 mV	7 mV

Reagents. *(a) Scintillators:*

(i) KL 353*, a dioxan-based liquid scintillator miscible with up to 12% of water. To render this scintillator thixotropic the finely ground silica powder Cabosil*was used.

(ii) Scintillation grade toluene* containing 15 g/l of Butyl-PBD†.

(b) Solubilizing agents:

(i) Hyamine hydroxide*, 1 M in methanol.

(ii) Soluene 100§.

(iii) NE 520‖.

Other chemicals used were of Analytical Reagent Grade.

Radiochemical materials. All were supplied by the Radiochemical Centre, Amersham, Buckinghamshire. Stock solutions were:

(i) Phosphorus-32, supplied as high specific activity orthophosphate in dilute hydrochloric acid (1 mCi/ml).

(ii) Carbon-14, supplied as $Na_2{}^{14}CO_3$ having a specific activity of 20 mCi/mM and a concentration of 1 mCi/ml (cat. No. CFA 1).

---

* Koch-Light Laboratories, Colnbrook, Bucks., England.
† 2-(4′-*t*-butylphenyl)-5-(4″-biphenylyl)-1,3,4-oxadiazole, Ciba Chemicals, Duxford, Cambs., England.
§ Packard Instrument Co. Ltd., York House, Empire Way, Wembley, Middlesex, England.
‖ Nuclear Enterprises Ltd., Edinburgh, Scotland.

Radiochemical standards were:
(i) Phosphorus-32, supplied as a standardized solution of $NaH_2{}^{32}PO_4$ in water.
(ii) Carbon-14, supplied as standardized n-hexadecane (carbon-14).

Dispensing techniques. Small quantities up to 1 ml were dispensed using Marburg micropipettes* with disposable tips. Quantities between 1 ml and 15 ml were dispensed with Aimer D-L Repeating Syringes†. The standard deviation on the reproducibility of these was within 1% of their calibrated volume.

Filtration equipment. Samples were vacuum filtered using Millipore § glass funnels and sinters. The glass funnels were coated with PTFE to minimize cross-contamination. Bacterial slimes, being mucilaginous, were filtered through Whatman GF/C discs; algae were filtered through 50 mm diameter 0.45 μm Millipore HA filters (cat. No. HAWP 04700).

Carbon-14 method for measuring primary production. The carbon-14 method for measuring primary production, developed by Steemann Nielsen,[2] involves adding a known quantity of $NaH^{14}CO_3$ to a water sample, followed by incubation and subsequent removal of the algae by membrane filtration for carbon-14 assay. Total carbon assimilation is calculated from the concentrations of carbon-14 fixed, carbon-14 added, and all forms of natural inorganic carbon dioxide present (measured by methods outlined by Golterman[3]). A small correction is necessary as carbon-14 is fixed at a slightly slower rate than natural carbon.[2,4] In the method used at the Water Pollution Research Laboratory a 1.5 ml aliquot containing between 2 and 6 μCi of sterile $NaH^{14}CO_3$ is added to the sample (160 ml in a round-bottom glass-stoppered bottle) using a repeating syringe. A replicate aliquot of the $NaH^{14}CO_3$ is diluted to 250 ml with 0.01 M Na_2CO_3 solution, and 0.5 ml aliquots are removed for calibration using the methods described later.

ASSAY OF LIQUID SAMPLES

Liquid samples were assayed using liquid scintillation techniques for carbon-14 and Cerenkov counting for phosphorus-32.

Cerenkov counting of phosphorus-32 in aqueous solution

Cerenkov radiation is produced as a bluish light when β-particles of energy greater than about 300 keV travel through a medium such as water. Cerenkov radiation forms only a small proportion (less than 1%) of the total energy losses and thus its use as an assay technique is limited to those radionuclides which are high energy β-emitters (phosphorus-32 has E_{max} of 1.71 MeV) and has only been possible since the development of automatic liquid scintillation counting equipment capable of detecting very low intensity light pulses.[5,6] As no scintillator is required and chemical (but not colour) quenching is absent, sample preparation is extremely simple and the technique is ideal for the assay of phosphorus-32 in aqueous solution.

Samples were assayed in 25 ml opaque polythene vials, which were found to yield a higher counting efficiency than low background glass vials (see Fig. 2), presumably owing to diffusion of the directional Cerenkov emissions. Efficiency of counting was very dependent on volume for both types of vials (Fig. 2), and in any series of samples it was therefore necessary to standardize the sample volume carefully. In most cases a total sample

* Eppendorff Co., imported by Anderman & Co., Tooley Street, London S.E.1, England.
† Aimer Products Ltd., 56–58 Rochester Place, Camden Road, London N.W.1, England.
§ Millipore (U.K.) Ltd., Heron House, 109 Wembley Hill Road, Wembley, Middlesex, England.

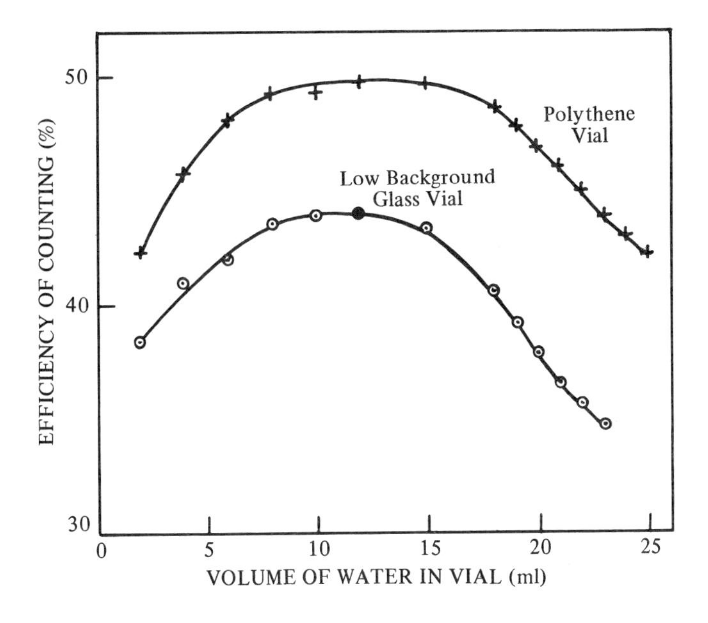

Fig. 2. Effect of sample volume on Cerenkov counting efficiency.

volume of 22 ml was used.

Counting efficiency for each series of samples was standardized by removing 2 ml aliquots for counting in a scintillator giving a known high and reproducible counting efficiency. For this purpose, KL 353 containing 6.3% ethyl alcohol, which raised the water capacity of the scintillator from 12% to over 20%, was used. Over the range 0 to 20% water the counting efficiency of this scintillator for phosphorus-32 was 98±0.5% relative to RCC standard phosphorus-32 solution. Efficiency of Cerenkov counting 22 ml replicate samples was found to be 44.2±0.2% by this means. Although this is much lower than the efficiency of liquid scintillation counting, the minimum detectable levels of activity are much less because the Cerenkov background activity is lower and the volume of the aqueous sample which can be counted is much larger. The comparison of performance is summarized in Table 3.

In experiments involving the assay of phosphorus-32 in natural waters by Cerenkov counting, various factors have to be considered, the most important of which are:

1. adsorption of phosphorus-32 onto vial walls;[7]
2. precipitation of $Ca_3(^{32}PO_4)_2$ from solution in hard natural waters;
3. microbial activity. (After sampling, phosphorus-32 can be rapidly removed from solution by micro-organisms in non-sterile samples; there may be uptake by organisms already present, or growth of micro-organisms on vial walls and in solution).

To control these factors, a stopping solution was added to the vials prior to sample collection; a solution containing EDTA to chelate Ca^{2+} ions, KH_2PO_4 to dilute $^{32}PO_4{}^{3-}$ and thus prevent its adsorption onto surfaces, and formalin to prevent uptake by micro-organisms was used for this. In most experiments sample volumes of 21 ml were added to

Table 3. Performance of counter for aqueous samples.

	Maximum sample volume (ml)	Counting efficiency (%)	Efficiency X volume	Background (c.p.s.)	Minimum detectable activity (nCi/l)[a]
Liquid Scintillation	4	98	3.92	1.3	0.70
Cerenkov	20	44.2	8.80	0.25	0.12

[a] 20 min counting period.

Table 4. Effectiveness of stopping solutions in Cerenkov counting of phosphorus-32.

Period from start	5 h	1 day	2 days	5 days
% activity retained in vial after washing	0.03±0.03	0.07±0.03	0.15±0.02	0.25±0.04
Activity in control vials (0 h = 100±2.6)	–	–	101±2.4	99±1.4

1 ml of stopping solution. Concentrations in the final stopped solution were: EDTA . 2Na, 2.74 g/l (to allow for a Ca^{2+} content of 300 mg/l in the sample); KH_2PO_4, 12.1 mg/l; and formalin, 1%. To determine the effectiveness of this solution, a series of vials was prepared containing these concentrations with a phosphorus-32 sample in local (Stevenage) hard water. At intervals of 5 h and 1, 2 and 5 days replicate samples were decanted and the vials were washed twice with tap water, filled with tap water, and recounted. Control samples without washing were counted at each interval. Results are shown in Table 4.

Carbon-14 assay by liquid scintillation

The method involves assay of 0.5 ml of a known dilution of the aqueous $Na_2{}^{14}CO_3$ inoculum by homogeneous liquid scintillation counting using either the internal 'channels ratio' method[1,8] or an internal standard to calibrate the efficiency of counting.

A channels ratio calibration was prepared by quenching samples containing known quantities of carbon-14 (as n-hexadecane) with increasing quantities of either acetone, water, or methyl acetate; Fig. 1 shows normal and quenched pulse spectra of carbon-14 in KL 353 and Fig. 3 shows the efficiency of counting in each of the channels C_I and C_{II} as a function of the ratio of the count rate in C_{II} to that in C_I. The curves were independent of the quenching or diluting agents, in agreement with the findings of Baillie[8] and Bush.[1] Efficiency in C_I is a linear function of channels ratios over a wide range and is represented by the equation:

$$\text{Efficiency} = 103.5 - 85.0 \times \text{channels ratio}$$

This equation together with background values and counting times is programmed into a desk calculator (IME Digicorder) to facilitate data processing.

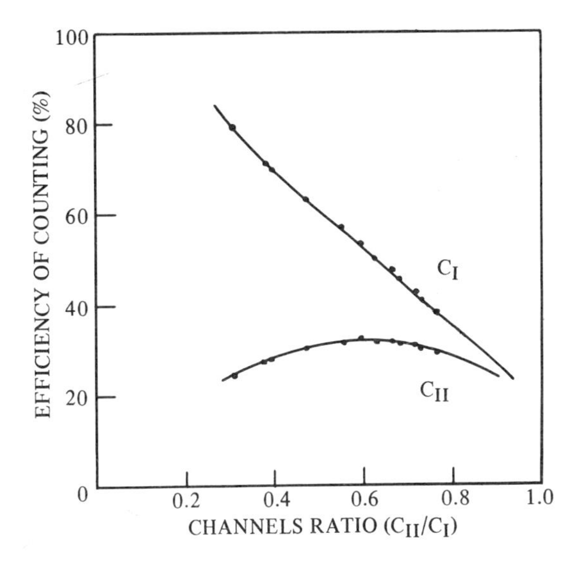

Fig. 3. Typical calibration of efficiency of counting against ratio of count rates in overlapping channels.

Table 5. Quenching of four similar samples of 10 ml KL 353 containing 0.5 ml water with standard n-hexadecane (carbon-14).

Sample	Efficiency of counting carbon-14 in aqueous solution (%)	Decrease in counting efficiency on adding 15 mg n-hexadecane (%)
1	67.7	4.2
2	72.0	2.5
3	69.5	1.3
4	68.5	2.2

It was found, using the channels ratio method, that small volumes of water and standard n-hexadecane, which cause little quenching in the dioxan scintillator when used separately, cause increased quenching when mixed. Table 5 shows that this change in overall efficiency is not easily quantified and can be greater than 4% (a relative change of 6%).

Thus, under these conditions n-hexadecane is not recommended as an internal standard for aqueous carbon-14 solutions, and the determination of counting efficiency by channels ratio gives more reproducible results. Table 6 shows the results of six determinations of a $NaH^{14}CO_3$ stock solution by this method.

LIQUID SCINTILLATION COUNTING OF BIOLOGICAL SAMPLES

Biological tissues have presented serious problems with respect to assay of radioactivity. The simplest method is to count the dried material using an end-window Geiger-Müller counter, but this suffers from inherent disadvantages such as self-absorption and geometry. Assay by liquid scintillation is a much preferred method and can be attempted

Table 6. Replicate determinations of the activity of a standard $NaH^{14}CO_3$ solution.

Sample	Weight of aliquot (mg)	Count rate (c.p.s.)	Counting efficiency by channels ratio (%)	Absolute activity of inoculum (disintegrations/s)
1A	597	319.5	63.0	2.12×10^5
1B	592	325.4	62.5	2.20×10^5
2A	594	315.6	62.7	2.12×10^5
2B	607	314.2	61.5	2.10×10^5
3A	595	317.9	62.5	2.14×10^5
3B	586	314.3	62.3	2.15×10^5

Mean activity of inoculum $(2.14 \pm 0.01) \times 10^5$ disintegrations/s.

by two general techniques:

1. solubilization of tissues, and
2. the production of suspensions (especially with unicellular algae etc.).

Tissue solubilization

Various solubilizing agents have been used in attempts to prepare toluene solutions of tissues, proteins etc. Hyamine hydroxide, a lipophilic quaternary nitrogen base, was the first such agent to be used.[9] It could dissolve various biological materials, even those with a high protein content, but was ineffective for dissolution of bacterial cell walls.[10] Bacterial slimes are frequently encountered in studies of river pollution and a method for their solubilization into scintillator solutions would have wide application. Screening tests were therefore carried out on a number of possible solubilizing agents including a recently marketed product 'Soluene 100', using a range of aquatic plant materials and bacterial slimes.

Materials were added in the proportion of 1 ml solubilizing agent to 100 mg fresh weight of tissue (washed, blotted, and air dried) or the corresponding dry weight (10 to 25 mg depending on the ratio of dry to fresh weight of the particular tissue). Formamide was included in the screening test in the proportion of 10 ml/100 mg fresh weight or corresponding dry weight because it has been found to solubilize certain bacterial cells.[11] The results are shown in Table 7.

Soluene 100 is obviously the best general solubilizing agent for the range of tissues used, and for the bacterial slime it was the only one to be at all effective.

A mixture of Soluene-dissolved tissues with scintillator resulted in a yellow colouration and a very high count rate in the sample regardless of the presence of any radioactive isotope. This count rate decreased on removal from light, but approached a constant value much higher than could be attributed to radioactive emission (see Fig. 4). This phenomenon of chemiluminescence* has been observed with hyamine solutions of biological tissues.[12] Part of the effect may be due to the presence of biological tissue (Debye and Edwards[13] found that proteins in alkaline solution exhibited chemiluminescence which they attributed to tyrosine and tryptophan), but a mixture of solubilizer and scintillator alone also results in some chemiluminescence. Dunn[14] noted the interaction of Butyl-

* Probably a combination of chemiluminescence and phosphorescence.

Table 7. Effect of solubilizing agents.

Tissue material	Solvent	Solution of fresh material (a)	(b)	Solution of dried material (a)	(b)
Filamentous alga (*Cladophora*)	NE 520	None	None	None	None
	Hyamine	+	+	+	+
	Soluene	+	++	+	+
	Formamide	None		None	
Rooted plant (*Potamogeton*) leaves	NE 520	None	None	None	None
	Hyamine	+	+	None	None
	Soluene	+++	+++	None	None
	Formamide	None		None	
Rooted plant (*Potamogeton*) stems	NE 520	None	None	None	None
	Hyamine	+++	+++	+	+
	Soluene	+++	++++	+	+
	Formamide	None		None	
Bacterial slime (sewage fungus)	NE 520	None	None	None	None
	Hyamine	+	+++	+	+
	Soluene	+++	++++	++++	++++
	Formamide	None		None	

Degrees of solution: + (little);
++ (approximately 50%);
+++ (almost complete);
++++ (complete).
Treatment (a) involved heating at 60° for 12 h.
Treatment (b) involved treatment (a) followed by addition of 0.1 ml concentrated hydrochloric acid; 10 ml of scintillator was then added and the mixture immersed in the ultrasonic bath for 20 min.
Treatment (b) was not applied to formamide.

PBD scintillator and Soluene 100 and concluded that their use in combination was not possible. However, it was found in the present work that complete instantaneous quenching of this chemiluminescence could be achieved by the addition of 0.1 ml concentrated hydrochloric acid. This resulted in a slight phase separation but there appeared to be no coprecipitation of phosphorus-32 when this occurred. When phosphorus-32 was added to the Soluene-scintillator samples recovery of radioactivity was 99±1%, whether the phosphorus-32 was added before or after admixture of the acid with scintillator.

Counting of high activity (several hundred c.p.s.) phosphorus-32 slime samples by the end-window Geiger-Müller counter followed by Soluene liquid scintillation counting showed both techniques to be very reproducible; assay by Geiger counter was 23.1±2.4% of that achieved by liquid scintillation (98±0.5% efficiency). However, at lower levels there was less reproducibility, presumably owing to errors in the Geiger counting (self-absorption of phosphorus-32 and also adsorption of phosphorus-32 onto the glass fibre discs onto which the slime was filtered). At very high activities the dead time of the Geiger counter (300 μs) had to be taken into account.

Assay of algal suspensions

Recent workers have used liquid scintillation assay to determine carbon-14 in algae

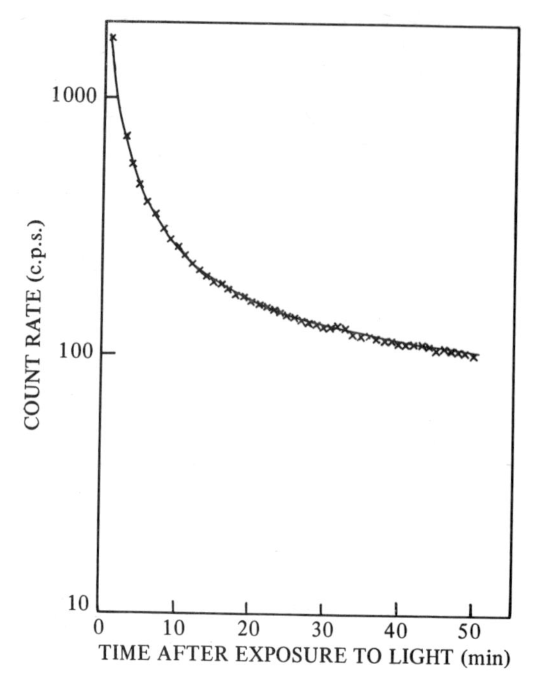

Fig. 4. Chemiluminescence and/or phosphorescence produced by Soluene, slime, and scintillator, and its partial decay on removal from light. True sample count rate after quenching with acid, 1.5 c.p.s.

on a membrane filter by two principal methods. In the first, the membrane is added to a counting vial containing scintillator (toluene based), which renders the membrane transparent. It has been claimed[15,16] that the algae are counted on the membrane with the same efficiency as carbon-14 in the liquid phase, but Pugh[17] showed that, in certain cases, this was not so, and that a correction for self-absorption, using the channels ratio, was required. In the second method, the algae are suspended directly in scintillator by dissolving the membrane. Krishnamoorthy and Vaswanathan[18] used acetone to dissolve the membranes which were then mixed with scintillator. Schindler[19] found that damp (but not dry) membrane filters were directly soluble in dioxan-based scintillators leaving the algae in suspension, but in his use of internal standardization with soluble standard toluene (carbon-14) it is not clear how the efficiency of counting of particulate carbon-14 is obtained.

Although it has been shown[20] that self-absorption can be neglected when assaying suspensions of algae, Hayes[21] showed that when self-absorption did occur, the lower energy region of the pulse spectrum was degraded first (*cf.* quenching where the higher energy region is degraded preferentially). Therefore it is to be expected that the channels ratios, outlined earlier, will be increased by quenching and decreased by self-absorption.

For the assay of phosphorus-32 and other high energy β-emitters in algae, the simple suspension technique of Schindler[18] can be recommended, using KL 353 rendered thixotropic with an equal volume of Cabosil. No self-absorption should occur and efficiencies of at least 95% should be obtained.

Table 8. Effects of shaking on the counting properties of algal suspensions.

Sample	Method of shaking	Change in count rate caused by shaking (%)
Microcystis suspensions	Mechanical	+3.3±0.5
Microcystis suspensions	Ultrasonic	+20.6±0.9
Soluble carbon-14 with non-radioactive *Microcystis* suspension	Ultrasonic	−1.9

To study the situation with carbon-14, a suspension of ^{14}C-labelled *Microcystis* (a blue-green alga forming microscopic gelatinous colonies) was formed in KL 353 as follows. The algae were filtered onto a membrane; this was then dissolved in 4 ml of non-quenching methyl acetate leaving a suspension of the colonies which was mixed with 10 ml of KL 353. To prevent settling (which is very marked with algae in liquid scintillators) the suspension was rendered thixotropic by the addition of 2 ml of 15% w/v Cabosil in KL 353. The suspension was counted in the two channels C_I and C_{II} before and after the addition of n-hexadecane internal standard. The channels ratio of the suspended carbon-14 at 0.67 was almost 0.10 higher than that of the soluble carbon-14, implying that quenching processes specific to suspended carbon-14 predominated over self-absorption.

Further samples of *Microcystis* colonies labelled with carbon-14 were suspended in scintillator and counted in the two channels before and after mechanical shaking, and after varying intervals of immersion in the ultrasonic cleaning bath. Table 8 shows that, in contrast to mechanical shaking, ultrasonic shaking had a marked effect on the count rate, raising this by about 20%.

The optimum time of shaking was about 25 min, after which there was no significant increase in count rate or efficiency. After this period it was found that the colonies had been broken up into individual cells of less than 5 μm diameter; these should have no self-absorption.[22] It was then confirmed, by the addition of internal standards, that soluble and particulate carbon-14 fractions were being counted with the same efficiencies; from Fig. 3, the values of the channels ratios were used to calculate the absolute activity of the algae, and the efficiency of counting as a function of time of shaking was plotted (Fig. 5). The slight decrease in solution counting efficiency shown in Table 8 presumably results from the increased opacity of the fine suspension.

Using the calculated absolute activity of the algae, the efficiency of algal suspension counting as a function of channels ratios was then plotted (see Fig. 6). This shows the slope of the line to be almost twice that for chemical quenching, and similar to the colour quenching curve of Bush,[1] and to that for algae attached to membranes.[17] It can be concluded that the loss in count rate and counting efficiency, and the anomalous channels ratios for these algal suspensions, are probably the result of very localized colour quenching, i.e. the very close proximity of a scintillation to a large 'black' particle of algae may result in absorption of up to half of the emitted light energy. This effect can be eliminated by ultrasonic shaking, and then the normal chemical quench correction curve can be used. Self-absorption from individual algal cells should be negligible; however, if its occurrence is suspected the effect can be decreased by discriminating against the lower pulses at the

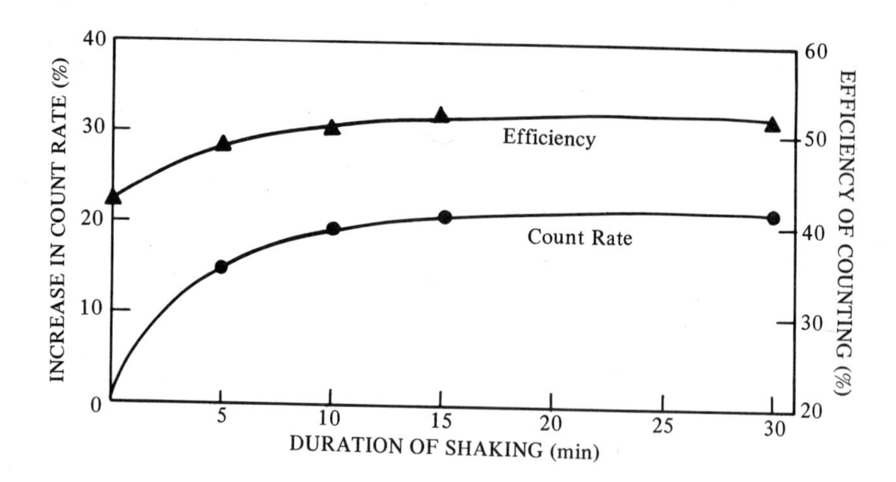

Fig. 5. Effect of ultrasonic shaking on the counting of carbon-14 in algal suspensions.

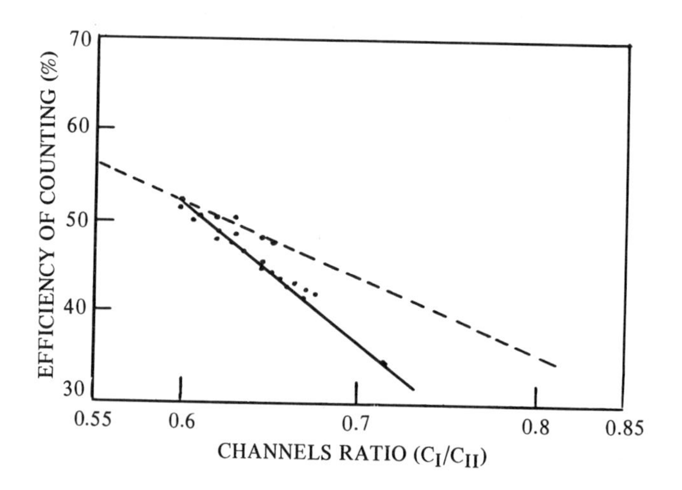

Fig. 6. Efficiency of counting of suspensions of *Microcystis* as a function of channels ratio. Solid line – particulate carbon-14 quench curve; broken line – chemical quench curve from Fig. 3.

expense of a slight decrease in counting efficiency.

For routine measurements, the dampened membrane is placed in a counting vial containing 10 ml of KL 353 rendered thixotropic with an equal volume of Cabosil. This is followed by immersion for 25 min in the ultrasonic bath. The vial is then assayed by counting the carbon-14 in the two channels.

APPLICATION AND CONCLUSIONS

Cerenkov counting of phosphorus-32 has been used to obtain quantitative estimates of biological slime growths. Initial uptake of phosphorus from water is primarily due to the micro-organisms present. In an organically polluted reach of water the micro-organisms are almost entirely attached heterotrophs (i.e. slime organisms) and thus uptake of phosphorus (or phosphorus-32) gives a measure of slime biomass. This method was used to measure slime biomass in a polluted effluent channel. Liquid scintillation techniques have been used to assay carbon-14 in primary productivity studies on a multi-purpose water-supply reservoir subject to periodic blooms of algae.

The techniques described should prove suitable for other similar studies of production and biomass. Cerenkov counting especially, with the advantages of high sensitivity and little sample preparation, seems well suited to studies in aquatic systems. Because of the key role of phosphorus in biological systems, phosphorus-32 lends itself to the study of a variety of problems, but chlorine-36 and krypton-85 are other available isotopes which should prove amenable to Cerenkov counting.

ACKNOWLEDGEMENTS

The authors wish to thank Miss Angela Smith for sample preparation and processing of results.

REFERENCES

1 E. T. Bush, *Anal. Chem.* **35**, 1024 (1963).
2 E. Steemann Nielsen, *J. Conseil, Conseil Perm. Intern. Exploration Mer* **18**, 117 (1952).
3 H. L. Golterman, *Methods for the Chemical Analysis of Fresh Waters,* Blackwell Scientific Publications, Oxford, 1969.
4 V. F. Raaen, G. A. Ropp and H. P. Raaen, *Carbon-14,* McGraw-Hill, New York, 1968.
5 R. P. Parker and R. H. Elrick, *Brit. J. Radiol.* **39**, 236 (1966).
6 H. Braunsbery and A. Guyver, *Anal. Biochem.* **10**, 86 (1965).
7 W. Hassenteufel, R. Jagitsh and F. F. Koczy, *Limnol. Oceanog.* **8**, 152 (1968).
8 L. A. Baillie, *Intern. J. Appl. Radiation Isotopes* **8**, 1 (1960).
9 J. M. Passman, N. S. Radin and J. A. D. Cooper, *Anal. Chem.* **28**, 484 (1956).
10 J. H. Hash, *Anal. Biochem.* **4**, 257 (1962).
11 A. Y. Neujahr and B. Ewaldsson, *Anal. Biochem.* **8**, 487 (1964).
12 R. J. Herberg, *Science* **128**, 199 (1958).
13 P. Debye and J. O. Edwards, *Science* **116**, 143 (1952).
14 A. Dunn, *Intern. J. Appl. Radiation Isotopes* **22**, 212 (1971).
15 D. A. Wolfe and C. L. Schelske, *J. Conseil, Conseil Perm. Intern. Exploration Mer* **31**, 31 (1967).
16 O. T. Lind and R. S. Campbell, *Limnol. Oceanog.* **14**, 787 (1969).
17 P. R. Pugh, *Limnol. Oceanog.* **15**, 652 (1970).
18 T. M. Krishnamoorthy and R. Vaswanathan, *Indian J. Exp. Biol.* **6**, 67 (1968).
19 D. W. Schindler, *Nature* **211**, 844 (1966).
20 J. D. Yardbrough, A. F. Finders and J. C. O'Kelley, *Intern. J. Appl. Radiation Isotopes* **17**, 453 (1966).
21 F. N. Hayes, *Nucleonics* **3**, 48 (1956).
22 F. Barreira and M. Laranjeira, *Intern. J. Appl. Radiation Isotopes* **2**, 145 (1957).

DISCUSSION

B. E. Gordon: How did you establish that the disappearance of $^{32}PO_4$ from the polluted channel was due only to slime uptake and not for example due to precipitation of salts?

E. J. C. Curtis: Samples of slimes, surface sediments and aquatic rooted plants were removed for assay from the whole length of the channel. From these data it was calculated that, of the total phosphorus-32 removed, less than 4% was present in rooted plants and sediments (which would include any precipitated salts).

Chapter 14

Application of Cerenkov Technique to Continuous Measurement of Radioactive Isotopes Isolated by an Automatic Analytical Process

J. Colomer, M. Cousigne and G. Metzger

Commissariat a l'Energie Atomique,
75.752-Paris 15, France

INTRODUCTION

In an attempt to improve the radioisotope analyses made in our laboratory on the products of fission mixtures, we attempted the replacement of the classical radiochemical methods (separation, then measurement) by automatic methods in which a more sophisticated analytical process is followed by on-line measurement utilizing Cerenkov counting.

In the classical methods which are often manual, it is necessary to determine the chemical separation efficiency. This difficult determination is made by different techniques such as gravimetry, absorption spectrometry, etc. and when radiometric measurement is made (β or γ technique) the counting rate so obtained must be corrected and one is forced to consider factors such as detection efficiency, self-absorption and coefficient of geometry, decay scheme and radioactive equilibrium, quenching, etc.

In order to avoid these disadvantages we are conducting research, the object of which is the production of a liquid phase radioanalyser in modular form, allowing the continuous analysis of radioisotopes as they become isolated.

Figure 1 shows a block diagram of the radioanalyser. The separation unit at uniform concentration is designed around an analytical process, the object of which is to isolate one or more chemical elements. The detection unit which uses Cerenkov detection analyses on-line the corresponding radioisotope and the measurement is performed versus a standard of activity of the same radioelement, treated under identical experimental conditions.

After a short examination of the analytical methods used in the separation unit, we describe the technique employed in the detection unit and give some examples of its application.

THE ANALYTICAL PROCESS: SEPARATION UNIT

The process is based on the non-active isotope dilution method in conjunction with the substoichiometric separation principle, the theory of which was presented by Ruzicka.[1,2]

The activity A of the radioelement to be measured can be determined from the separated activities (a) of the sample solution and (a_0) of a standard solution of the same

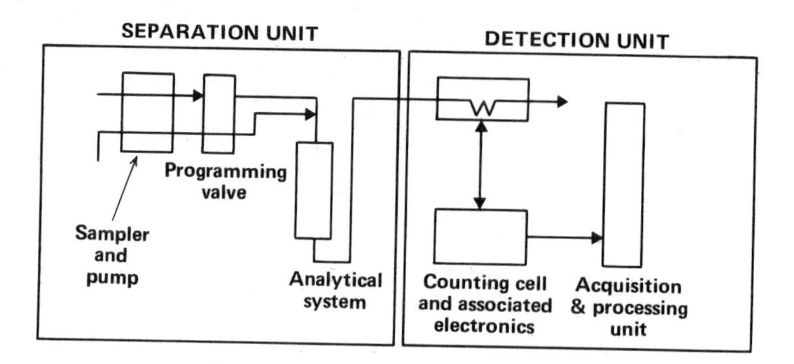

Fig. 1. Block diagram of the radioanalyser.

radioelement by the expression:

$$A = A_0 \frac{a}{a_0}$$

A_0 is the known initial activity of the standard solution.

This expression can be used and the analysis is then possible provided a number of conditions are satisfied:

1. The amount of carrier of the standard of activity must be minimal.
2. Exactly the same amount of isotopic carrier must be added to the two solutions.
3. A good isotopic exchange must be realized between the carrier and the corresponding radioelements.
4. For a correct evaluation of a and a_0, an identical amount of the element in these two solutions must be isolated (the substoichiometric separation principle).

This method need not take account of the determination of chemical separation efficiency nor of the absolute determination of activity; it consists of a simple relative measurement. The separation processes used were:

i) The substoichiometric separation with continuous solvent extraction.[3,4]
ii) The displacement at uniform concentration through cation exchange resins.[5,6]

The dilution and substoichiometric separation are automatically made by the separation unit; it consists of a sampler, proportioning pump (Technicon), manifold, mixing coil, separation trap or ion exchange column.

RADIOMETRIC DETECTION: DETECTION UNIT

The problem, after continuous separation of the isotope to be measured at uniform concentration, was to perform an immediate radiometric measurement in order to profit from the presence of these isotopes without its parents and daughters; we did not need to take into account filiation and radioactive equilibrium. Therefore, we decided to adopt on-line measurement to avoid interference by decay products.

Choice of this method

The continuous measurement of β activity in fluid streams in chromatographic effluents is a routine and useful technique in biological investigations;[7-9] the principle and methods are applicable to any flowing stream such as that which comes out of the separation unit.

The various cells designed until now, belong to three groups:

Plastic scintillator cells. In this type of cell, the first to be described in the literature, the stream to be measured is circulated through a cell of plastic scintillator formed like a helix or a flat coil (SPF, NE 102 or NE 814).

Cells packed with suspended fluor particles. The most popular system is a cell filled with small anthracene crystals; the fluid flows in the spaces between the solid particles of fluor. Cells of this type are constructed of inert materials with low background and high light transmission.

Homogeneous counting systems. The fluid stream is continually mixed with a liquid scintillator before passing through the cell, made in low background material.

Note that the first and second types of cells cannot be used when the fluid stream is an organic solvent such as that used in continuous solvent extraction techniques. The third type of cell implies that the flowing solution be mixable with the scintillator; one is faced with exactly the same problems as in the discrete counting of aqueous samples by means of liquid scintillators, and chemical quenching is possible.

Since Belcher's first work,[10] an important advance in methods of radioassaying solutions of β-emitting radionuclides has been achieved. This is also due to the development of liquid scintillation counting equipment, designed to detect very low intensity light pulses.[11-14]

Among all the classic continuous counting processes, the use of the Cerenkov effect seemed to hold the most promise, so we decided to adopt this method, but if necessary we shall use γ detection or a homogeneous counting system.

The Cerenkov technique has been chosen for its simplicity, absence of chemical quenching, low background, relative selectivity as the result of energy threshold, and its possible application whatever the nature of the fluid stream and the possibility of reusing samples for subsequent investigations.

Detection unit

The β-counting equipment consists of a detection cell and an electronic unit with data acquisition accessories. Flow cells designed for the continuous counting of β-emitters should satisfy a series of conditions:

1. Their geometry must provide the highest possible collection of light by the photomultiplier.
2. The effective volume of the cell should be large enough for a good counting rate of the fluid stream, but to avoid loss of resolution this volume should remain compatible with the rate of change in radioactivity of the effluent stream.

A tubular cell of very low volume is used (500 μl), formed like a W to obtain an instantaneous and correct evaluation of activity in the fluid stream at the output of the separation unit (Fig. 2). This cell may be constructed of glass or, more desirably, of quartz; the aqueous solution or the organic solvent which carries the radioelement is the radiative medium as much as are the walls of the cell.

The detection cell is placed in the median plane perpendicular to the axis of the two

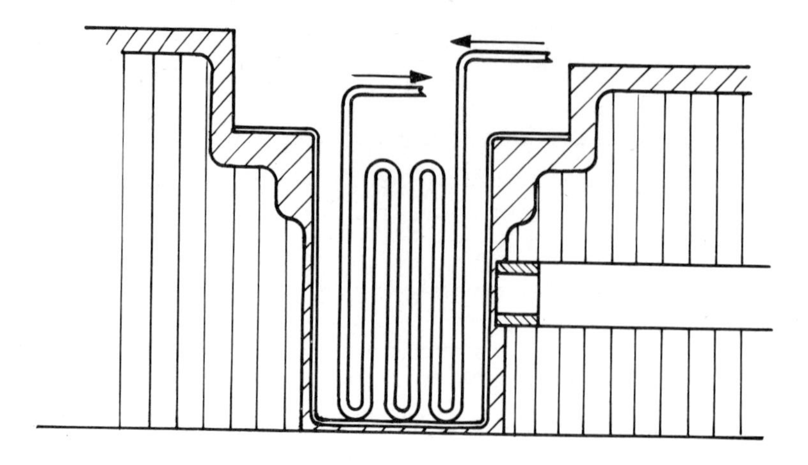

Fig. 2. Cerenkov detection: the continuous flow cell placed in the sample holder of Intertechnique SL 20.

photomultipliers of an Intertechnique liquid scintillation spectrometer model SL 20. This counting equipment is very simple to set up and to operate, and is economically priced, with good performance and reliability. The tubing used for inlets and outlets may easily act as a light pipe and so sufficiently long connecting lines made of opaque material such as black Teflon are used.

The highly directional nature of Cerenkov emission affects the light collection efficiency of a two photomultiplier system. It is useful to frost the walls of the cell to increase slightly the detection efficiency.

The relative selectivity of Cerenkov detection as the result of energy thresholds (~700 keV for water stream experimentally determined in our detection unit) recommended application of that technique for various complex counting problems; for instance, several radioisotopes corresponding to a radioelement obtained after chemical separation.

Weak β-emitters and α-particles are not counted. Counting efficiency need not be ascertained because the measurement is performed versus a standard of activity of the same radioelement, treated under the same experimental conditions.

One should also be aware of the fact that the efficiency of a flow cell is only a partial measure of its performance and that the sensitivity attained will depend on its volume and the flow rate of the liquid. Some detection efficiencies using standard solutions have been determined; the effective volume of the cell was determined by weighing and the efficiency was established by circulating a solution of known specific activity through the detector. The Cerenkov counting efficiencies for aqueous samples so determined are given in Table 1.

It should be noted here that for a fixed gain of the amplifier, the detection efficiency will greatly depend on a great number of parameters such as the form of the β spectrum of the nuclide, the geometric arrangement of the counting cell and the photomultipliers (light collection) refractive index of the cell and flow stream, optical properties of the solution (colour quenching), and the electronic setting of the sum and coincidence circuits.

We hope to increase the experimentally determined efficiency shown in Table 1 by changing the cell.

Table 1. Cerenkov counting efficiencies experimentally determined for some aqueous samples.

Nuclide	E_{max} (MeV)	Counting efficiency (% of disintegration)
Thallium-204	0.776 (98%)	2
Molybdenum-99	1.18 (83%) 0.80 (3%) 0.41 (14%)	7
Barium-140	1.02 (60%) 0.48 (40%)	10
Strontium-89	1.463	17
Praseodymium-144	2.996 (97.8%) 2.30 (1.2%) 0.807 (1.0%)	34

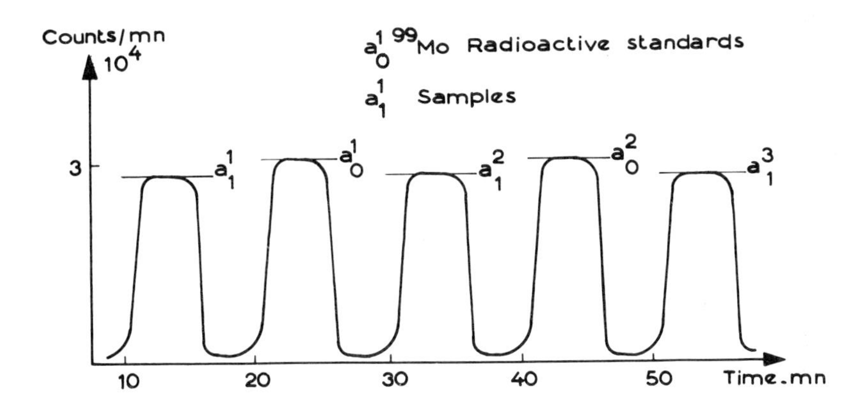

Fig. 3. Graphic recording of activity levels of analysis of molybdenum-99.

Data readout

The display of level of counting is performed on-line with a graphic recorder Servo-trace Sefram, by means of a ratemeter Intertechnique model DA 101, so verification of performance is possible during the course of analysis. This graphic information is in the form of a series of activity levels corresponding to either sample or standard of activity (see Fig. 3). The liquid scintillation spectrometer model SL 20 is also specially equipped with a Teletype for data print out; the same information which is printed is also punched on tape. The punch tape output is an ideal interface for using the instrument with an off-line digital computer which makes a complete statistical treatment of the data.

The arithmetic means of several measurements (5 to 6 counting rates in general) determine the uniform level of activity corresponding to the sample or a standard isolated

Table 2. Sensitivity data.

Isotope separated from a mixture	Sensitivity (μCi/ml)
Molybdenum-99	10^{-3}
Strontium-89	2.10^{-3}
Barium-140	10^{-2}
Yttrium-91	2.10^{-3}

at uniform concentration. The ratio of two successive levels (one for the sample, the other for the standard of activity) gives us the ratio $a{:}a_0$ of the general expression:

$$A = A_0 \frac{a}{a_0}$$ (see p. 182)

It should be noted here that for such a comparative measurement done in a single detection cell, we need not take account of geometry, counting efficiency, branching decay or colour quenching; all these parameters are similar for sample and corresponding standard activities. Likewise, we are able to ignore on the one hand, radioactive decay, and on the other hand, radioanalyser drift, by making alternate measurements on sample and standard in short time intervals.

APPLICATIONS

Cerenkov counting was applied to automatic or semi-automatic analysis of molybdenum-99, strontium-89, barium-140 and yttrium-91. The analytical process of molybdenum-99 is based on continuous substoichiometric separation with solvent extraction and that of strontium-89, barium-140 and yttrium-91 is based on continuous displacement at uniform concentration through ion exchange resins.[15]

The reproducibility of the analysis is good; taking the example of molybdenum-99, we obtain after the complete analytical process a standard deviation of 0.3% at the level of 30000 c.p.m.; for strontium-89 and barium-140 we obtain 0.5% for the same counting rate. The maximum error on a ratio of activity is 1% at a confidence level of 95%. The total accuracy of the analysis depends overall on the standard of activity.

The sensitivity is now low: this is not on account of the Cerenkov counting efficiency but is due to the use of a flow cell of very low volume and the necessity for a great number of successive counting intervals in short periods of time to produce an instantaneous and correct evaluation of activity in the effluent stream. Some data of sensitivity are given in Table 2.

It should be noted here that these automatic analyses with on-line detection are more rapid (4 to 6 times) than the classic radiochemical method.

CONCLUSIONS

The on-line detection by Cerenkov counting of radioactive isotopes as they are isolated is a practicable and useful technique in our automatic analytical process based on the substoichiometric separation in conjunction with non-active isotope dilution.

In comparison with classic continuous measurements of activity in fluid streams by scintillation methods, the Cerenkov counting provides a great number of advantages, i.e. simplicity, easy handling of samples, possibility of using organic solvents and reusing samples for subsequent investigations.

The measurement is performed versus a standard activity of the same radioelement treated under the same experimental conditions; it consists of a simple comparison between two count rates whose ratio is equal to the specific activity ratio of the two flowing streams which is in turn identical to the ratio of the initial activities. We need not take into account several factors which one is forced to consider when making absolute measurements or a measurement on a classically isolated sample.

The time of analysis of a sample has been decreased, the sensitivity of analysis is now low but repetitive performance is good.

This process of radiochemical analysis may be extended to the measurement of trace impurities using radioisotope dilution.

ACKNOWLEDGEMENTS

The authors wish to acknowledge the help given by Mrs. M. Lenagard, Miss N. Wimphen and Mr. L. Roux for technical assistance in the experiments; the Laboratoire de Métrologie de la Radioactivité (L.M.R.) of the C.E.N.-Saclay for having supplied the radioactive standards and to Intertechnique for fabricating the interface for adapting the teletype with punch tape to their model SL 20 counter.

REFERENCES

1 J. Ruzicka and J. Stary, in *Substoichiometry in Radiochemical Analysis* (Ed. M. Williams), Pergamon Press, Oxford, 1968.
2 J. Stary and J. Ruzicka, *Talanta* **18**, 1 (1971).
3 J. Ruzicka and M. Williams, *Talanta* **12**, 967 (1965).
4 J. Ruzicka and J. Stary, *Talanta* **8**, 228 (1961).
5 B. Tremillon, *Les Séparations par Résines Echangeuses d'Ions,* Gauthiers-Villars, Paris, 1965.
6 J. Colomer, M. Cousigne and G. Mezger, *Le Déplacement et ses Possibilités d'Application à l'Analyse de Radioélément par Dilution Isotopique,* Rapport CEA R 4100, 1971.
7 K. Piez, *Nuclear Chicago Tech. Bull.* **15** (March 1964).
8 E. Rapkin, *Bulletin Pickernuclear* **11**, 6L (May 1967).
9 E. Schram, in *The Current Status of Liquid Scintillation Counting* (Ed. E. D. Bransome), Grune and Stratton, New York, 1970.
10 E. M. Belcher, *Proc. Roy. Soc. (London)* **A216**, 90 (1953).
11 K. von Haberer, *Atomwirtschaft* **10**, 36 (1965).
12 M. M. Ross, *Anal. Chem.* **41**, 1260 (1969).
13 R. P. Parker, *Digitechniques Tech. Rev.*, Intertechnique, Paris, 1969.
14 R. P. Parker and R. H. Elrick, in *The Current Status of Liquid Scintillation Counting* (Ed. E. D. Bransome), Grune and Stratton, New York, 1970, p. 110.
15 J. Colomer, M. Cousigne and G. Metzger, *International Symposium on Rapid Methods for Measurement of Radioactivity in the Environment,* Neuherberg, 1971, I.A.E.A., SM 148/10, Vienna.

SECTION IV

METHODS OF SAMPLE PREPARATION OF ORGANIC MATERIALS INCLUDING BIOLOGICAL SYSTEMS

Chapter 15

Sample Preparation Techniques in Biochemistry with Particular Reference to Heterogeneous Systems

B. W. Fox

Paterson Laboratories, Christie Hospital and Holt Radium Institute, Manchester, England

INTRODUCTION

It is generally agreed that for an accurate radioassay of any β-emitter, homogeneous scintillation counting techniques are essential. For labelled tissues, the most accurate method is undoubtedly the combustion technique; however, the combustion procedure can be impracticable for very large numbers of samples, e.g. from chromatographic columns, and from electrophoretic and centrifugation procedures. In order to achieve complete degradation of many biological samples, often mixed with salts, metal ions and macromolecules, it is often more convenient to employ complexing or degradative procedures. Where specific components are needed to be measured, such as a labelled macromolecule, lipid, or other solvent-extractable material, the pre-processing may be simplified by extraction or precipitation procedures, followed by disc or other heterogeneous counting methods.

Analytical procedures using paper, thin layer, or column chromatography, polyacrylamide or agarose electrophoresis, caesium chloride or sucrose gradient centrifugation, each benefit from specific short cuts which enable reproducible counting to be undertaken rapidly within each method.

I therefore propose to a) discuss some general aspects of the use of solubilizers, disc counting and suspension techniques; b) to describe some of the techniques found to be useful in analytical and preparative procedures involving chromatography, electrophoretic and centrifugation methods; c) to describe in greater detail some aspects of the technique of colloid scintillation counting.

SOLUBILIZERS

I do not intend to review the field of solubilizers, as this has been covered by Rapkin[1] and others in several excellent reviews.

One of the problems associated with the use of the quaternary ammonium hydroxides such as Hyamine-10X, Soluene-100 or NCS is the phenomenon of chemiluminescence, especially noticeable with Soluene-100. The phenomenon is clearly associated with the presence of peroxides in the scintillating solvents, and dioxane is a notorious solvent in this respect. Herberg[2] noticed that the count rate due to chemiluminescence appeared to be inversely related to the viscosity of the solution and also observed that it increased on exposure to light. Winkelman and Slater[3] showed that benzoyl peroxide, normally employed for colour bleaching, tended to produce an enormous count rate in the presence of Hyamine-10X or NCS. The recognition that peroxides were the cause of the trouble led to the use of hydrochloric acid or, more recently, 10% ascorbic acid (McClendon *et al.*)[4] to act as peroxide scavengers. We have found that the chemiluminescence due to peroxides in Dioxane–naphthalene–phosphor mixtures can be largely overcome by keeping the scintillant solution over granulated zinc, but this only seems to work with sodium hydroxide-induced chemiluminescence and not with Soluene-100. Winkelman and Slater[3] did find that hydrochloric acid reduced the count rate to low levels but higher than the minimum reached by natural decay of the chemiluminescence. It seems, therefore, that hydrochloric acid does not destroy the chemiluminescent reaction but simply slows it down to tolerable rates. Reducing the temperature of counting from 20°C to 8°C simply reduces chemiluminescence by 'one third', presumably again reducing the rate of reaction. For low activity counts, therefore, it may be necessary to allow decay, without hydrochloric acid treatment, for over a week. I am sure there is scope for the production of a solid peroxide scavenger that can be added to scintillant mixtures.

Both internal standard and channels ratio techniques are suitable for quench correction. The external standard ratio should be used with extreme care since it is not clear whether a solubilizer produces a true solution or whether it possesses some of the properties of an heterogeneous, colloidal system. Many commercial preparations are now being introduced by liquid scintillation firms, and I would like to appeal very strongly to avoid the use of secret recipes for these agents so that such factors as the advisability of using an appropriate quench correction procedure can be assessed by the user. It is not clear for example, if surfactants are used as well as quaternary bases, but without this knowledge the user could use processed information derived from automated systems employing external standard ratios which may have no meaning in colloid systems.

In our experience, most tissue samples can be radio-assayed very conveniently by using the perchloric acid–hydrogen peroxide degradation technique in the scintillation vial suggested by Mahin and Lofberg,[5] provided the mixture is heated at not more than 75°C and for longer than 30 min and the product blended into a toluene-based scintillant with 2-methoxyethanol.

I would now like to discuss certain aspects of heterogeneous or two-phase systems. The two phases may be either solid–liquid or liquid–liquid.

Solid–liquid

The most efficient system for weak β-emitters is to use the liquid phase as fluor, although early attempts to use the solid phase as fluor have been tried, e.g. Pilot B beads.[6]

Since the development of counting β activity on filter paper in the early 1960's, it has been realized that glass fibre and membrane filters are especially efficient supports for precipitates, either as a pulp suspension or as discs. Discs are easy to use and need not employ more toluene-based scintillant than is necessary to wet the paper. Furthermore, as Pinter *et al.*[7] have shown, up to twenty-five discs can be counted in a single vial, with no reduction in counting efficiency. On a standard disc of GF/C Whatman, 24 mm diameter, about 0.2 ml of aqueous solution may be dried. Twenty-five discs would thus represent a total of 5 ml and may therefore make it possible to obtain significant counts from low specific activity material by simply adding more discs (see Fig. 1).

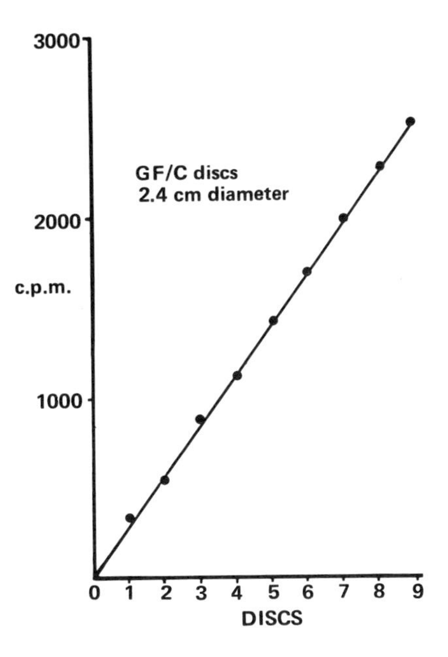

Fig. 1. Additive effect of counting glass fibre discs in the same vial. Each disc had 0.2 ml of tritiated thymidine solution dried on it.

A word about cost. A simple brass punch to cut out 24 mm circles from larger diameter glass fibre discs can reduce costs appreciably.

Gill[8] has examined the efficiency of a number of different materials as solid supports and has concluded that glass fibre is superior. Even scotch tape has been used recently to determine some parameters of the kinetics of epidermal proliferation of labelled hairless mice by stripping off the surface layers and counting adhering material in a toluene phosphor.[9]

The use of ion exchange paper, in particular DEAE in the form of discs was also described by Furlong *et al.*,[10] Breitman,[11] and Loftfield and Eigner[12] to distinguish charged from uncharged species in enzyme systems. It became clear that many assays could be undertaken, such as thymidine kinase,[11] hydroxamate formation,[12] galactose-1-phosphate from galactose,[13] adenosine diphosphate glucose pyrophosphorylase,[14] adenosine diphosphate creatine phosphototransferase[15] and glycerol kinase and hexokinase.[16] As an example, the thymidine kinase system first described by Breitman allowed 98% removal of thymidine and allowed 100% retention of the monophosphate. However, a study of this system has recently been undertaken by Roberts and Tovey[17] who have shown that the conditions of salt elution and paper drying are very critical for some enzyme assays.

Thin layer chromatography

Thin layer chromatography has been widely used and the fumed silica product Cab-O-Sil has been successfully employed to improve the poor counting efficiency caused by geometry artifacts by preventing wall absorption of trace quantities of radioactive materials, thereby allowing nearly 4π counting rather than 2π counting.[18] When Cerenkov counting is used, however, e.g. with phosphorus-32, it is important to realize that Cab-O-Sil can introduce abnormal quenching artifacts as reported by Wiebe *et al.*[19] recently. Polyacrylamide gels have also been used extensively, and methods of solubilization of these have been recently reviewed by Paus.[20] He concluded that gels up to 5% can be conveniently solubilized by heating to 60 to 65°C in Soluene-100, BBS-3 or NCS. However, it is worth noting that Soluene-100 apparently interacts with PBD fluors and considerable counting losses can occur.[21]

Liquid–liquid

One of the earliest accounts of the use of emulsions in liquid scintillation counting was by Meade and Stiglitz[22] when they showed that the counting efficiency of body fluids and tissues could be improved by using the detergent Triton X-100 in conjunction with Hyamine-10X. Erdtmann and Herrmann[23] appreciated the possibility of exploiting emulsions for the measurement of β-emission which was later examined by Patterson and Greene[24] who suggested empirically derived compositions of Triton X-100 and toluene. The first systematic study of the system was undertaken by van der Laarse,[25] who whilst examining ground water samples, also described the appearance of all possible combinations of the three components by plotting on a triangular plotting system such as is used in phase composition research. The diagram is not strictly speaking a phase diagram since the whole system in all its areas is probably two-phase; however, we will refer to the diagrams as phase diagrams. They also employed a *figure of merit* to describe the efficiency, being the product of the absolute efficiency and the volume of aqueous phase used. The so-called *merit value* which I shall adopt is a product of the percentage efficiency and the percentage of aqueous solution present, based on a total volume of 10 ml. The value is, of course, based on the instrument used and to compare values between instruments it is necessary to divide by a comparative factor. I have not made use of this factor, and all measurements which I shall show were conducted on a Beckman LS 200 at ambient temperature operation, with a reference toluene standard efficiency of 45%.

The system consists of a range of transparency from crystal clear to completely opaque and also a range of viscosity from that of toluene to a thick gel which retains its position on inversion of the vial. These are separately assessed for each point of the diagram on an empirical scale of 0 to 4.

The clear areas represent submicroscopic colloidal droplets from 5 nm to 0.1 μm whereas the opaque areas are true emulsions with a droplet range from 0.1 μm to probably greater than 1 μm. The mean range of β-particles from tritium in matter of unit density is approximately 0.56 μm or, better still, 50% is absorbed within a thickness of 0.4 μm (half value layer). The Triton X-100 apparently forms an organized double layer, in which the phenolic ends are associated within the droplet and the hydrophilic ethoxy chains extend out into the aqueous layer, interlacing with other chains arranged in the opposite direction.[26] The final efficiency of counting will then depend on a) the relative volumes of fluor-containing organic phase to the β-emitter containing aqueous phase, and b) the size of the fluor-containing droplet in relation to the path length of the β-emitter.

In practice, these parameters need not be assessed, since the areas of interest can be determined for an individual solution by determining the merit value for each point within the diagram using a polar tritium standard and the stability of the counts of each sample measured repeatedly over a reasonable counting period, 24 to 48 h. It is not possible to predict in detail the appearance or stability of the various parts of this diagram, but after constructing quite a number of these, several general points of interest emerge.

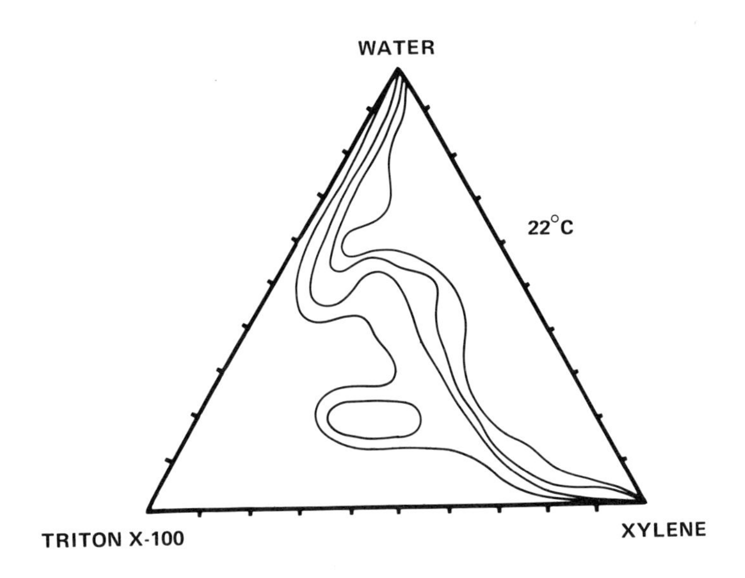

Fig. 2. The phase appearance of all combinations of water, Triton X-100 and xylene at 22°C. The areas outlined range from completely transparent (bottom left) to completely opaque (right hand edge).

To construct a diagram, a series of 36 vials are arranged in a vial box in a triangular pattern and starting from each corner, rows of successively increasing volumes of the three components are dispensed so that the total volume in each vial is 10 ml. The toluene component contains 0.5 ml of a concentrated fluor solution (2.67 g PPO, 0.067 g POPOP in 50 ml toluene) the rest of the volume being pure toluene. The vials are shaken thoroughly and a visual assessment of transparency and viscosity is made (see Fig. 2). The backgrounds are assessed and the vials spiked with 10 μl standard tritiated water, shaken and recounted several times by recycling over a period of 24 h. Comparison of these counts enables one to assess counting stability. The merit values are ascertained for the 2nd or 3rd cycle, plotted on the diagram and contours drawn corresponding to increments of merit value of 100 (see Fig. 3). A point can then be selected which combines high merit value with good counting stability in a transparent or opalescent region. The optimal value is then obtained by making up several samples of this mixture with varying proportions of concentrated fluor–toluene solution, to ascertain the highest merit value. All these stages were conducted for a series of solutions used in biochemical work.[27]

Turner[28] has pointed out that with the toluene : Triton X-100 (2: 1): water system, a discontinuity in appearance and efficiency occurs as water is added. This discontinuity may be predicted on examining the diagram for toluene: Triton-X: water (Fig. 4). The use of a different detergent, e.g. Triton N-101, one of the nonyl-phenol series shows a slightly different phase diagram (see Fig. 5), but has fewer merit value contours than that of the X-100 diagram. Incidently, BBS-3: water: toluene is also different, but optimal values are lower than those obtained with the Triton X-100: water: toluene system (see Fig. 6). The value of this detergent, however, is said to be with high salt conditions, and these are currently being assessed in this way. The appropriate optimal concentrations and merit values are summarized in Table 1. An example of the unpredictability of the phase diagram

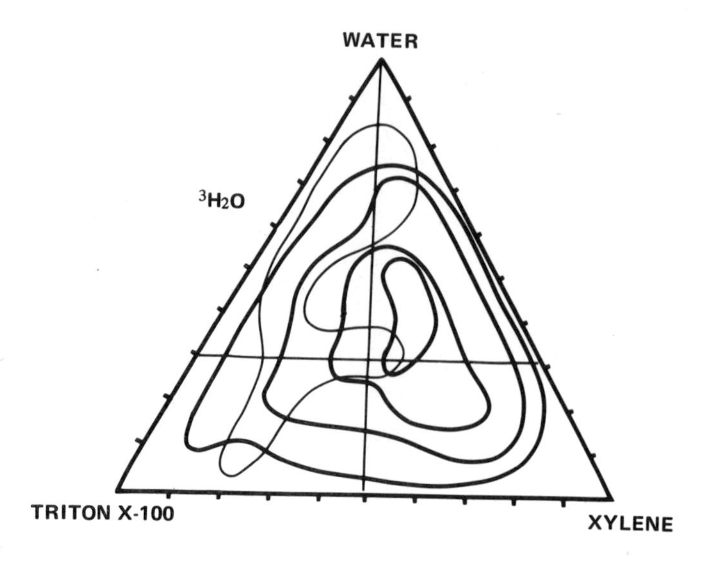

Fig. 3. The merit value diagram for the water: Triton X-100: xylene system. Each contour (thick line) represents an increment of 100 starting from the outside. The fine line encloses the area of comparative counting stability and the point of intersection of the vertical and horizontal lines represents the final point chosen for counting. The final composition of the scintillant used will therefore be the ratio indicated by the base line. The ratio of this mixture to the aqueous solution involved would then be indicated by the vertical line.

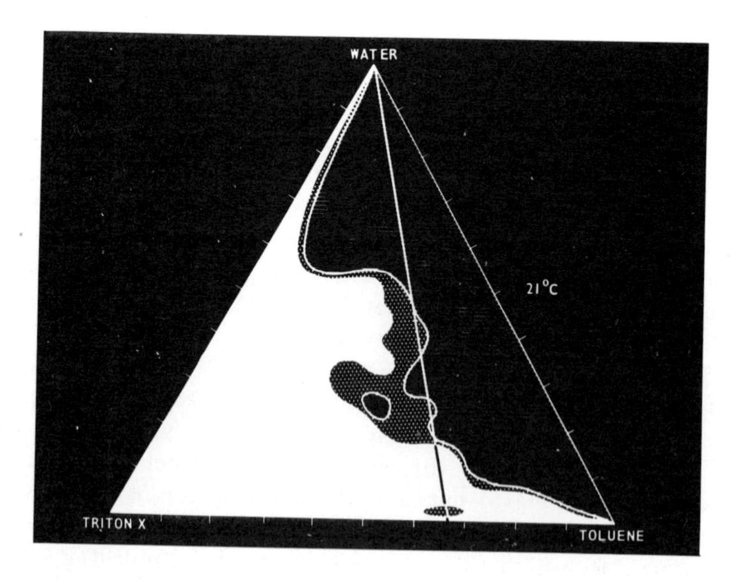

Fig. 4. A 'phase diagram' for the system water: Triton X-100: toluene. (See legends to Figs. 2 and 3 for interpretation).

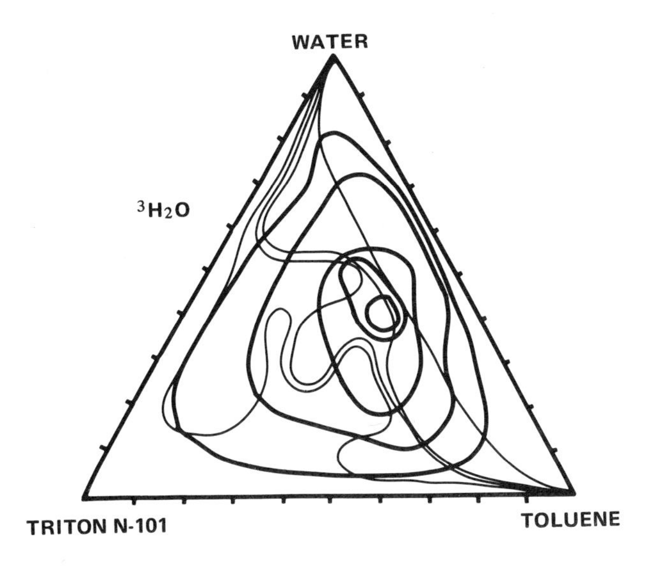

Fig. 5. A combined phase (fine lines) and merit value plot (thick lines) of the water: Triton N-101 : toluene system. (See legends to Figs. 2 and 3 for interpretation).

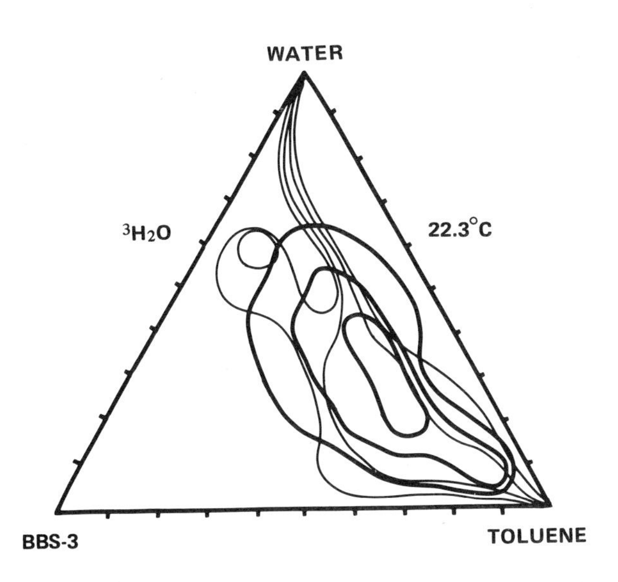

Fig. 6. A combined phase/merit value diagram for the Biosolve (BBS-3): water: toluene system, using PPO as fluor. (See legends to Figs. 2 and 3 for interpretation).

is shown by the 5% sucrose: toluene: Triton X-100 system.[27]

The most efficient region is within one of the 'islands' of clarity. The diagrams for comparable concentrations of fructose, glucose and maltose are quite different.

If sucrose gradients are being conducted, 0.5 ml fractions can be diluted with water for neutral gradients or an appropriate dilute hydrochloric acid solution for alkaline sucrose,

Table 1. Surfactant/solvent mixtures for tritiated water.

Surfactant	Solvent	Phosphor Mix Phosphor	Phosphor Mix Water	Tritium Counts Merit Value	Tritium Counts Efficiency	Fluor
Triton N-101 (2)	Toluene (1)	6	4	510	12.5	PPO (opaque)
Triton N-101 (4)	Xylene (3)	7	3	378	12.6	PPO bis MSB
Triton X-100 (13)	Xylene (17)	7	3	409	13.6	PPO
BBS-3 (2)	Toluene (3)	5	5	273	5.5	PPO
Triton X-100 (5)	Benzyl alcohol (2)	7	3	130	4.3	PPO
Triton X-100 (1)	Toluene (1)	6	4	560	14.0	PPO

to bring the total volume to 5 ml. 5 ml of a toluene: Triton X-100 (2: 3): phosphor mixture is then added and shaken. If the experiment involves the use of direct cell lysis to examine labelled DNA and its precursors, the TCA precipitable DNA can be assessed by glass fibre disc methods and the data subtracted from a duplicate Triton-X: toluene system to give the distribution of acid soluble oligonucleotides on the sucrose gradient. With caesium chloride, a similar dilution of 0.2 ml fractions with 1.8 ml distilled water and after optical density measurement, is added to 8 ml of toluene: Triton X-100 phosphor (1 : 1) containing PPO and can then be counted with a merit value of 264.

Another feature of the colloid system can be seen in assaying chromatography eluates involving gradients. In the particular case of 1 M to 3 M hydrochloric acid elution of nucleotides on chloride anion exchangers, optimal counting conditions can be found for both 1 M and 3 M hydrochloric acid solutions (see Fig. 7), but these are different. However, one can superimpose the merit diagrams and locate a counting area where a constant merit value can be obtained (see Fig. 8) so that no quench correction will be necessary, although the merit values are slightly less than optimal. A similar estimate may be made from the ammonium formate system for separating nucleic acid bases (see Table 2). A further advantage of the colloid system is, of course, in the counting of complex biological solutions such as cell media, urine and plasma. The diagram for Fischer's medium containing 20% serum is shown in Fig. 9, and for urine, Fig. 10 demonstrates the kind of stability profile obtainable. These are summarized in Table 3.

Quench correction

The colloid scintillation system should be regarded in the same way as any heterogeneous system. The only satisfactory way in which quench correction can be undertaken is by spiking with a standard of comparable polarity to that of the β-emitter. If there is any doubt, standards with two extremes of solubility, namely tritiated water and tritiated toluene can be compared and a region where a common merit value occurs is selected. Channels ratio measurements follow very closely those of internal standards in those systems which have been studied but it is reasonable to expect that divergences may occur. Indeed, Turner[28] has pointed out some small divergences in the less stable parts of the diagram. External standard ratios have very wide divergences in this system and should not

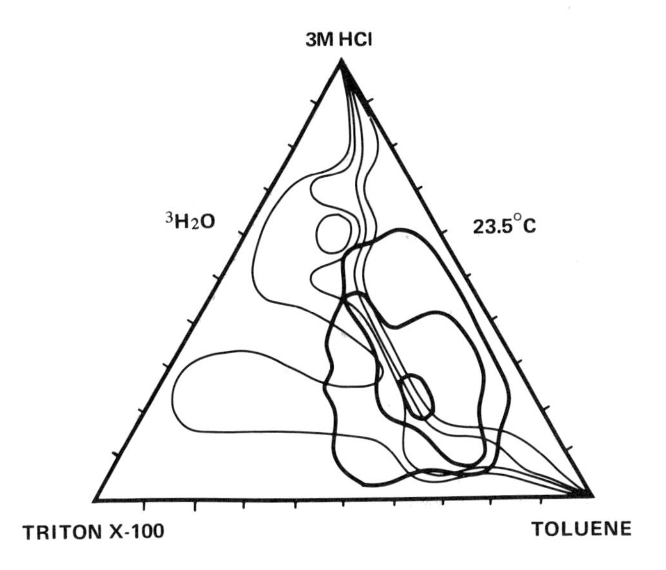

Fig. 7. A combined phase/merit value diagram for the system 3 M HCl: Triton X-100: toluene, using PPO as fluor. (See legends to Figs. 2 and 3 for interpretation).

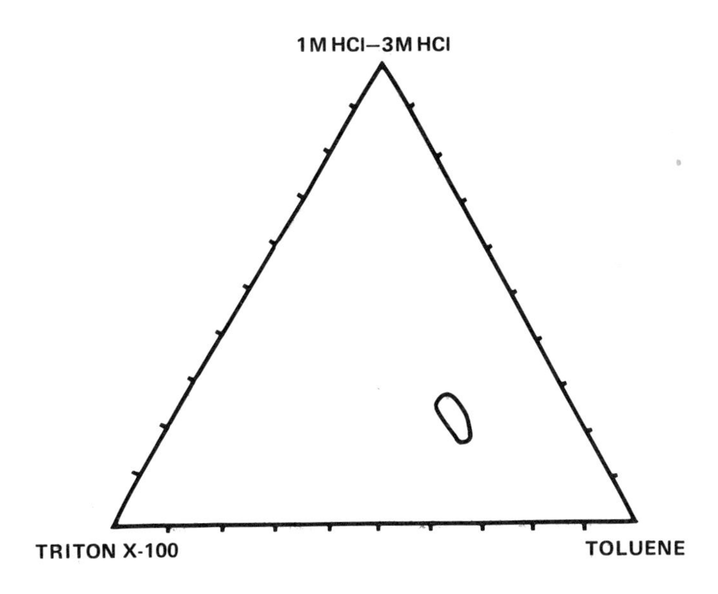

Fig. 8. The result of combining the merit values for 1 M and 3 M HCl: Triton X-100: toluene systems. The contour encloses that area where the merit value remains constant.

be used. It is possible that important information on colloid structure can be obtained from these measurements but they are useless for quench correction. I cannot stress too highly the importance of knowing whether one is dealing with a homogeneous or liquid–liquid two-phase system when one employs the newer gelling or solubilizing agents.

Table 2. Column chromatography.

Aqueous solution	Solvent composition Triton X-100	Toluene	Phosphor mix Phosphor	Aqueous	Tritium counts Merit value	Efficiency
Ammonium formate 1 M	17	23	7	3	502	16.7
Ammonium formate 0.03 M	2	3	5	5	504	10.1
Gradient 0.03 M→1.0 M	1	1	7	3	392→377	13.1→12.6
Hydrochloric acid 1 M	2	5	7	3	461	15.3
Hydrochloric acid 3 M	2.3	5	8	2	339	16.9
Gradient 1 M→3 M	1	2	7.5	2.5	321→326	12.9
Formic acid 0.1 N	6	11	8.5	1.5	320	21.3

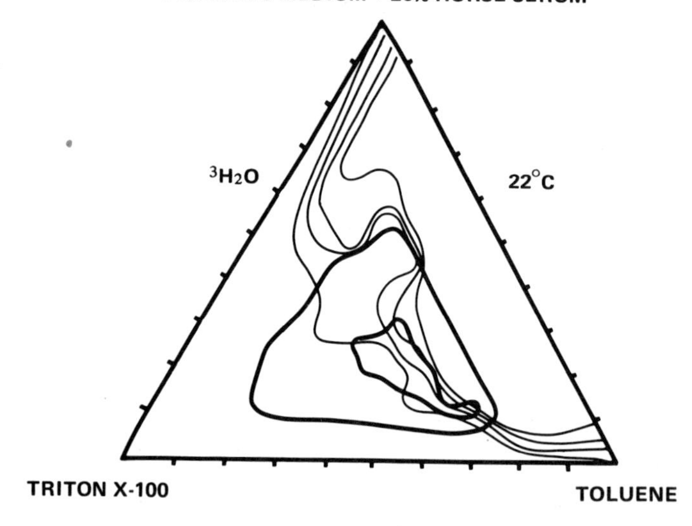

Fig. 9. A combined phase/merit value diagram for the Fischer's Medium containing 20% horse serum, Triton X-100 and toluene. (See legends to Figs. 2 and 3 for interpretation).

CONCLUSION

In conclusion, I hope I have shown that heterogeneous systems play an extremely important part in modern biochemical procedure and the scope for their further exploitation is considerable. I am sure that the potentialities of the colloid system have only just been realized and that with a more sophisticated appreciation of the physical properties

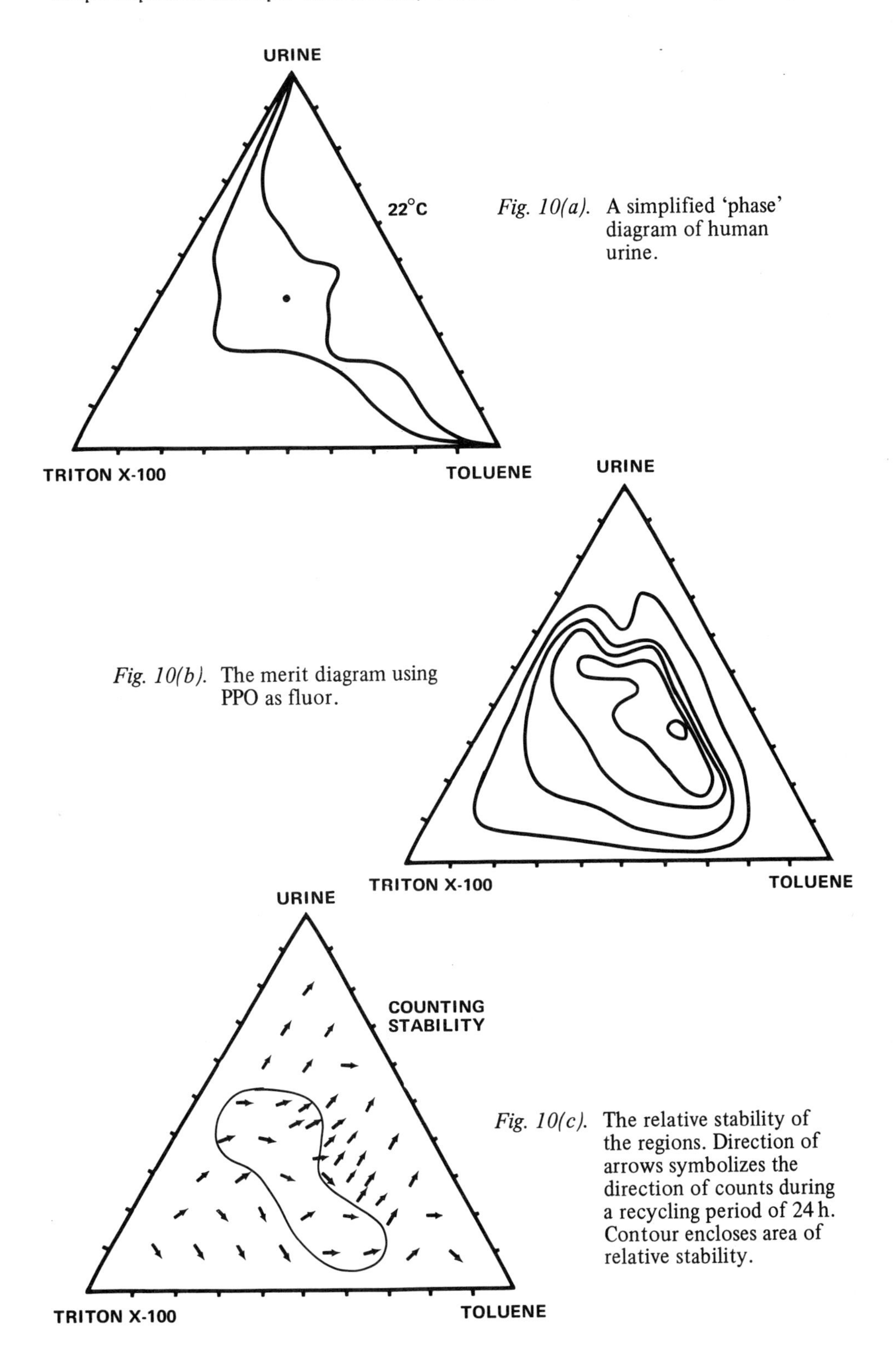

Fig. 10(a). A simplified 'phase' diagram of human urine.

Fig. 10(b). The merit diagram using PPO as fluor.

Fig. 10(c). The relative stability of the regions. Direction of arrows symbolizes the direction of counts during a recycling period of 24 h. Contour encloses area of relative stability.

Table 3. Media and biological samples.

Solution	Solvent composition		Phosphor mix		Tritium counts	
	Triton X-100	Toluene	Phosphor	Aqueous	Merit value	Efficiency
Fischer's medium + 20% serum	7	9	8	2	217	10.9
TYG micro-biological medium	1	1	8	2	480	24.2
Human urine	1	1	6	4	439	11.0
Human plasma	4	7	8	2	177	8.9

of the colloidal particles in relation to the very wide range of surfactants available, new techniques of assay of soft β-emitters in complex media could be developed.

REFERENCES

1 E. Rapkin, *Intern. J. Appl. Radiation Isotopes* **15**, 69 (1964).
2 R. J. Herberg, *Science* **128**, 199 (1958).
3 J. Winkelman and G. Slater, *Anal. Biochem.* **20**, 365 (1967).
4 D. McClendon, M. P. Neary, M. Galassi and W. Stephens, in *Organic Scintillators and Liquid Scintillation Counting* (Ed. D. L. Horrocks and C. T. Peng), Academic Press, London and New York, 1971, p. 587.
5 D. T. Mahin and R. T. Lofberg, *Anal. Biochem.* **16**, 500 (1966).
6 D. Steinberg, *Nature* **182**, 740 (1958).
7 K. G. Pinter, J. G. Hamilton and O. N. Miller, *Anal. Biochem.* **5**, 458 (1963).
8 D. M. Gill, *Intern. J. Appl. Radiation Isotopes* **18**, 393 (1967).
9 O. H. Iverson (personal communication).
10 N. B. Furlong, *Anal. Biochem.* **5**, 515 (1963).
11 T. R. Breitman, *Biochim. Biophys. Acta* **67**, 153 (1963).
12 R. B. Loftfield and E. H. Eigner, *Biochim. Biophys. Acta* **72**, 373 (1963).
13 M. R. Atkinson and A. W. Murray, *Biochem. J.* **94**, 64 (1965).
14 H. P. Ghosh and J. Preiss, *J. Biol. Chem.* **241**, 4491 (1966).
15 J. F. Morrison and W. W. Cleland, *J. Biol. Chem.* **241**, 673 (1966).
16 E. A. Newsholme, J. Robinson and K. Taylor, *Biochim. Biophys. Acta* **132**, 338 (1967).
17 R. M. Roberts and K. C. Tovey, *Anal. Biochem.* **34**, 582 (1970).
18 F. A. Blanchard and I. T. Takahashi, *Anal. Chem.* **33**, 975 (1961).
19 L. I. Wiebe, A. A. Noujaim and C. Ediss, *Intern. J. Appl. Radiation Isotopes* **22**, 463 (1971).
20 P. N. Paus, *Anal. Biochem.* **42**, 372 (1971).
21 A. Dunn, *Intern. J. Appl. Radiation Isotopes* **22**, 212 (1971).
22 R. C. Meade and R. Stiglitz, *Intern. J. Appl. Radiation Isotopes* **13**, 11 (1962).
23 G. Erdtmann and G. Herrmann, *Radiochim. Acta* **1**, 98 (1963).
24 M. S. Patterson and R. C. Greene, *Anal. Chem.* **37**, 854 (1965).

25 J. D. Van der Laarse, *Intern. J. Appl. Radiation Isotopes* **18**, 485 (1967).
26 R. G. Barradas and F. M. Kummerle, *J. Electroanal. Chem.* **11**, 128 (1966).
27 B. W. Fox, *Intern. J. Appl. Radiation Isotopes* **19**, 717 (1968).
28 J. C. Turner, *Intern. J. Appl. Radiation Isotopes* **19**, 557 (1968)

DISCUSSION

B. Scales: I would like to re-enforce the warnings given by Dr. Fox about the uncritical use of Triton X-100. I have two slides which illustrate the peculiarities of the toluene: Triton X-100: water system, and show that gross errors can be incurred in the calculation of d.p.m. if these emulsion systems are used without due care and attention.
Figure 11 shows two identical series of 14 vials made up by adding increasing volumes of water to a fixed volume (10.0 ml) of a 2: 1 toluene: Triton X-100 system containing 0.6% butyl PBD. Two important points to note are a) the general decrease in opacity (and increase in viscosity) as the volume of water increases, and b) the presence of two regions of instability; the first, which occurs in this example with the addition of about 0.25 ml water probably corresponds to the small pool of instability noted in Fig. 4 of Dr. Fox's paper, the water droplets visibly setting out on the base of the vial. The second occurs with the addition of from 2 to 3 ml water and corresponds to the larger central area of instability in that figure.
Figure 12 illustrates the count rates of toluene-soluble and a toluene-insoluble material in these systems at 18°C and 2 to 4°C. Most samples were made up in triplicate and counted for 10 min intervals over a 3-day period. The points show the average of these counts, and the bar lines illustrate the maximum drift in observed count rate (or calculated disintegration rate) over the full counting period. With ^{14}C-hexadecane at 18°C, the pattern is similar to that reported by Turner (*Intern. J. Appl. Radiation Isotopes* **19**, 557 (1968)). A marked decrease in count rate occurs with the addition of 2.5 to 3.0 ml water, and this only returns to normal when the system is stabilized by the further addition of water. Under refrigerated conditions, the same sample showed gross instability of count rate with 4.0 ml water. With higher water concentrations, the count rate was more stable but was drastically reduced.
When the patterns are examined using a water-soluble material (^{14}C-sodium succinate), a marked drop in count rate occurs with the addition of 0.25 ml water. Thereafter, a steady count rate is observed at 18°C, but at 2 to 4°C it falls rapidly when more than 4.0 ml water is present.
Any attempt to calculate d.p.m. of the ^{14}C-sodium succinate using the counting efficiency derived from a ^{14}C-hexadecane spike of the same samples results in the graphs within the lower axes. Instead of obtaining a constant d.p.m. value (4800 d.p.m. expected) at all water concentrations, the results can range from 2200 up to 5700 d.p.m. depending on the system used.
In metabolism studies, labelled materials of unknown structure present in biological samples are frequently quantitated using Triton X-100 systems. The lipophilic characters of the various labelled chemical entities present, will, by virtue of the normal detoxification mechanisms, often be very different from each other, and from that of the parent compound. For these reasons, the incautious use of internal standardization procedures in emulsion counting, whether carried out with lipophilic or hydrophilic substances, can give very misleading results.

B. W. Fox: I agree with these comments. The only really satisfactory standard would be

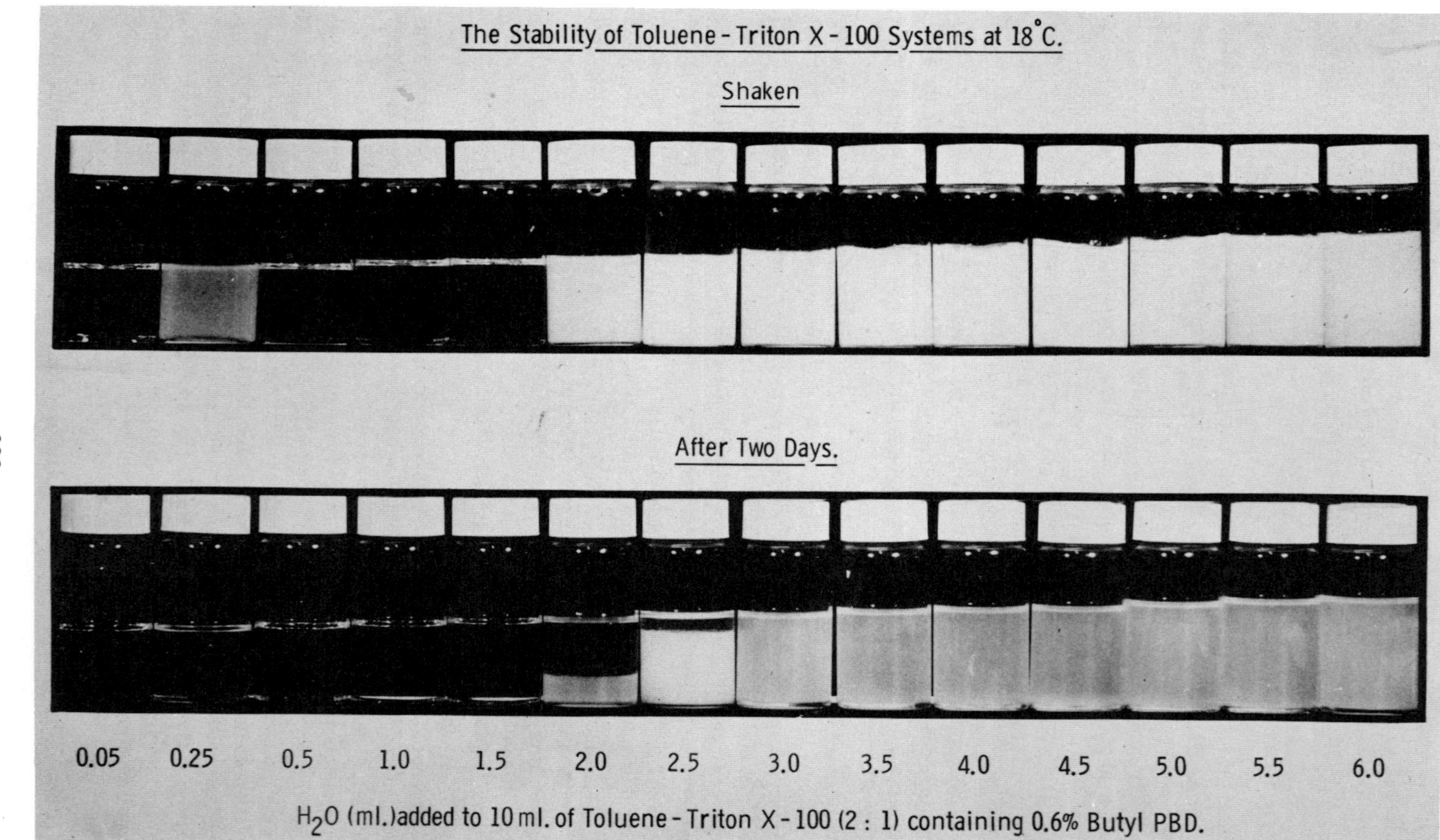

Fig. 11. The stability of toluene : Triton X-100 systems at 18°C.

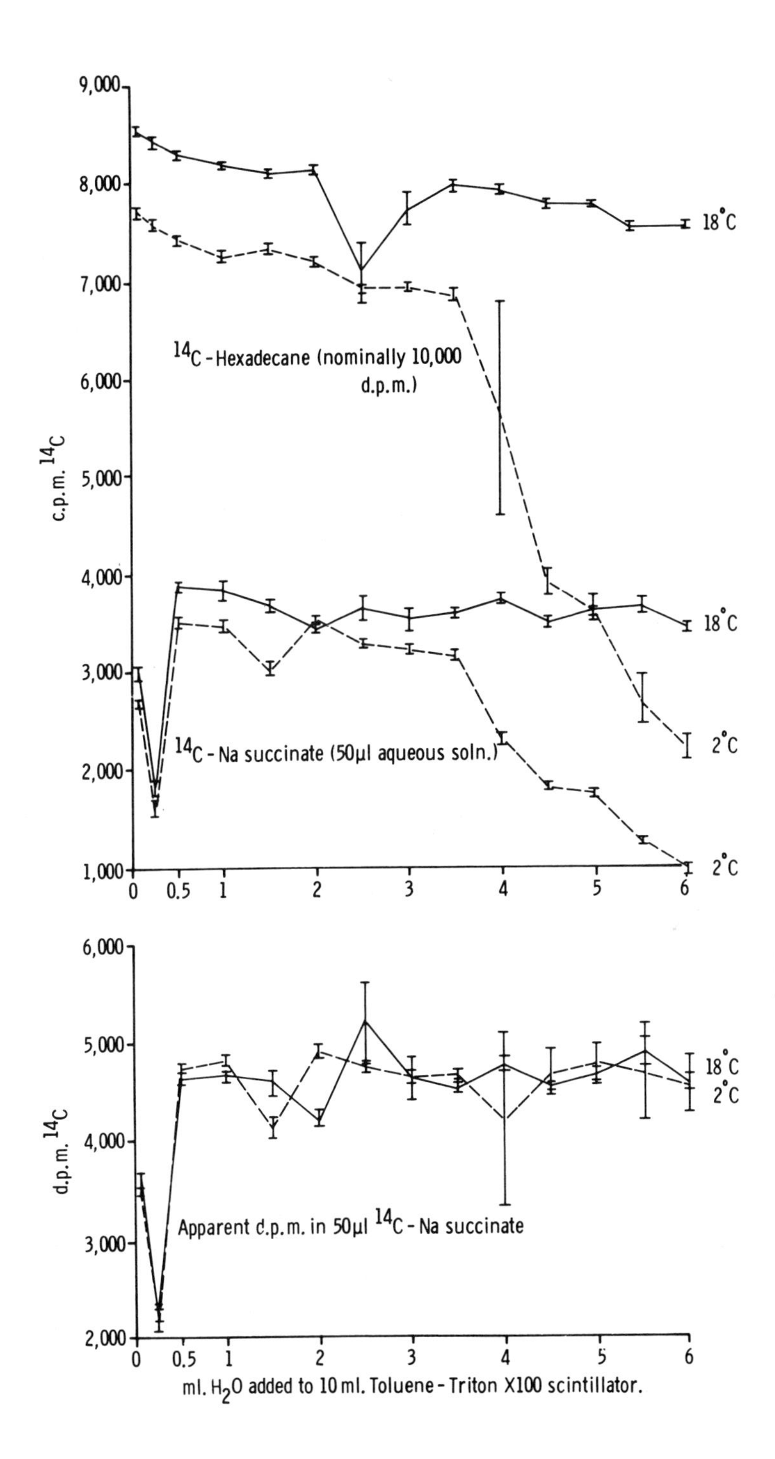

Fig. 12. Variations in carbon-14 counting efficiency of lipophilic or hydrophilic substances in toluene: Triton X-100 systems.

one which is made up of the species being studied. In biochemical work, most of the species of interest are water-soluble and hence tritiated water and ^{14}C-sodium succinate would be a reasonable compromise to be used in colloid counting. I published merit diagrams comparing tritiated water and tritiated toluene in *Intern. J. Appl. Radiation Isotopes* **19**, 717 (1968), and it can be seen from these curves that the major discrepancies occur in low surfactant and hence unstable regions, and not the regions recommended for counting. Your comments underline the necessity to adhere to the recommended composition.

D. Bowyer: In view of the apparent ease of determination of counting efficiency by means of the external standard method it would seem important to try to find external standard counting conditions suitable for use with heterogeneous systems; in this respect the energy of the external standard is important.
Which external standard sources were used in the experiments which demonstrated that the method was unreliable for heterogeneous systems?

B. W. Fox: In the initial studies, the relatively weak caesium-137 (20 μCi) was used as external standard, and this gave meaningless ratios. In other machines with stronger sources a figure is obtained, but I feel that this is in general unreliable due to the unknown relationship of droplet to environment producing conditions not unlike those in any other form of heterogeneous system. What is the external standard measuring? The efficiency of the droplet or the efficiency of the environment? The net result is some complex function of both. It is possible that providing the micelle is very small in size, the system may approach that of a homogeneous solution, but at this moment I would prefer to avoid using the external standard as a means of quench correction in this system.

F. Battig: Contrary to your observations, we find excellent agreement of external standard ratios with tritium efficiency if using soluble regions of mixtures of biological fluids like urine, bile, plasma and serum in Instagel.

B. W. Fox: I think you have been fortunate. In our experience the external standard ratio is too low to be meaningful. In the case of high activity external standard sources, it is possible that data from a single composition within the phase diagram may be meaningful, say for colour quenching, but it is dangerous to apply it when crossing the phase diagram.

H. Dobbs: I would like to thank Dr. Fox for presenting the plenary lecture in this session. I would like to comment on the high quality of his slides. I found them aesthetically pleasing and am sure they reflect the excellence of his work. As always, his presentation was both eloquent and informative. I am sure that the systems he has described give accurate results when used correctly. However, I would like to sound a word of warning. Many of the people who use liquid scintillation counting techniques have little knowledge of the fundamental processes. Once they have been introduced to a method of sample preparation they stick to it slavishly regardless of the possible pitfalls. I am concerned that the methods used with great success in Dr. Fox's laboratories may be misused in other laboratories. Careless or ignorant application of Dr. Fox's elegant techniques to unknown samples could lead to gross errors in the measured counting efficiency.

Chapter 16

Liquid Scintillation Counting of Biological Macromolecules: Extraction from Aqueous Solution and from Glass Fibre Filters

P. N. Paus

Department of Microbiology, University of Oslo, Blindern, Oslo 3, Norway

INTRODUCTION

Modern liquid scintillation spectrometers enable one to measure simultaneously, with accuracy and great sensitivity, the presence of different isotopes, provided that samples have been prepared correctly. The difference between non-polar isotope standards and polar biological macromolecules introduces special difficulties. The measurement of the radioactivity in these macromolecules has been performed in several ways, most of which have disadvantages.

With filter precipitation techniques energy absorption is a great danger, and makes accurate double labelling work impossible as the weak β-emitters are affected more than the strong ones,[1] and the quench spectra are different from those obtained in solution. Combustion techniques have so far been too time-consuming to attain any great importance. Direct counting of aqueous solutions may necessitate the use of scintillation liquids with great water capacity. These have lower counting efficiencies than the pure toluene scintillation liquids, and most of them are quite expensive. When molecules with polar groups are counted in aqueous solution, the possibility of unstable count rates must be considered.[2]

Counting in aqueous solution

Figure 1 shows the decline in count rate with time obtained when ^{14}C-labelled RNA, dissolved in 50 μl buffer, was counted in 15 ml toluene–2-methoxyethanol (85 : 15 v/v) scintillation liquid in a spectrometer with discriminators set for counting of tritium and carbon-14 according to the simultaneous equation method.[3] Similar results were obtained when proteins (cold TCA-precipitate from monkey serum) were counted, although the reduction in count rate was somewhat slower. The reduction was even greater when ^{3}H-labelled RNA was counted, the final count rate being some 40% of the initial. After 5 to 36 h the rates became stable. Manual shaking of the vial restored the initial count rates, followed by a second decline. The more hydrophobic the scintillation liquid, the faster and more pronounced was the drop. External and internal count rates were constant, and no precipitation, phase separation or colour alteration could be detected by the naked eye. As the scintillation liquid was unaffected during the observation period, the reason

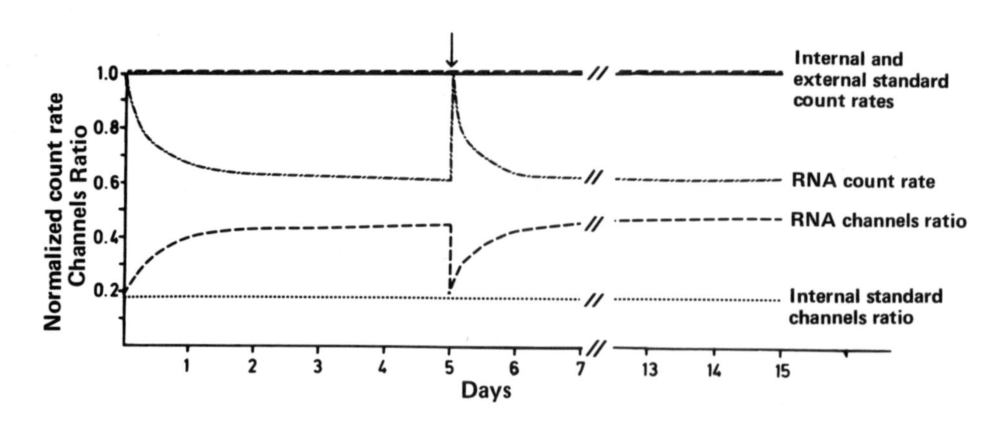

Fig. 1. Count rates and channels ratios as functions of time. 50 μg ^{14}C-labelled RNA in 50 μl 0.1 M sodium acetate, pH 6, dissolved in 15 ml toluene–2-methoxyethanol (85:15 v/v) scintillation liquid. Channels ratio: c.p.m. tritium channel/c.p.m. carbon-14 channel. Arrow indicates manual shaking.

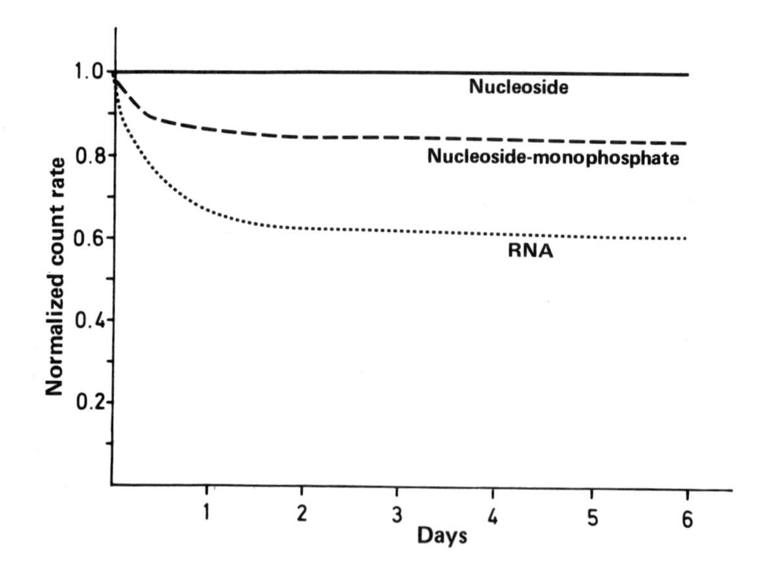

Fig. 2. Count rates of RNA, AMP and thymidine as functions of time. Each substance was dissolved in 50 μl 0.1 M sodium acetate, pH 6, before addition of 15 ml scintillation liquid.

for the unstability had to be sought in the radioactive molecules themselves (see Fig. 2). A nucleoside-monophosphate (AMP) showed a less pronounced fall in count rate than RNA, and a nucleoside (thymidine) had a constant count rate.

The polar groups thus seem to be important for the instability of the count rates. With increasing amounts of water/μg RNA, the drop became faster and more pronounced.

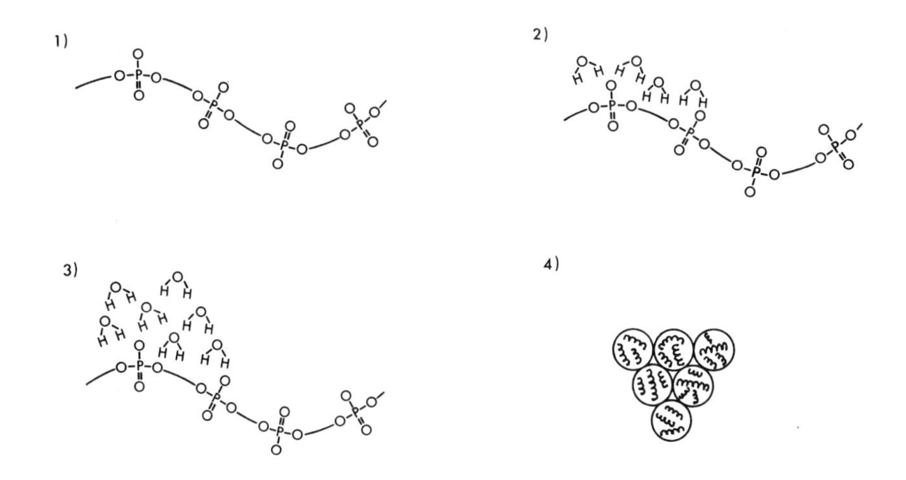

Fig. 3. Hypothesis for selective quenching of polar macromolecules.

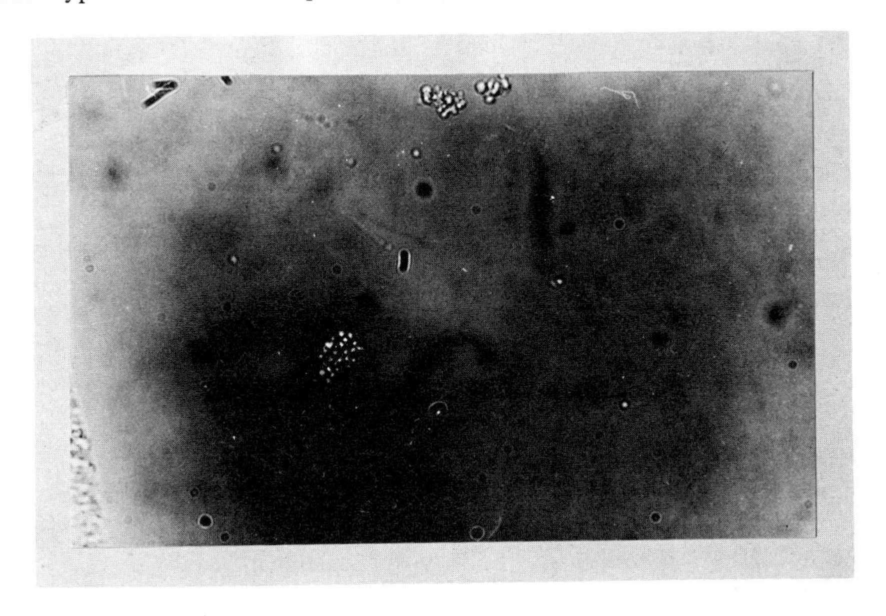

Fig. 4. Microscopy of the scintillation liquid after addition of 50 μg RNA dissolved in 50 μl 0.1 M sodium acetate buffer pH 6 (see text).

The quenching must be caused by weak forces, as it was abolished by manual shaking. Selective association of the aqueous molecules to the polar phosphate groups of RNA, caused by the dipole moments of the water molecules, seems probable. The polar molecules will thus be surrounded by growing aqueous spheres, interfering with the transfer of β-energy (see Fig. 3). The spheres could be seen under a microscope (see Fig. 4), having apparent diameters of 0.015 to 0.15 mm. They were absent when RNA was absent. On addition of hyamine, the count rates were stabilized (see Fig. 5), the quaternary ammonium

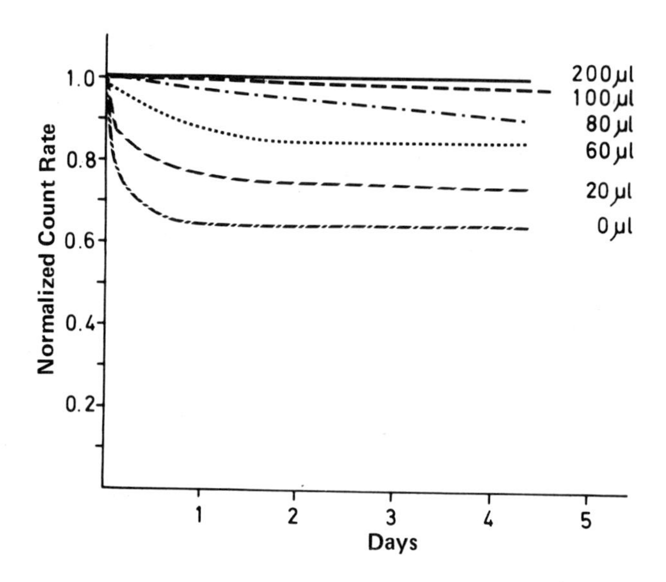

Fig. 5. Stabilization of count rates by addition of hyamine. 50 µg RNA was dissolved in 50 µl 0.1 M sodium acetate buffer, pH 6, and hyamine added in the indicated amounts before addition of scintillation liquid.

groups probably becoming associated with the acid groups of the nucleic acids, forming a hydrophobic coat around the molecules. However, stabilization depended upon the scintillation liquid and the buffer used. Some detergents, such as Triton X-100, tended to break the bonds and some buffer ions blocked the dissociable groups of the macromolecules or of hyamine.

Based on these experiments, the following methods were elaborated. Originally designed for counting of nucleic acids in aqueous solution,[4] they probably have a wider field of application.

METHODS

Extraction from aqueous solution

By shaking the aqueous solution containing the radioactive molecules with a mixture of N.C.S. or Soluene-100 in n-pentanol, 1 : 4 v/v, or in n-hexanol, 4 : 6 v/v, the quaternary ammonium bases will form low quenching hydrophobic complexes with the molecules having sufficient dissociable acid groups. Upon phase separation, these complexes will be found in the organic, upper phase, which can easily be pipetted off and counted in a pure toluene scintillation liquid.

Using 1 ml extraction liquid, 1 ml of the Soluene–n-pentanol and 0.95 to 0.97 ml of the N.C.S.–n-pentanol mixtures could be recovered when 2 ml 0.04 M acetate buffer was extracted. N.C.S. count rates were higher than expected, compensating for the slightly reduced volume (see Table 1). Hyamine, being dissolved in methanol, did not give full phase separation. Increasing the buffer volume reduced the organic phase volume after phase separation. With 10 ml buffer, only about 0.7 ml Soluene–n-pentanol was recovered. Evidently, with 2 ml buffer, the volume of extraction liquid dissolved in the aqueous phase exactly balanced the amount of buffer dissolved in the organic phase. RNA was quantita-

Table 1. RNA extraction and counting with different complexing agents in pentanol (1 : 4 v/v).[a]

Extraction liquid	Extracted from	No. of parallels	Observed count rates in per cent of calculated (±SD)	
			After 1 h	After 14 days
Soluene/pentanol	0.04 M Na acetate pH 6.0	5	96±3%	94±4%
Soluene/pentanol	0.1 M Na acetate pH 6.0	5	101±2%	98±3%
N.C.S./pentanol	0.04 M Na acetate pH 6.0	5	91±9%	88±9%
N.C.S./pentanol	0.1 M Na acetate pH 6.0	5	105±1%	102±1%
Hyamine/pentanol	0.04 M Na acetate pH 6.0	5	80±1%	65±9%
Tetraethylammonium hydroxide/pentanol	0.04 M Na acetate pH 6.0	3	0	0
Diethyldihexylamine/pentanol	0.04 M Na acetate pH 6.0	3	0	0
Diethyldihexylamine/pentanol	0.1 M Na acetate pH 6.0	3	0	0

[a] A toluene–2-methoxyethanol (85 : 15 v/v) scintillation liquid was used.

Table 2. Effect of Soluene in different solvents (1:4 v/v).[a]

Extraction liquid	No. of parallels	Observed count rates in per cent of calculated (±SD)	
		After 1 h	After 14 days
Soluene/pentanol	5	101±2%	98±3%
Soluene/hexanol	5	96±1%	96±1%
Soluene/heptanol	3	7±1%	–
Soluene/benzyl alcohol	3	11±3%	–
Soluene/cyclohexane	3	12±1.5%	–
Soluene/heptane	3	16±1.5%	–
Soluene/benzene	3	0	–
Soluene/toluene	3	0	–

[a] RNA was extracted from 0.1 M sodium acetate pH 6.0 and counted in a toluene–2-methoxyethanol (85:15 v/v) scintillation liquid.

tively extracted from buffer volumes up to 7 to 8 ml, rising count rates compensating for the decreasing volume of the organic phase. Soluene–n-hexanol permitted nearly complete phase separation with complete extraction of the same volumes. The other solvents tried were too hydrophobic to allow quantitative extraction (see Table 2).

The ratio of the quaternary ammonium base to the aliphatic alcohol was adjusted according to the ionic strength of the solution. A high ratio permitted extraction from solutions of high ionic strength. When the ratio was too high, however, phase separation did not take place. RNA was quantitatively extracted from solutions with ionic strengths between 0.02 and 0.12 by a 20% v/v solution of Soluene in pentanol, and a 80% v/v solution extracted RNA quantitatively from solutions of ionic strengths as high as 0.6.

Complete extraction of RNA was obtained over the entire pH range tested (3.25 to 9.5), as expected from the dissociation constants of RNA and Soluene.

Ions that blocked the complexing groups depressed the extraction (e.g. RNA extraction was depressed by magnesium, and a high chloride content unbalanced by strong cations depressed extraction generally). Sucrose, EDTA and SDS in the usual concentrations did not affect the extraction beyond their effect on the ionic strength of the solution.

To avoid photo- and chemiluminescence, sufficient acid should be added to the scintillation liquid to give a final acidic pH in the counting vial, causing the spurious counts to disappear during the time needed for sample temperature equilibration in the spectrometer. Pure toluene scintillation liquids became milky white on addition of the extraction liquid. This cleared in a few hours as water collected in small droplets at the bottom of the vial. Tritium count rates rose from 98% to 100% of the expected rate during this time. Carbon-14 count rates were unaffected. Scintillation liquids containing Triton X-100 gave unstable count rates.

Some buffer ions interfered with the complexes, causing unstable count rates. In these cases, extremely hydrophobic or hydrophilic scintillation liquids should be used. In the former, the buffer is not incorporated into the scintillation liquid. In the latter, the thermodynamic equilibrium favours stability. We find that 0.8 to 0.9 ml can routinely be pipetted off with a constriction pipette from 1 ml upper phase. When 0.9 ml was pipetted off, 53% tritium efficiency and 93% carbon-14 efficiency were obtained by balance point counting in a closed channel system. Appropriate corrections should be made for the size of the aliquot. The efficiencies of the pure scintillation liquid were 55% and 94% respectively.

The method is fast. A hundred fractions may be extracted and transferred to the scintillation vials in 1 to 1½ h, excluding the time needed for phase separation.

Beside nucleic acids, quantitative extractions of monkey salivary α-amylase (pI about 6.0 and 8.0), of cytochrome-C (pI 10.6), of a crude total TCA precipitate from serum in monkeys labelled with radioactive amino acids, and of soluble HeLa cell protein were obtained from 0.05 M sodium chloride in 0.01 M sodium acetate buffer pH 6. Smaller molecules, like adenosine monophosphate and orotic acid, were only partially extracted.

Obviously, the method outlined has its limitations. The need for unblocked complexing groups is absolute. Conditions found for one kind of biological macromolecule may therefore be invalid for another. Consequently, in every case, the recovery of isotope should be checked under the actual conditions used. On the other hand, extraction is insensitive to volume changes over quite a wide range. Dilution with water or salt solution may extend the limits for complete extraction. In the case of complete extraction, the method provides high counting efficiencies and the possibility of performing multiple labelling experiments.

Extraction from glass fibre filters

When complete extraction is not possible, precipitation with collection of the precipitates on glass fibre discs may be preferable. Precipitated macromolecules with sufficient acid groups may then be brought into solution by quaternary ammonium bases. The procedure should not be used with nitro-cellulose filters, as these give a severely quenching yellow colour upon incubation.

Routinely, the dried filters are incubated in closed scintillation vials with 1 ml 20% v/v Soluene in toluene at 50°C. Then 15 ml pure toluene scintillation liquid is added, the vial shaken until the filter floats freely, and counted as soon as sample temperature equilibrium is obtained in the spectrometer. To avoid photo- and chemiluminescence enough concentrated acetic acid is added to the scintillation liquid to acidify the samples.

Using this method, 98 to 99% of rat liver total RNA and DNA was brought into solution after incubation for 1 to 2 h. Elution was checked by removal of the filters from the vials after counting, and recounting of the vials and filters separately. 3 to 5 h were needed for an equal elution of soluble HeLa cell protein (labelled with radioactive amino acids for 48 h). Recently, a similar method has been published by Birnboim[5] using 25% v/v N.C.S. in toluene for counting ^{3}H-labelled RNA.

Increase of the Soluene content beyond 20% reduced the counting efficiency, as Soluene had a marked quenching effect compared with toluene. Less than 20% Soluene reduced the elution. Increase in incubation time of up to 24 h did not increase the count rates. The glass fibre paper disintegrated when incubated in 50% Soluene for more than 12 h.

High counting efficiencies with stable count rates were obtained. Direct balance point counting gave 52% tritium efficiency and 93% carbon-14 efficiency, while the eluted molecules were counted with 53% tritium efficiency and 93% carbon-14 efficiency as judged by the channels ratio method.

If quantitative direct extraction is possible, it is probably the method of choice, because it is faster, and because errors due to incomplete precipitation or absorption to the filter during filtration are avoided. With both methods, however, a complete account of disintegrations should be made. Even a slightly reduced recovery conveys the danger of selection because of different polarity of the molecules.

CONCLUSIONS

Molecules with polar groups may show unstable count rates when counted in aqueous solution. RNA count rates may be stabilized by addition of hyamine, presumably because of the formation of RNA–quaternary ammonium base complexes.

RNA, DNA and many proteins in aqueous solution may be extracted into an organic phase, suitable for liquid scintillation counting, by shaking with a mixture of Soluene-100 or N.C.S. in n-pentanol or n-hexanol.

RNA, DNA and many proteins may be eluted from glass fibre filters by incubation with quaternary ammonium base solubilizers.

The two latter methods abolish energy absorption by filters or precipitates, give high counting efficiencies, and allow double labelling experiments to be performed with greater accuracy.

REFERENCES

1 F. N. Hayes, B. S. Rogers and W. H. Langham, *Nucleonics* **14**, No. 3, 48 (1956).

2 P. N. Paus, *Anal. Biochem.* **31**, 502 (1969).

3 G. T. Okita, J. J. Kabara, F. Richardson and G. V. LeRoy, *Nucleonics* **15**, No. 6, 111 (1957).

4 P. N. Paus, *Anal. Biochem.* **38**, 364 (1970).

5 H. C. Birnboim, *Anal. Biochem.* **37**, 178 (1970).

DISCUSSION

B. W. Fox: Have you tried separating double and single stranded DNA by this procedure?

P. N. Paus: No. There might be differences in extraction under conditions when extraction is not complete (*cf.* RNA).

B. W. Fox: Have you noticed any differential extraction of RNA species since the use of quaternary bases is well known in counter current separation procedures?

P. N. Paus: Differential extraction has only been obtained with magnesium present in the aqueous phase. Ribosomal RNA extraction was then more depressed than rapidly labelled RNA.

B. W. Fox: Comment: Lithium is normally used in the aqueous phase for RNA separation; it would be interesting to try this ion in your system.

H. E. Dobbs: Dr. Paus has described a method, using liquid reagents, for preparing biological macromolecules for liquid scintillation counting. When selecting the papers for this session, we included a paper in which the preparation of biological samples for liquid scintillation counting by combustion techniques was reviewed. Unfortunately that paper was withdrawn. Suitable combustion techniques can be applied to all biological samples which are not easily soluble in liquid chemical reagents.

Chapter 17

The Estimation of Small Quantities of Carbon-14 Labelled Adenine Nucleotides following their Separation by Ion Exchange Paper Chromatography

G. C. Carney

Bristol Polytechnic, England

The technique of polarography combined with the analysis of adenine nucleotides in a mitochondrial suspension provides a powerful analytical approach to certain aspects of oxidative phosphorylation. The use of the oxygen electrode alone is a well established technique. Its sensitivity, however, has tended to far outstrip the methods available for conveniently estimating the efficiency of the phosphorylation process. The present communication has the purpose of emphasizing the fact that the separation and estimation of the nucleotides derived from trace amounts of ADP is well within the scope of the teaching as well as the research laboratory.

A typical experiment for studying the efficiency of oxidative phosphorylation proceeds as follows. Samples to be analysed are withdrawn from the incubation medium containing the respiring mitochondria in volumes of 3 to 5 μl. The samples are then spotted on strips of DEAE 81 (OH form) ion exchange paper. For convenience, after being spotted the strips are cut from a previously ruled sheet. Ascending chromatography is then carried out using 20% formic acid at room temperature.

This type of chromatography results in a very tight binding of the nucleotides to the paper. The enzymatic reactions are immediately halted, thus doing away with the need for perchloric acid or TCA treatment. Following chromatography the strips are dried and cut into fragments about 1 cm square which are placed serially in separate vials containing the fluor solution. Immersion in this solution results in very little radioactivity leaving the paper.

The data obtained from the group vials give the total counts above background for the regions of the original chromatogram in which the ATP, ADP and AMP have become distributed. Each locus may have its activity spread over three to five vials depending on the concentration. This enables the activity due to each nucleotide to be calculated as a percentage of the total activity. The regions between the separated components have an activity which is the same as, or very close to, background.

Samples may be withdrawn at frequent intervals during an experiment. The open cell of the Gilson 'Oxygraph' lends itself to this procedure in that small samples may be obtained very conveniently and without disturbing the oxygen uptake recording. This procedure is illustrated in Fig. 1. The labelled compound is added through the same

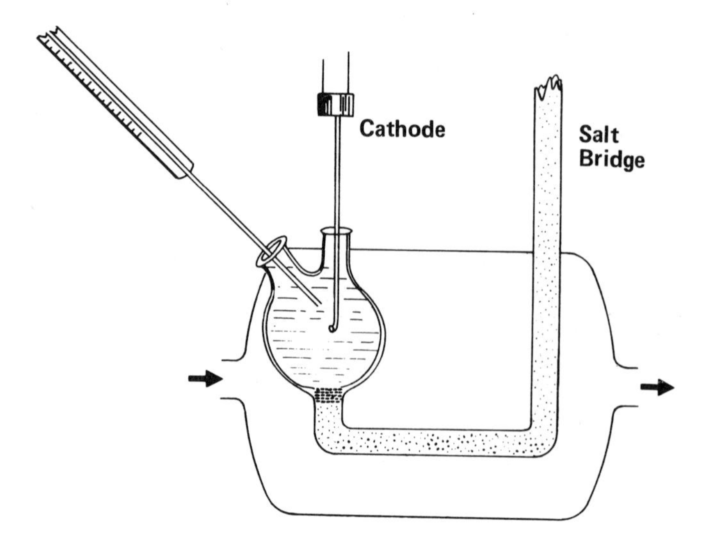

Fig. 1. The Gilson 'Oxygraph' cell showing the use of a Hamilton syringe for withdrawing small volumes. Large arrows indicate direction of flow in water jacket.

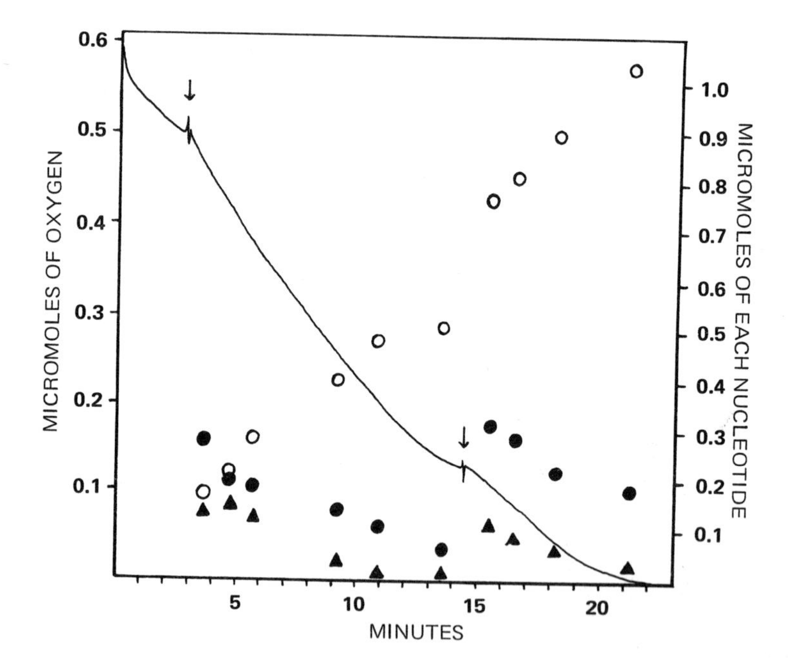

Fig. 2. The respiratory response to ADP in mitochondria isolated from the turtle liver. Arrows indicate times when ADP was added, 0.6 μmol in each case. ○ = ATP, ● = ADP and ▲ = AMP. Total nucleotide concentration established at 260 nm. Substrate, succinate. Mitochondria present equivalent to 1 mg protein. (From experiment conducted with N. Grimes, University of Bowling Green, Ohio).

opening. Its volume is larger and the mixing procedure slightly deflects the oxygen recording. The combined data are shown in Fig. 2. In this case consecutive additions of ADP have been made. The reduced rate of oxygen uptake as the ADP is used up is clearly seen. Also shown is the renewed stimulation of respiration when fresh ADP was added.

Apart from the oxygen relationship, the study of the balance between the concentrations of the three nucleotides is a fertile field of research. Data quite similar to those in the present report (but obtained by means of the luciferase assay) have been published by Godinot *et al.*[1] A result which is sometimes surprising is the intense activity of myokinase and ATPase revealed in certain preparations, for example the embryo chick liver[2] and housefly sarcosomes.[3]

The use of the oxygen electrode in teaching the theory of oxidative phosphorylation in the laboratory has been adequately described.[4] The combined approach presently described has also been very successful in practice. Students have been able to conduct oxidative phosphorylation experiments in a way which they have found to be particularly interesting. They have also obtained valuable experience in the technique of liquid scintillation counting and in the statistical handling of the data. The low activity of the isotopes made possible by this counting method is an additional asset in the teaching laboratory from the point of view of safety and economy.

REFERENCES

1 C. Godinot *et al., Eur. J. Biochem.* **8**, 385 (1969).
2 A. M. Holm, *Thesis,* University of Bowling Green, Ohio, U.S.A., 1968, p. 59.
3 G. C. Carney, *Life Sci.* **8**, 435 (1969).
4 J. M. Foster, *Bioscience* **19**, 541 (1969).

DISCUSSION

H. Dobbs: Comment: There is a one word answer to the questions raised in Dr. Carney's paper; it is 'combustion'. In our laboratories we have yet to find a sample which will not succumb to this technique. Quantitative measurements of the distribution of radioactivity on paper chromatograms can be made by cutting the chromatogram into strips and burning the individual pieces of paper. We have successfully applied combustion techniques to the assay of thin layer chromatograms. We scraped the areas of the plate (containing highly quenching materials adsorbed on alumina) into rice paper cachets. The alumina was mixed with powdered cellulose. Each cachet was then burned in the normal manner. After introduction of the liquid scintillator into the combustion flask the alumina was allowed to settle before an aliquot of the liquid scintillator was removed for counting.

Chapter 18

Sample Preparation for Tritium Counting in the Application of the Digoxin Radioimmunoassay Technique to Lysed Blood

A. P. Phillips and C. A. Sambrook

Home Office Central Research Establishment, Aldermaston, Berkshire, England

INTRODUCTION

The cardiac glycosides, among which digoxin is almost exclusively prescribed in Great Britain, are extremely effective in the treatment of congestive heart failure and supraventricular tachyarrhythmias. Unfortunately there is a narrow margin between therapeutic and toxic doses, and since toxicity comprises aggravation of these cardiac conditions it was formerly impossible to distinguish between insufficient and over-dosage. To emphasize the seriousness of the situation, a recent survey revealed that toxicity results in 8 to 20% of hospital patients taking digoxin, with a subsequent mortality of 7 to 50% (Baller *et al.*).[1] The situation may improve as a result of the introduction of a radioimmunoassay (Smith *et al.*)[2] sufficiently sensitive to measure therapeutic digoxin levels in plasma, particularly now that the technique is available as a commercial kit.

This study on scintillation counting was a preliminary to the adaptation of the digoxin radioimmunoassay method to the analysis of whole blood taken at post-mortem examination. The high degree of haemolysis in post-mortem specimens poses a problem of colour quenching. The facility to examine such samples should be useful in forensic science, where overdose fatalities are sometimes encountered. Furthermore, by providing a retrospective measure to the clinician it may help to increase our understanding of the delicate balance between therapy and toxicity.

EXPERIMENTAL

Materials

The Triton X-100 used initially was purchased from Lennig Chemicals Ltd. Toluene, PPO and POPOP were the scintillation grades of Koch-Light Laboratories Ltd. Naphthalene, n-pentanol and Cellosolve were the laboratory reagent grades of B.D.H.; ethanol was their technical grade. Scintillation grade hyamine hydroxide and NE 250 solubilizer were purchased from Nuclear Enterprises Ltd., as were the scintillation cocktails NE 233, NE 216 and NE 250 (based respectively on toluene, xylene and dioxan).

Lanoxitest kits for the radioimmunoassay of digoxin were from Wellcome Reagents Ltd. 'Analar' hydrogen peroxide, 100 volumes 30% w/v, was purchased from B.D.H.

Table 1. Relative performance of 10 ml scintillator cocktails with 0.5 ml N.H.S. supernatant.

SCINTILLATOR COMPOSITION							
Toluene/ POPOP/PPO[a]	NE 233	NE 216	NE 250	Naphthalene		c.p.m. as percentage of NE 250 standard	new pence/ 10^3 c.p.m.
–	50%	–	–	–	50% Triton X-100	48.4[b]	2.0
60%	–	–	–	–	40% Cellosolve	35.7	3.1
52%	–	–	–	8%	40% Cellosolve	38.9	2.8
60%	–	–	–	–	40% Ethanol	36.2	2.5
52%	–	–	–	8%	40% Ethanol	51.2	2.2
80%	–	–	–	–	20% NE 520	42.6	14.0
72%	–	–	–	8%	20% NE 520	42.2	14.0
–	80%	–	–	–	20% NE 520	50.6	13.8
–	85%	–	–	–	5% Hyamine, 10% n-pentanol	57.7	4.3
85%	–	–	–	–	5% Hyamine, 10% n-pentanol	35.6	6.3
77%	–	–	–	8%	5% Hyamine, 10% n-pentanol	45.5	4.8
–	–	100%	–	–	– –	< 1	–
–	–	–	100%	–	– –	121.0	3.0
5 ml of NE 250						100.0	1.8

[a] 0.4% PPO, 0.01% POPOP
[b] Counted with the sample changer temperature at 8.5°C.

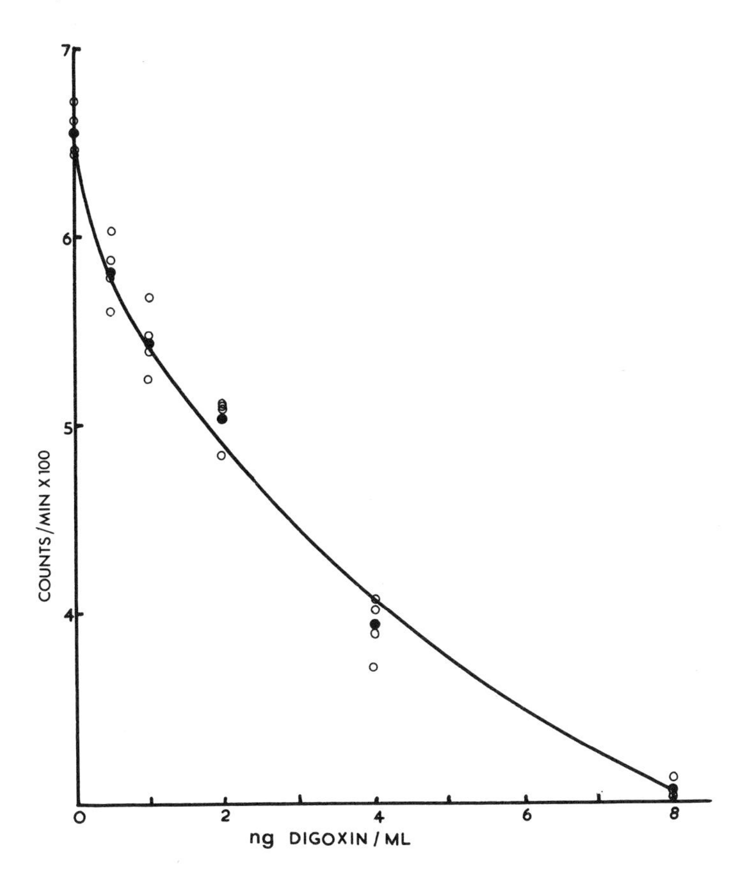

Fig. 1. Radioimmunoassay standard curve on lysed blood, using the Japax bleach technique.

Japax bleach is manufactured by Newlands Bros. and Mumford Ltd., 324 Harrow Road, London W9.

Counting methods

Scintillation counting of samples described in Table 1 and Fig. 1 was carried out in an Intertechnique SL 40 model, by kind permission of the Medical Research Council. The sample changer chamber in this instrument was at 12°C. The remaining data were derived in an Intertechnique SL 31, with the changer chamber at 8.5°C. 2×10^4 counts were collected routinely from each specimen.

Since the specific activity of the ^{3}H-digoxin was unknown, relative counting efficiencies are presented.

Radioimmunoassay method

The principle of the assay is that the reaction of a limiting quantity of antidigoxin serum with a quantity of ^{3}H-digoxin is inhibited by digoxin contributed by a test plasma, the degree of inhibition being related to the performance of a series of known digoxin solutions expressed as a standard curve. After the reaction has reached quasi-equilibrium, any ^{3}H-digoxin not bound to antibody is adsorbed onto activated charcoal, and the ^{3}H-digoxin-antibody in the resulting supernatant counted in a liquid scintillation spectrometer.

Concentrated buffer, unlabelled and tritiated digoxin, anti-digoxin serum, and activated charcoal are provided in the Lanoxitest kit. The reaction mixture prescribed by the manufacturers consists of 0.1 ml plasma mixed with ^{3}H-digoxin and antiserum in a total of 0.8 ml. Normal horse serum (N.H.S.) is supplied for use in place of plasma in blank and standard tubes. After incubation, 0.2 ml of the suspension of charcoal are added, the mixture incubated further and then clarified by centrifugation in a bench centrifuge. 0.5 ml portions of the final supernatant were routinely counted for tritium in this laboratory, assays being performed at least in triplicate. When several scintillation systems were under comparison, a single large reaction mixture was utilized, and 0.5 ml portions of the supernatant dispensed to individual counting vials. These reaction mixtures contained either N.H.S., outdated blood bank whole blood lysed by a single freezing/thawing cycle, or post-mortem blood believed to be digoxin-free.

Choice of a scintillator for N.H.S. supernatants

A variety of scintillator cocktails were examined for use with N.H.S. supernatants. The latter displayed only a faint straw colouration, and thus colour quenching was not expected to be a serious problem. Samples were loaded into the scintillation counter immediately after brief shaking, with the exception of Triton emulsions which were first allowed to stand in a 55°C oven for 5 min. The relative performance of the systems may be seen in Table 1. Count rate has been expressed as a percentage of that obtained with 5 ml of NE 250, the system chosen as a reasonable compromise of efficiency, economy and ease of use.

Samples prepared in NE 250 were allowed to equilibrate for several hours before counting. Even so, count rates continued to rise, there being a further increase of about 5% by 24 h, and thus where great accuracy was required samples were not counted until 24 h after preparation.

Scintillation counting of haemolysed supernatants

Replacement of N.H.S. by lysed whole blood in the radioimmunoassay resulted in serious quenching, presumably due to the pronounced redness of the supernatant and subsequent scintillation mixture. In 5 ml NE 250 the count rate was reduced to 30% of that obtained with samples prepared in parallel with N.H.S., and in 10 ml NE 250 a relative count of 54% was achieved.

Three inexpensive decolourants were considered, hydrogen peroxide, chlorine water and a commercial hypochlorite bleach, 'Japax'. Since the standard 5 ml of NE 250 was rather intolerant of a significant increase in aqueous content over the normal 0.5 ml (an additional 0.5 ml of distilled water reducing the count rate by 20%), the minimal necessary volumes of the decolourants were investigated. Reasonable volumes of chlorine water had little effect even after prolonged incubation. The maximal bleaching effect was obtained with 0.20 ml of hydrogen peroxide or 0.225 ml of Japax within a few minutes at room temperature, the resulting fluid being a rather deeper straw colour than N.H.S. supernatants. As a precaution the pH of samples was checked, and it was necessary to add 0.025 ml of 3 M acetic acid to adjust the Japax-treated solutions to about pH 6.5. Supernatants decolourized with hydrogen peroxide displayed marked inter-sample variation, with replicas differing by up to 70%, and this method was considered no further. The supernatants bleached with Japax counted reproducibly at 77% of the efficiency of comparable N.H.S. supernatants; in 10 ml NE 250 they counted at 95%. Normal background count rates were registered in the absence of ^{3}H-digoxin.

Once bleached haemolysed samples had been allowed to equilibrate with the temperature of the scintillation spectrometer, count rates were stable for at least 72 h. The rapid equilibration contrasts with the 24 h period necessary for N.H.S. supernatants counted directly in NE 250 and is probably a consequence of the increase in homogeneity conferred by the bleaching process. In fact, Japax-treated samples in NE 250 formed emulsions which separated into two clear phases on standing, the aqueous phase retaining the yellow colour of the bleached haemolysed supernatant. In contrast, untreated N.H.S. supernatants in NE 250 resulted in a flocculent precipitate under a single clear liquid phase. Treatment of N.H.S. supernatants with Japax bleach before counting in NE 250 resulted in a bilayer system with the rapid equilibration found for haemolysed supernatants.

As a final test, a standard curve was prepared from radioimmunoassays based on lysed whole blood, using the Japax treatment. The results are shown in Fig. 1. The criteria for a successful assay were in general satisfied. Replicas agreed well, with coefficients of variation between the various sets of quadruplicates varying from 1.9 to 4.7%.

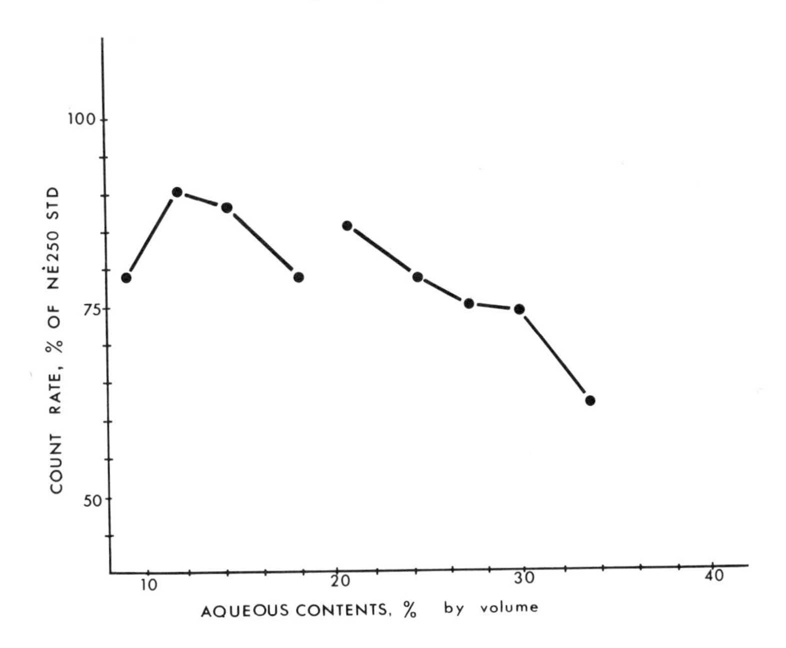

Fig. 2. Counting performance of N.H.S. supernatants in Triton emulsions with varying water content.

Further examination of Triton emulsion systems

Because of the advantages accredited to Triton emulsion counting of tritium in the recent literature (Williams and Florkowski,[3] van der Laarse,[4] Williams,[5] Turner[6]) this type of system was reconsidered. First it was necessary to ascertain to what extent the relatively poor efficiency encountered in the 1 : 1 Triton X-100 : NE 233 mix (Table 1) had been due to the technical quality of the Triton used. Substitution of scintillation grade Triton (Intertechnique Ltd.) in this system, in fact, only increased the efficiency

Table 2. The counting of post-mortem supernatants.

Bleach (ml)	3 M Acetic acid (ml)	Additional water (ml)	NE 250 (ml)	Triton/toluene (ml)	c.p.m. (mean of Triplicates)
–	–	–	5.0	–	420
–	–	–	10.0	–	707
–	–	0.5	–	4.5	61
0.225	0.025	0.25	–	4.5	580
0.225	0.025	–	5.0	–	743

to 56% of that in 5 ml NE 250. Increasing the water content in stages up to 23% by volume only served to reduce the count rate, finally to 49% of the NE 250 figure.

However, such a high Triton content has not been found to be ideal for efficient counting[5] and so a cocktail was prepared comprising 2:1 analar toluene:scintillation grade Triton with 0.7% PPO and 0.035% POPOP. The aqueous contribution of 0.5 ml N.H.S. supernatant was augmented to varying degrees by the addition of distilled water in a total vial content of 5.5 ml. As seen in Fig. 2, the resulting discontinuous curve is very similar to those reported previously.[3,5] The maximum count rate encountered was 90% of the NE 250 system at 12% aqueous content.

Of the mixtures shown in Fig. 2, that with an aqueous content of 18.2% was selected for a detailed stability study. Repeated counting of such samples showed that the count rate remained within the ±2 SD limits for 48 h after initial temperature equilibration. This system was applied to Japax-bleached supernatants based on actual post-mortem blood. As may be seen in Table 2, bleached samples counted well in the emulsion, the efficiency relative to the bleached NE 250 system (77%) being almost identical to the relative performance of untreated N.H.S. supernatants in Triton and NE 250 (Fig. 2). There was no evidence of chemiluminescence. The Triton system was particularly susceptible to colour quenching by untreated post-mortem supernatants.

It is noteworthy that our calculated cost of 1.1 new pence/10^3 counts for N.H.S. supernatants counted in 4.5 ml of this Triton/toluene cocktail is better than any of the costs listed in Table 1.

REFERENCES

1 G. A. Baller, T. W. Smith, W. H. Abelmann, H. Haber and W. B. Wood, *New Engl. J. Med.* **284**, 989 (1971).

2 T. W. Smith, V. P. Butler and E. Haber, *New Engl. J. Med.* **281**, 1212 (1969).

3 P. H. Williams and T. Florkowski, Proceedings of the Monaco Symposium on Liquid Scintillation Counting, 1967, p. 703.

4 J. D. van der Laarse, *Intern. J. Appl. Radiation Isotopes* **18**, 485 (1967).

5 P. H. Williams, *Intern. J. Appl. Radiation Isotopes* **19**, 377 (1968).

6 J. C. Turner, *Intern. J. Appl. Radiation Isotopes* **20**, 499 (1969).

DISCUSSION

H. Dobbs: Comment: At Reckitt and Colman we put Harpic in our lavatories, but I have never thought of putting it in our counting bottles!

Chapter 19

Dynamics of Quenching by Plasma from Patients with Various Pathological Conditions

J. Assailly, C. Bader,* J.-L. Funck-Brentano and D. Pavel

Unité de Recherches sur l'Application des Radio-Eléments à l'Etude des Maladies Métaboliques, INSERM U.90, Hôpital Necker, Paris, France

INTRODUCTION

In the last few years, techniques have been described which permit the direct incorporation of plasma into organic scintillation mixtures by using tensioactive agents.[1-3] Many studies have been perfected on the physical characteristics of these heterogeneous systems. A great number of these studies emphasize the difficulties encountered with sample stability,[4,5] and with the measurement of detection efficiency[6,7] when using the classical methods of quenching correction with these systems.[8-10] Perhaps, due to these difficulties, the use of tensioactive agents in clinical laboratories working with β-emitters has not been well accepted and clinical assays on patients with these nuclides are usually accomplished with complex physical or chemical treatment of plasma samples.

Clinical tests with β-emitters are shown in Table 1 together with the corresponding preparative method most currently employed for each test.

Our aim has been to perfect a reliable method for the preparation of plasma samples while benefiting from the inherent simplicity obtained through the use of tensioactive agents. We have developed, for clinical studies, a simple method for counting untreated plasma in conjunction with tensioactive agents.

In the first part of this communication, the methods used for direct plasma sample preparation, counting efficiency measurement and quench correction are described. In the second part, we discuss our results which show that at least for clinical studies, counting efficiencies vary widely. It is shown that this quenching is due to the colour changes in plasma samples. Finally, the practical consequences of counting and the calculation of d.p.m. from c.p.m. in labelled plasma samples are discussed.

MATERIALS

All these studies have been conducted with an SL 40† liquid scintillation counter, equipped with an 'on-line' computer. The coefficients of the quenching curves were com-

---

* Attaché de Recherches.

† Intertechnique, 78-Plaisir, France.

Table 1. Clinical studies with the aid of β-radioisotopes.

Designation	Tracer	Molecule	Plasma processing	Minimal number of plasma samples required
Whole body water	Tritium	Water	Plasma protein precipitation. Combustion	2
Extracellular (sulphate) space	Sulphur-35	SO_4^{--}	Precipitation, isolation with ion exchange resin	4
Calcium metabolism	Calcium-45	Ca^{++}	Complex	6
Iron metabolism	Iron-55	Fe^{++}	Complex	8

puted with a PDP 10 computer.§ The various products used for sample preparation were:

1. toluene
2. Triton X-100®*
3. PPO†
4. 2,5-POPOP†

For quenching studies we have used haemoglobin or Hank's Liquid which is red at neutral pH and yellow at acid pH. For chemical quenching studies we have made use of pyridine which is water-soluble. The plasma samples were collected from patients suffering from various diseases which may be described here as:

1. nephrotic syndrome
2. hepatic syndrome

The various studies currently performed with β-tracers in our laboratories are represented in Table 2.

METHODS

Plasma sample preparation

The method used for plasma sample preparation is the same one as we have previously described[11] and is schematically represented in Table 3. This is a direct incorporation in a toluene–Triton mixture but with a dilution of plasma in physiological saline. This procedure results in a better stability of the plasma components and permits detection efficiency determination with the aid of the 'pseudo-internal' method as described below.

Counting efficiency measurement

The preparative method results in an heterogeneous system currently designated in the scientific literature as an 'emulsion'. As has been shown by many authors, it is not

* Registered Trademark, Rohm and Haas, U.S.A.
† PPO = 2,5-diphenyl oxazole; POPOP = *p*-bis (2,5-phenyl oxazolyl) benzene.

Table 2. Clinical studies with β-radiotracers in our laboratory.[a]

		Tracer	Molecule
Compartmental studies	Total body water	Tritium	H_2O
	Extracellular space	Sulphur-35	SO_4^{--}
Metabolic studies	Calcium metabolism	Calcium-45	Ca^+
	Azathioprine	Sulphur-35	Imuran®[b]

[a] I.N.S.E.R.M. U.90 (J.-L. Funck-Brentano) and Clinique Nephrologique (J. Hamburger) Hôpital Necker, Paris.
[b] Imuran®: Burroughs Wellcome, U.S.A.

Table 3. Direct plasma sample preparation.

NaCl 9 g/l	2 ml	For plasma dilution
+		
Plasma	1 ml	
+		
S.E.M.	12 ml	Cooling (ice bath) Plasma + NaCl solution + S.E.M. Counting at 4°C
S.E.M.:	Scintillator emulsifying mixture Composition: toluene scintillator solution PPO 4 g/l POPOP 100 mg/l	

This solution is mixed with Triton X-100® in the proportion of 1 : 1 toluene : Triton X-100.

possible with those systems to make use of the internal standard method for counting efficiency measurement. This method when applied to heterogeneous systems results in two values corresponding to the various phases constituting the system. Thus, we have used a method previously described by us as 'pseudo-internal' standardization.[11] This method is schematically represented in Table 4.

Two samples are prepared – the first one, as described above; the second, with the addition of a radioactive standard solution of known activity diluted in physiological saline. Thus, we obtained two samples with exactly the same physical structure and phase properties. The computation of detection efficiency is classical and is briefly presented in Table 4. The maximum counting efficiency is designated numerically as 1.00.

Quenching studies

For studying quenching, we have made use of two quenching factors:

1. B/C – ratio of the count rate in two channels B and C

Table 4.
Counting efficiency measurement by 'pseudo-internal' standardization.

Mixing order	I Plasma sample (P_I)	II P_I + radioactive standard of known activity
1	NaCl 9 g/l (2 ml)	Radioactive standard
2	Plasma P_I (1 ml)	1 ml plasma P_I solution
3	S.E.M. (12 ml)	12 ml S.E.M.

Counting efficiency computation $E(P_I)$ of the plasma sample (P_I).

N_I: Counting rate of the plasma sample P_I (c.p.m.)
N_{II}: Counting rate of the standardized plasma (c.p.m.)
D_{st}: Activity in d.p.m. of the radioactive standard solution
$D(P_I)$: Activity in d.p.m. of the plasma sample P_I

$$E(P_I) = \frac{N_{II} - N_I}{D_{st}}$$

Activity in d.p.m. of the plasma sample P_I

$$D(P_I) = \frac{N_I}{E(P_I)}$$

2. E_1/E_2 – ratio of the count rate in two channels E_1 and E_2 of an external γ-emitter (caesium-137).

The discriminator levels are shown in Fig. 1. Using the SL 40 computer facilities, it is possible to determine simultaneously these two factors.

RESULTS

Detection efficiency

Counting efficiencies of the plasma samples labelled with tritium in the case of whole body water measurements, sulphur-35 for extracellular space and azathioprine metabolism and calcium-45 for metabolic studies are respectively:

1. For tritium, a great number of samples have detection efficiencies between 0.08 and 0.14. A few samples have efficiencies inferior or superior to these values. Finally, the extreme values observed were 0.02 and 0.18.
2. For ^{35}S-labelled plasma a great number of samples have detection efficiencies between 0.55 and 0.75. A few samples have values inferior or superior to these values. The extreme values observed for sulphur-35 detection efficiencies were 0.38 to 0.83.
3. For radioactive calcium-45 a great number of plasma samples have detection efficiencies between 0.63 and 0.83. A few samples have detection efficiencies inferior or superior to these values. Extreme values for calcium-45 detection efficiencies in plasma samples were 0.40 and 0.88.

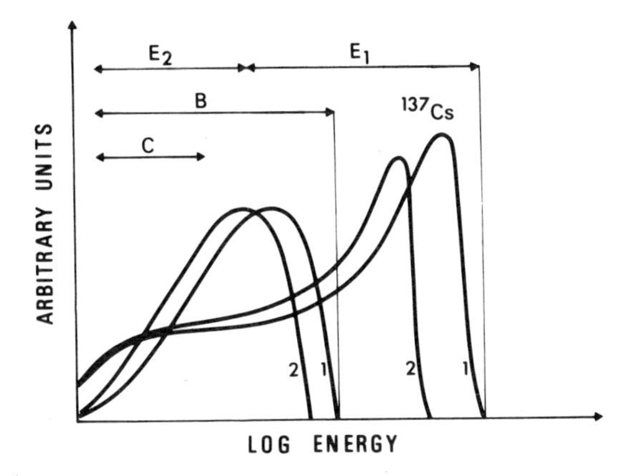

Fig. 1. Discriminator levels for quenching studies.

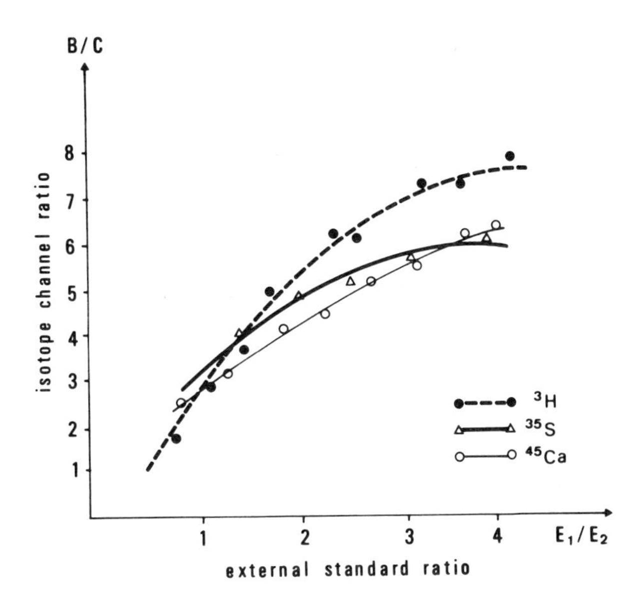

Fig. 2. Relationship between isotope channels ratio and external standard ratio.

Dynamic quench range of plasma samples

The results of counting efficiency measurement show that quenching is variable in plasma samples. Thus, we must choose a method for quench correction. The two classical methods are isotopes channels ratio and external standard channels ratio. A number of difficulties have been reported when using these two methods when applied to heterogeneous systems similar to the one used in this study.[8-10] Bush[10] recently pointed out that in those cases it is possible to check the validity of external standard ratio by studying the relationship between the factors B/C and E_1/E_2. We have made use of this method and by

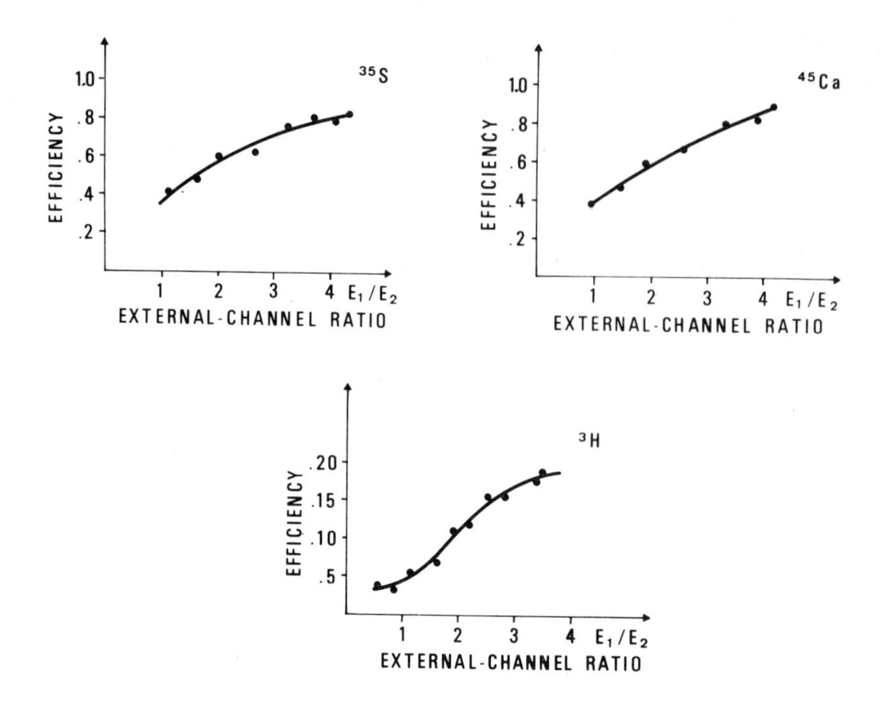

Fig. 3. Quenching dynamics of ^{3}H-, ^{35}S- and ^{45}Ca-labelled plasma samples as a function of the external standard channels ratio.

using the computer facilities of the SL 40 apparatus, we have obtained simultaneously these two factors for samples which have different counting efficiencies as measured by our 'pseudo-internal' method.

The results are shown in Fig. 2. There is a fixed relationship between the two quenching factors for the three isotopes used in our clinical studies. Thus, it is possible at least with our preparation system to make use of one or the other of these two quenching factors. Due to the easy measurement of the E_1/E_2 quenching factor and its better statistical accuracy we decided to use this factor to correct for quenching.

Curves of efficiency versus E_1/E_2 of ^{3}H-, ^{35}S- and ^{45}Ca-labelled plasma samples are represented in Fig. 3.

These quenching curves were obtained from clinical samples by measuring their counting efficiency and quenching factor. However, in view of the fact that the characteristics of our samples change after three days it is difficult to check long term instrument stability. Also we must add to our existing curve new values resulting from samples falling outside the curve limits. Computation parameters are modified according to these data. Thus, it was necessary to study the mode of quenching. In practice it is current to use a chemical quenching agent, for example carbon tetrachloride, water, acetone and so on. However, due to the various colours of the plasma samples which were green, yellowish or more or less red, it was expected that the quenching was of optical origin. At the last International Symposium on Liquid Scintillation Counting, Neary and Budd[12] stressed that the dynamics of quenching due to physical i.e. colour effects were different from those obtained with a chemical agent, at least with β-emitters having energies superior to tritium

Table 5. Preparation of samples for studying the origin of quenching.

Series	Added substances	Sample number 1	2	3	4	5	6	7	8	9
I Physical	Haemoglobin solution (μl) (100 mg/ml)	0	2	5	10	20	50	100	200	500
	Hank's Liquid pH 7.4 and 6.8 (μl)	0	50	100	200	350	500	1000	1500	2000
II Chemical	Pyridine (μl)	0	20	50	100	300	500	1000	1500	–

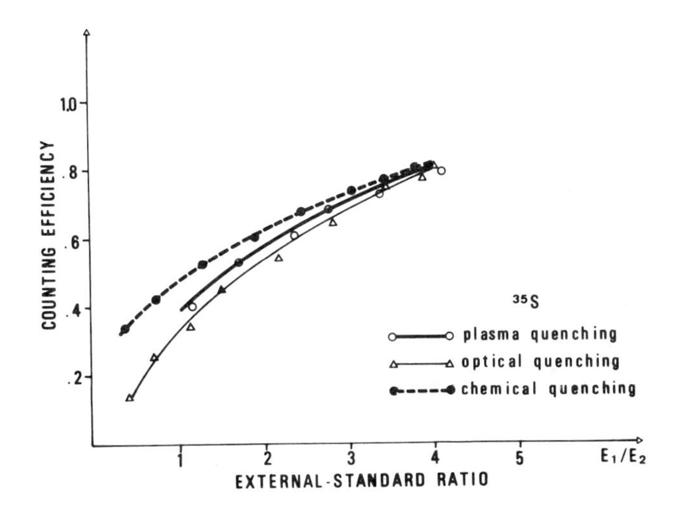

Fig. 4. Plasma quenching in comparison of optical and chemical quenching curves.

(E_{max} = 18 keV). Thus, we have checked our hypothesis concerning the physical origin of plasma quenching. To do this, two series of samples were prepared as described on p. 224 but with physiological saline in place of plasma and containing two different quenching substances (see Table 5).

The first series was optically quenched by different volumes of haemoglobin solution at fixed concentration or Hank's Liquid at various pH levels. The second series was chemically quenched with different volumes of pyridine. We have chosen pyridine as the chemical quencher for its high water solubility. The results are shown in Fig. 4.

One may observe that the quenching in plasma samples is mostly of physical origin. For tritium the colour and chemical curves are identical as Neary and Budd have demonstrated,[12] but are not represented here. Thus, it is possible when counting labelled plasma samples to use a physical colour quench curve to obtain reliable results.

Table 6. Substances which may be found in human plasma and which interact with the liquid scintillation process by physical (i.e. colour) or chemical quenching.

Substances	Quenching effect	Wavelength maximum absorption (mμ)[a]
Haemoglobin	Colour	275–600
Bilirubin	Colour	400–520
pH (Hydronium ion concentration)	Chemical	
Proteins	Chemical	
pO_2 (Oxygen partial pressure)	Chemical	

[a] Fluorescence spectra of PPO and POPOP: λ_{max} = 380 and 420.

COMMENTS

In this study we have shown that when measuring labelled plasma samples by liquid scintillation counting, counting efficiencies vary, at least in the case of plasma samples for clinical studies. The origin of this variation is colour quenching induced by the different colours of the plasma samples. A number of molecules which have absorption spectra overlapping the emission spectra of the fluorescent scintillators may be found in human plasma. Other molecules which can interact with the process of energy transfer between solvent molecules are also present in human plasma. Table 6 shows a few examples of these molecules. These factors are present in all plasma samples, and most quenching substances which significantly influence counting efficiencies interact by optical absorption processes.

PRACTICAL CONSEQUENCES

The practical consequences of counting and the calculation of d.p.m. from c.p.m. in labelled plasma samples, in particular in the field of clinical studies, may be summarized as follows:

1. Detection efficiency does not remain constant. On the contrary, there may be a large range of efficiencies due to quenching.
2. This quenching is principally of physical origin (colour effects) and may be corrected by using a correction curve developed from a physical quenching agent.
3. It is possible to use the isotope channels ratio or the external standard ratio method to correct for quenching.

Finally, it is hoped that this study will contribute to the application of β-tracers in the field of clinical investigation and biological research.

ACKNOWLEDGEMENTS

The authors are indebted to Mr. Palais (Intertechnique, France) for his interest and helpful suggestions for the completion of this work.

REFERENCES

1 A. E. Whyman, *Intern. J. Appl. Radiation Isotopes* **21**, 81 (1970).

2 A. Nadarajah, B. Leese and G. F. Joplin, *Intern. J. Appl. Radiation Isotopes* **20**, 733 (1969).
3 C. B. Oxby and P. A. Kirby, *Intern. J. Appl. Radiation Isotopes* **19**, 151 (1968).
4 R. H. Benson, *Anal. Chem.* **38**, 1353 (1966).
5 J. D. van der Laarse, *Intern. J. Appl. Radiation Isotopes* **18**, 485 (1967).
6 R. Lieberman, *Intern. J. Appl. Radiation Isotopes* **21**, 319 (1970).
7 P. H. Williams, *Intern. J. Appl. Radiation Isotopes* **19**, 377 (1968).
8 P. H. Springell, *Intern. J. Appl. Radiation Isotopes* **20**, 743 (1969).
9 H. E. Dobbs, *Intern. J. Appl. Radiation Isotopes* **19**, 155 (1968).
10 E. T. Bush, *Intern. J. Appl. Radiation Isotopes* **19**, 447 (1968).
11 C. Bader, J. Assailly, J. Chanard, J.-L. Funck-Brentano, *12e Table Ronde sur l'Exploration Fonctionnelle par les Isotopes Radioactifs,* Strasbourg, 1970.
12 M. P. Neary and A. L. Budd, in *The Current Status of Liquid Scintillation Counting* (Ed. E. D. Bransome), Grune and Stratton, New York, 1970.

DISCUSSION

H. Dobbs: At the risk of being labelled as a 'combustion obsessive' I would point out that plasma can be assayed very easily using combustion techniques. In our laboratories small volumes of plasma are evaporated to dryness in vacuum desiccators. The resultant dry powder is placed in rice paper cachets and burnt. This gives us very accurate and reproducible results. It also eliminates the possible inaccuracies due to surface adsorption which may be encountered when measuring tritiated compounds.

Chapter 20

Factors influencing the Detection of Incorporated ^{3}H-Thymidine in Biological Material

W. A. Cope and J. A. Double

Wellcome Laboratories of Experimental Pathology, St. Mary's Hospital Medical School, London W.2, England.

INTRODUCTION

DNA synthesis is commonly measured by pulsing with labelled thymidine which is readily available labelled with either tritium or carbon-14. With the high counting efficiencies now found in modern liquid scintillation spectrometers, for convenience and economic reasons ^{3}H-thymidine is becoming more widely used than ^{14}C-thymidine.

It is often assumed that tritium can simply be substituted for carbon-14 in techniques originally developed for use with the latter. However we have found that this change in usage may give rise to problems. In particular, counting of incorporated tritium may be subject to considerable inaccuracy when methods of sample preparation are used that were developed in work with carbon-14.

METHODS

Mouse ADJ PC6/A plasma cell tumour cells[1] were used as a convenient source of biological material in these experiments. The tumour was grown in BALB/c female mice. Labelled cells were prepared by pulsing a suspension of cells at a concentration of 10^6/ml in a medium consisting of 70% medium 199 (British Drug Houses) and 30% horse serum (Burroughs Wellcome No. 2), for an hour with either ^{3}H-thymidine (5 MeT), 5 Ci/mM at 1.0 μCi/ml, or with ^{14}C-2-thymidine, 42 mCi/mM at 0.05 μCi/ml. (Labelled thymidine was obtained from the Radiochemical Centre, Amersham). Labelled DNA was isolated from solid tumours by the methods of Kirby,[2] tumours being taken one hour after the mice had been injected intraperitoneally with 100 μCi/kg ^{3}H-thymidine.

The radioactivity bound within the labelled cells or DNA was assayed by various methods:

1. **Assay on filter discs.** 0.5 ml aliquots of suspension containing 0.5×10^6 cells were loaded onto Whatman glass fibre filter discs, (GF/A, 2.5 cm) contained in a Millipore 30-sample manifold. Each disc was then washed with approximately 10 ml aliquots of saline, 5% trichloroacetic acid (TCA), and ethanol in turn. The discs were then removed and placed in glass scintillation vials along with 10 ml scintillator.
2. **Perchloric acid extraction.** The total label incorporated into the cells was determined

by centrifuging a 5.0 ml aliquot containing 5×10^6 cells, washing with saline and 5% TCA, and extracting the pellets twice with 5.0 ml of 0.5 N perchloric acid at 70°C for 30 min. 0.5 ml of the extracts was then assayed for radioactivity by mixing with 10 ml scintillator.

3. **Tetraethylammonium hydroxide (TEH) extractions.** TEH extractions of either TCA pellets or of cells on filter discs were carried out by using two 5.0 ml extractions at 50°C overnight. The radioactivity was then assayed in 0.5 ml aliquots of the extracts.

The radioactivity of the samples was measured in an Intertechnique ABAC SL 40 liquid scintillation spectrometer. The counter was set to accumulate 10 000 counts, i.e. with an inherent 1% counting error. The results were expressed as disintegrations/min (d.p.m.)/0.5×10^6 cells. Each point on the graphs is the mean of five samples; the standard deviations are too small to show on the scales used.

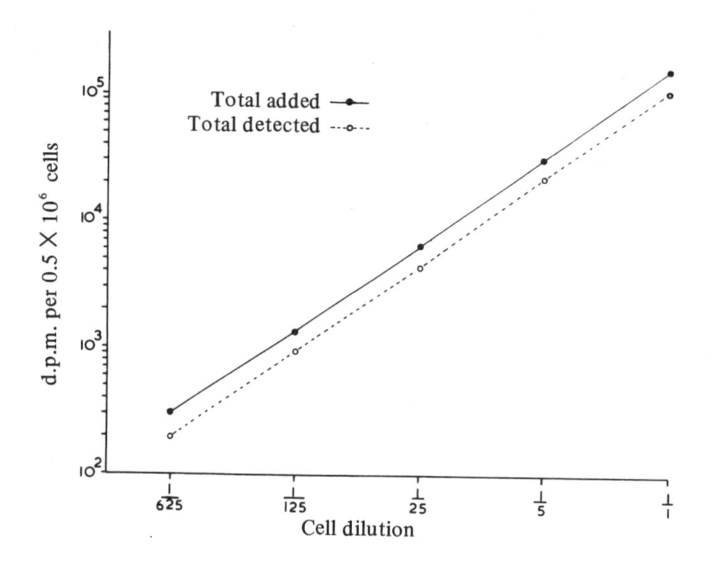

Fig. 1. Radioactivity on filter discs. Cells were labelled with ^{3}H-thymidine and serially diluted with unlabelled cells. The radioactivity incorporated was estimated by perchloric acid hydrolysis of a 5% TCA pellet ('Total added') and by the filter disc technique by direct measurement ('Total detected'). Each point in this and subsequent figures represents the mean of five samples. The standard errors are too small to plot.

RESULTS

Effect of sample preparation on amount of activity detected

In a series of experiments cells were labelled with ^{3}H-thymidine and serially diluted with unlabelled cells. Aliquots were assayed for tritium incorporated into DNA using different methods of sample preparation. Figure 1 shows the results of experiments in which the incorporated radioactivity was measured either on filter discs or after 0.5 N perchloric acid hydrolysis of a 5% TCA pellet. It is clear that there is a considerable discrepancy between the amounts of radioactivity detected by the two methods, activity after perchloric extraction being about 50% higher than that detected on the discs. This

Table 1. Labelled cells on filter discs. Aliquots of cells, labelled either with ^{3}H-thymidine or ^{14}C-thymidine were divided into three groups. The first was precipitated with 5% TCA and the second two loaded onto filter discs. The 5% TCA pellets were extracted with 0.5 N perchloric acid and were considered to be equivalent to the material added ('Amount added') to the discs. One group of discs was counted directly ('Amount detected'), the others were extracted with 0.5 N perchloric acid ('Amount extracted').

Isotope	Amount added	Amount detected	Amount extracted
Tritium	30716	20125	29172
Carbon-14	5072	4998	5017

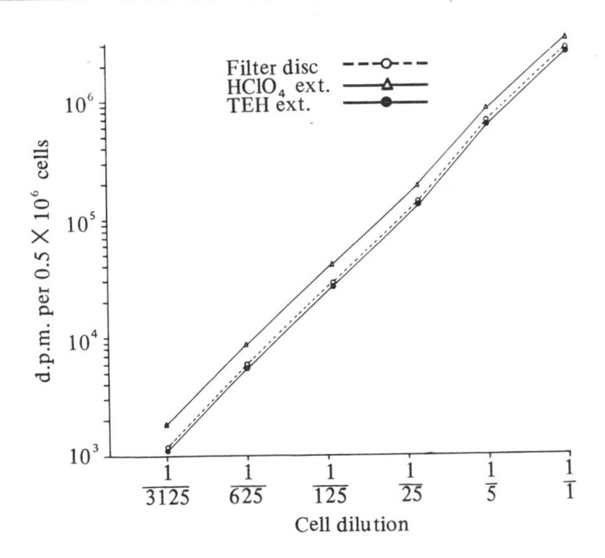

Fig. 2. Radioactivity on filter discs. Cells were labelled with ^{3}H-thymidine and serially diluted with unlabelled cells. The radioactivity incorporated was measured by the filter disc technique, either by direct measurement ('Filter disc'), or after perchloric acid extraction of the disc ('$HClO_4$ extraction') 2 X 5.0 ml 0.5 N at 70°C for 30 min or after TEH extraction of the material on the disc ('TEH extraction') 2 X 5.0 ml at 50°C overnight.

discrepancy could be the result of many factors, loss through the disc being the most obvious. However, Table 1 shows the results of an experiment in which aliquots of labelled cells were either loaded onto a disc or precipitated with TCA and the discs were then either counted directly or extracted with perchloric acid. It is clearly seen that the amount extracted agrees with the amount loaded onto the disc showing that little activity is lost through the filter.

Figure 2 shows the results of an experiment in which serial dilutions of labelled cells were processed on filter discs which were then divided into three groups. The first was counted directly, the second extracted with tetraethylammonium hydroxide (TEH) and the third with 0.5 N perchloric acid. The radioactivity measured in the TEH extract was almost the same as that measured on the discs. However with 0.5 N perchloric acid extraction there was a 50% increase in the amount of radioactivity measured.

Table 2. Labelled DNA on filter discs. Aliquots of ^{3}H-labelled DNA or ^{14}C-labelled DNA in 2% NaCl were loaded onto filter discs and, after precipitation and washing, the radioactivity on the discs was measured ('Amount detected'). Further aliquots were hydrolysed in 0.5 N perchloric acid and assayed for radioactivity to determine the total amount of activity added to the disc ('Amount added'). Other discs were extracted with perchloric acid and the radioactivity of the extracts measured ('Amount extracted').

Isotope	Amount added	Amount detected	Amount extracted
Tritium	147974	72589	138648
Carbon-14	5277	5090	5171

Table 3. Isotope ratios of labelled DNA on filter discs. Samples of double labelled DNA and single labelled DNA were dissolved in 2% NaCl. Aliquots of double labelled DNA and mixtures of single labelled DNA were assayed for radioactivity after perchloric acid hydrolysis ('Amount added'), or after precipitation and washing on filter discs ('Amount detected'). Samples of DNA precipitated on the discs were also extracted with perchloric acid and the radioactivity of the extracts measured ('Amount extracted'). The isotope ratios tritium : carbon-14 of the samples were then calculated.

Sample		Amount added	Amount detected	Amount extracted
^{3}H/^{14}C	^{3}H	34820	20492	33539
dual	^{14}C	3972	3945	3929
label	Ratio	8.76	5.19	8.54
^{3}H	^{3}H	70777	52784	63510
plus	^{14}C	2474	2564	2371
^{14}C	Ratio	28.6	21.0	26.8

In further experiments DNA that had been labelled with either ^{3}H-thymidine or ^{14}C-thymidine was assayed for radioactivity either directly on filter discs, in perchloric acid extracts of filter discs or after perchloric acid hydrolysis of the native material. Results are shown in Table 2. It is clear that with tritium labelling there is considerable loss of activity when DNA is measured on the filter discs but 94 to 98% of the total activity is recoverable by perchloric extraction of the discs, again showing that very little of the loss is due to passage of material through the disc. With the ^{14}C-labelled DNA there is no significant variation with sample preparation in the amounts of radioactivity detected.

These results suggest that in experiments using both tritium and carbon-14 the method of sample preparation could influence measurements of isotope ratio if tritium were used to label DNA. Table 3 shows the results of such an experiment using doubly labelled DNA and mixtures of ^{3}H- and ^{14}C-labelled DNA. Aliquots were assayed for radioactivity of filter discs and after perchloric extraction of the precipitated material on the discs. The results show that the ratio of tritium to carbon-14 appears to be much greater after perchloric acid hydrolysis.

The influence of sample preparation on the apparent activity of labelled DNA is further illustrated in Table 4. Radioactivity of equal aliquots of labelled DNA was measured after different methods of sample preparation. It is seen that with ^{3}H-labelled DNA

Table 4. Effect of sample preparation on activity detected. Samples of ^{3}H-labelled and ^{14}C-labelled DNA in 2% NaCl were assayed for radioactivity using different methods of sample preparation. Perchloric acid hydrolysis was carried out using 0.5 N acid at 70°C for 2 h. TEH solubilization was effected by using 25% TEH at 50°C overnight. Biosolv 111, Hyamine and Soluene were used according to the manufacturers' instructions.

	Tritium (d.p.m.)	Carbon-14 (d.p.m.)
Perchloric acid	34050	11050
TEH	20500	10311
Biosolv 111	20470	10165
Hyamine	31810	11248
Soluene	20710	11010

Table 5. Effect of sample preparation on DNA specific activity. Aliquots of labelled DNA at 0.1 mg/ml in 2% NaCl were made up and the specific activities in terms of d.p.m./μg DNA measured after different methods of a sample preparation. Perchloric acid hydrolysis was carried out using 0.5 N acid at 70°C for 2 h. TEH solubilization was effected using 25% solution at 50°C overnight. N.C.S., Soluene and Hyamine were used according to the manufacturers' instructions.

Agent	Specific activity	Percentage perchloric acid
0.5 N Perchloric acid	929	100
TEH	734	79
N.C.S.	723	78
Soluene	718	77
Hyamine	845	91

activity detected depends greatly on the method of extraction and that maximum activity is only found after perchloric acid hydrolysis. With the carbon-14 label, measurement of activity is unaffected by sample preparation. Table 5 shows the results of a similar experiment in which the radioactivity of ^{3}H-DNA is expressed in terms of specific activity (d.p.m./μg). Again this depends on the method of sample preparation and was maximal after perchloric acid hydrolysis.

From the previous experiments it is clear that the maximum count rate from ^{3}H-labelled DNA is only found after perchloric acid hydrolysis. A similar result is found after hydrolysis with DNase. It is well known that both procedures result in the production of small molecular weight fragments, bases, phosphates and sugar degradation products with perchloric acid and mononucleotides with DNase. Figure 3 shows results of an experiment in which equal amounts of DNA were either hydrolysed in perchloric acid or solubilized in TEH and then fractionated on a G50 Sephadex column using 2% sodium chloride as the eluent. Optical density measurement of the fractions shows a considerable difference in distribution of molecular weight of the fragments produced. Perchloric acid produces small molecular weight fragments, the peak corresponding to that of the thymidine marker, whereas the TEH solubilization produces a large range of molecular weight fragments,

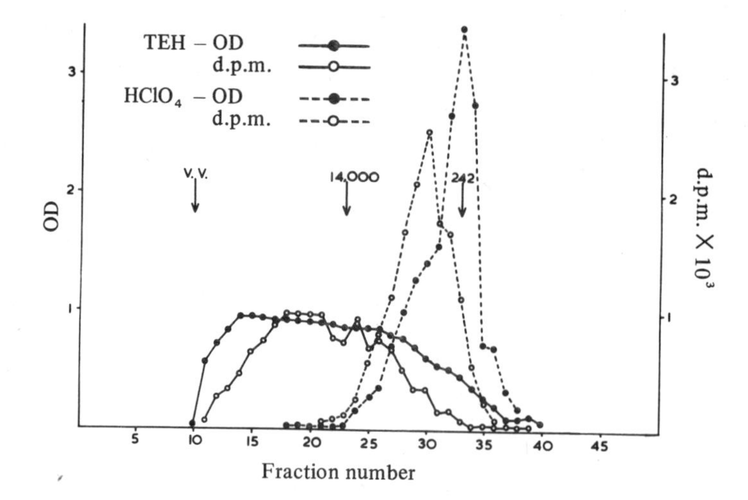

Fig. 3. Sephadex G50 fractionation of ^{3}H-DNA after solubilization in TEH or hydrolysis in 0.5 N perchloric acid. ^{3}H-DNA in 2% NaCl at 10 mg/ml was either solubilized in TEH or hydrolysed in 0.5 N perchloric acid. 1.0 ml of the resulting solutions was loaded on a 30 X 2.5 cm column of G50 Sephadex. 2% NaCl was used as the eluent, and 100-drop fractions were collected (approximately 5.0 ml). The optical density and radioactivity of these fractions were then measured and plotted as shown. Native DNA was used as a marker to find the void volume. Lysosyme (molecular weight 14 000) and thymidine (molecular weight 242) were also used as markers.

many of which are larger than the lysosyme marker (molecular weight 14 000). (This and the previous results suggest that molecular weight may influence resultant count rate of ^{3}H-labelled DNA). A possible explanation of this phenomenon is self absorption of the low energy β-particles within the larger DNA polymer. The total d.p.m. eluted from the column with the TEH solubilized DNA was approximately 55% of the total after perchloric acid hydrolysis which is in good agreement with the previous experiments.

Effect of quenching

It could be argued that the discrepancies shown when ^{3}H-labelled DNA is counted under different conditions were due to errors in quench correction in the systems used. The original quench curve had been made up using chloroform but the experiment summarized in Table 6 shows that all the other agents fall well within the limits of experimental error and could not possibly account for discrepancies of the magnitude found. In this experiment a series of ^{3}H-hexadecane standards in scintillation counting fluid were assayed and then reassayed after the addition of various quenching agents. Some deviation is observed, but some must be expected since the counter was set to accumulate 10 000 counts, i.e. with an inherent 1% error, and since samples are counted at about 25% efficiency multiplication of counts/min (c.p.m.) to d.p.m. would increase this error fourfold.

Figure 4 shows the chloroform quench curve used for the work. The curve was unaltered by the presence or absence of glass fibre filter discs, nor did the discs influence the d.p.m. output. The results of the external standard quench corrections were shown to be accurate by internal standardization. We do not suggest that external or internal stan-

Table 6. Accuracy of quench correction. ^{3}H-hexadecane was made up in scintillation counting fluid and aliquots were assayed for radioactivity. Various quenching agents were then added to the samples at a final concentration of 2%, and the samples were then reassayed for radioactivity. The error in d.p.m. measurement before and after quenching was then calculated.

Quencher (2%)	d.p.m. before quench	d.p.m. after quench	% Error
0.5 N Perchloric acid	9492	9767	+2.90
TEH	9281	9466	+1.78
5% TCA	9110	9167	+0.63
N.C.S.	9526	9492	−0.36
Water	9301	9503	+2.17
0.1 N NaOH	9401	9089	−3.41

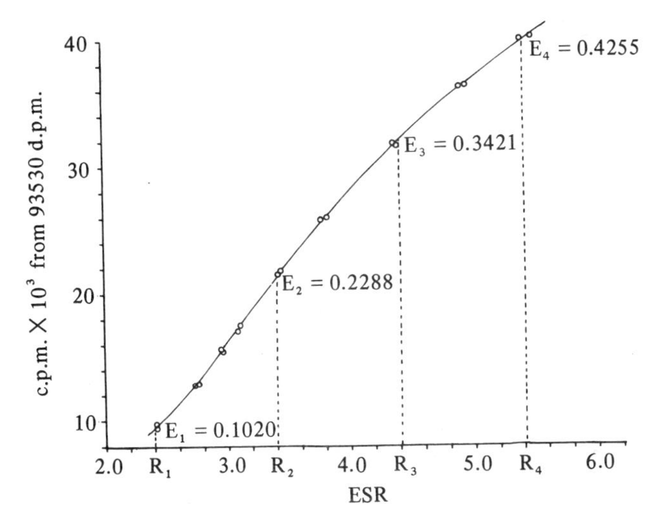

Fig. 4. Tritium quench curve – external standard ratio versus efficiency (chloroform). ^{3}H-hexadecane (Amersham) was used as a standard, diluted with scintillator to give 93530 d.p.m./vial. Chloroform was used as a quenching agent, in final concentrations ranging from 0.1 to 10% (v/v). Duplicate vials were counted and the mean c.p.m. plotted against the mean external standard ratio (ESR). The curve follows a third order polynomial function and the coefficients to the equation $y = ax^3 + bx^2 + cx + d$ (where y = efficiency and x = ESR) were calculated from the coordinates R_1E_1, R_2E_2, R_3E_3 and R_4E_4, These coefficients were then fed into the computer of the Intertechnique ABAC SL 40 where they were used in the conversion of c.p.m. to d.p.m.

dardization corrects for quenching occurring in the filter discs but only for contamination of the scintillator by water etc.

Figure 5 shows an absolute activity quench curve. This assumes that the effects, if any, produced by the presence of the filter disc are constant in all samples. The absolute activity of aliquots of ^{3}H-labelled cells was determined by perchloric acid hydrolysis of an equivalent 5% TCA pellet. The apparent activity in terms of c.p.m. detected in the

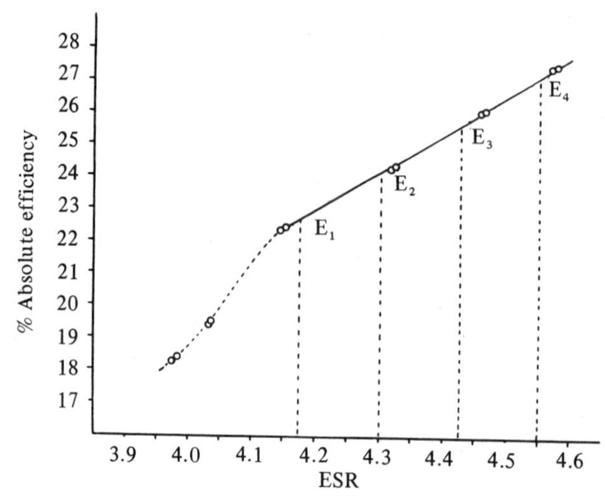

Fig. 5. Absolute activity quench curve – water quenching. The absolute activity present in aliquots of ^{3}H-labelled cells was determined by perchloric acid hydrolysis of a 5% TCA pellet. The apparent activity, in terms of c.p.m., detected in corresponding aliquots of cells on filter discs was also measured and, from the two measurements, an efficiency figure was calculated that would allow for the loss of energy due to self absorption and any loss through the disc. Water was used as a quenching agent since it was the agent most likely to be left in the filter disc technique. Water was added in final concentrations ranging from 0 to 5% (v/v) as shown by the six pairs of points on the graph, the 0% concentration being represented by the top, right hand pair of points and the 5% concentration by the bottom, left hand pair. Over the working range, i.e. 0 to 3% the curve is linear. The coefficients to the equation $y = ax^3 + bx^2 + cx + d$ (where y = absolute efficiency and x = ESR) were calculated from the coordinates, R_1E_1, R_2E_2, R_3E_3 and R_4E_4. As the curve is linear, a and $b = 0$.

corresponding aliquots of cells in filter discs was also measured and from the two measurements an efficiency figure was calculated that would allow for loss of energy due to self absorption and any loss through the disc. Water was used as a quenching agent since it was the agent most likely to be left in the filter disc technique.

CONCLUSIONS

These findings show that sample preparation can influence the activity detected when measuring ^{3}H-thymidine labelled DNA. This does not invalidate techniques using ^{3}H-thymidine providing they are internally controlled and absolute values are not required. In experiments using different methods of sample preparation caution should be applied in determining absolute values. The same caution must also be applied in systems using tritium and carbon-14 together, where DNA is labelled with tritium and carbon-14 used to determine another parameter.

ACKNOWLEDGEMENTS

This work was supported by grants from the Cancer Research Campaign, Leukaemia Research Fund, and the Nuffield Foundation. We wish to thank Miss J. W. Babbage for technical assistance, and Dr. M. C. Berenbaum for advice in preparation of the paper.

REFERENCES

1 M. Potter and C. L. Robertson, *J. Nat. Cancer Inst.* **25**, 847 (1960).
2 K. S. Kirby, *Biochim. Biophys. Acta* **55**, 545 (1962); *Biochem. J.* **96**, 226 (1965); K. S. Kirby and J. H. Parish, *B.E.C.C. Ann. Rept.* (1967).

DISCUSSION

P. Stanley: I would like to raise three points.
Firstly, I believe the problem you have encountered is due to the very low β-energy associated with tritium. Both β-absorption and quenching of this isotope diminishes its counting efficiency. In addition biological samples on filters tend to absorb water thus aggravating the situation. It is also difficult to obtain precipitates on filters in a reproducible manner and to ensure that those precipitates are thinly and evenly spread.
^{3}H-thymidine has been shown in many systems to label macromolecules other than DNA, for instance RNA and lipid. Is the DNA you purify free from these components?
Is your ^{3}H-thymidine free from contaminating breakdown products of the compound? Was it rechromatographed to check this before use?

J. A. Double: We have suggested in the text that the problem we encountered was due to the very low β-energy associated with tritium, that energy losses occurred within the DNA polymer, and that external counting conditions, i.e. within the scintillation system have little effect on the resulting observed efficiency. The effect of water quenching was shown in Fig. 5. It is clearly seen that only when unrealistic amounts of water are added is there a significant reduction in efficiency. From our results, where in all instances count rates from five identical samples are within 5% of one another, it is unlikely that the spreading of the precipitates has any great influence on the resultant efficiency.
The DNA used was highly purified and definitely free from RNA and lipids. However, it is possible that if these were present and labelled, the same phenomenon of self absorption of the low energy β-particles might still be observed.
The ^{3}H-thymidine used was free from contaminating breakdown products as shown by TLC. However, such products would not label DNA and any unbound label would be removed by the extraction and purification procedures.

M. Grant: Are you sure that your DNA preparations are free from other materials such as lipids? Since you are working with mouse tumour cells which unlike many of the bacterial systems probably contain the enzymes necessary to degrade thymidine, one might expect to find the label from thymidine in a variety of cellular components.

J. A. Double: The DNA samples in these experiments were very clean.

D. E. Bowyer: High temperature digestion of samples may lead to a loss of isotope as volatile materials such as ^3H_2O and $^{14}CO_2$.

J. A. Double: The digestions were carried out in sealed tubes, and it was highly unlikely that any such losses occurred.

P. Johnson: I would like to follow up Dr. Bowyer's point about possible loss of volatile materials with high temperature digestion procedures. A very simple answer to this is to modify a standard scintillation vial as shown in Fig. 6, in which a snug fitting polythene shive has a central well to accommodate a gas chromatograph septum. By drilling a central hole in the vial cap and in the polythene shive, samples and reagent can conveniently be withdrawn or added, but the system is sufficiently gas tight to allow high tem-

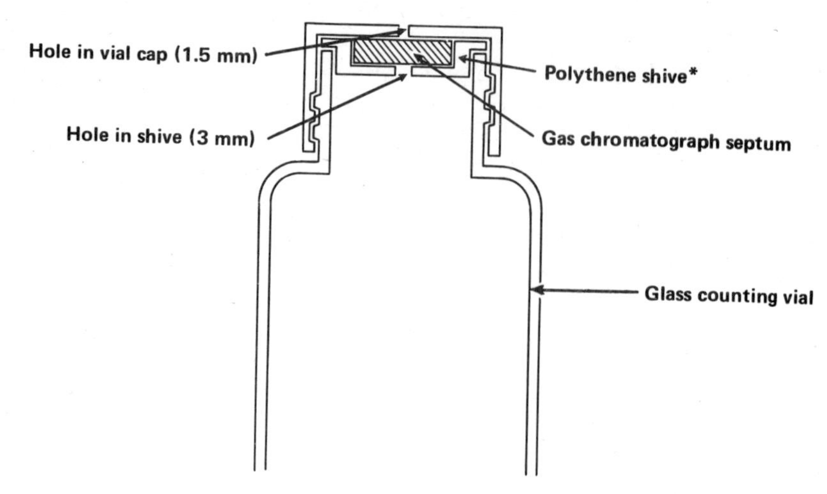

Fig. 6. Modified scintillation vial.

perature digestion without losses of volatile material such as $^{14}CO_2$.

J. A. Double: From the answer to the previous question, I do not think that the complexities you suggest are necessary.

P. Johnson: Comment: Except that sealed tubes do not allow convenient repetitive withdrawal of samples or addition of reagents; neither are tubes compatible with direct scintillation counting, so that further transfers are necessary.

B. E. Gordon: Special comment to all research workers: It must be pointed out to those doing biological and biochemical research that there is a major problem associated with liquid scintillation counting of materials not truly soluble in the counting solvent. The problem is that losses due to absorption or self absorption cannot be corrected for. Thus when counting a protein, a degraded protein, or any metabolite of unknown solubility, one may achieve highly reproducible but severely inaccurate results. This arises from the adsorption of a fixed amount of the labelled materials on any available surface.

The use of auxiliary reagents which raise the counting rate (e.g. solubilizers) is also suspect since there is no test to determine when all the material has been desorbed. Furthermore, the argument that after the count rate has stabilized one has useful data is invalid for the same reasons.

The situation is usually worse with ^{3}H-labelled materials than with carbon-14, but carbon-14 is subject to the same uncertainties. Therefore, it is essential that one establish, beyond doubt, that the counting system employed in work reported is free of such errors. This can only be done by calibrating the method against a series of combustion analyses and counting the $^{14}CO_2$ and/or $^{3}H_2O$.

The comment that one must calibrate the combustion method is easily answered by combusting standards. There is no merit whatsoever in the argument that samples high in salts, metals, etc. cannot be combusted to yield quantitative recovery of carbon and hydrogen. There are a number of variations in the combustion method in the literature which can take care of refractory materials such as inorganics.

* Amplex Appliances Ltd., Tylney Road, Bromley, Kent.

Finally, it should be stressed that, because of the great tendency of such high polymers to adsorb, absorb, or coagulate small changes in the composition of the samples (e.g. molecular weight of a protein) or in the counting solution can change a quantitative system to a non-quantitative one. Thus one must be exceedingly careful to preserve the conditions of a proven (i.e. by combustion) counting technique.
I believe that referees of papers submitted to journals should view critically the point of accuracy of counting data and should in the future reject those which do not demonstrate, unequivocally, this accuracy.

P. Johnson: Comment: The question of precision versus accuracy is an old one and the two are sometimes confused. With regard to our own work, I am inclined to agree with the comments of Dr. Dobbs earlier in this meeting (pp. 215, 231) about burning a sample if other methods are difficult. We also routinely attempt some cross check on our other methods by using a combusion technique on random samples but no method is foolproof.

B. Scales: I agree wholeheartedly with Dr. Gordon's comments, but how do you check the efficiency of your combustions? Some samples are undoubtedly difficult to combust; whilst it is not advisable to simply place an aliquot of labelled materials on the surface of the sample prior to combustion, it is impossible to go to the other extreme and prepare, as a standard, a sample of animal tissue which contains the various radioactive materials dispersed in the same manner as would occur after dosing the animal with labelled compound. So very often a compromise has to be reached. Whether you use a commercially available automatic combustion system, a semi-automatic oxygen flask technique, or a combustion furnace, the efficiency of combustion of some types of samples leaves much to be desired.

B. E. Gordon: I am still at a loss to understand the difficulty of quantitatively combusting an organic material (particularly a biological sample) to carbon dioxide and water. A quartz combustion tube with appropriate packing has worked for decades on the most refractory samples (polynuclear aromatics, carbon on silica or alumina particles, and metallic carbides). The question of dispersion of the labelled materials in a tissue sample is irrelevant – it all goes to carbon dioxide and water.

SECTION V
DATA PROCESSING

Chapter 21

Acquisition and Handling of Liquid Scintillation Counting Data

J. L. Spratt

Department of Pharmacology, College of Medicine, University of Iowa, Iowa, U.S.A.

INTRODUCTION

Since all previous speakers have necessarily been discussing data handling and since all of the speakers following are specifically committed to the subject, I find myself in an unusual situation. Likewise, in contrast to the lecturers in other sessions, I am not a developer of liquid scintillation spectrometers nor do I devote a substantial part of my effort to investigating the basic chemistry and physics of the liquid scintillation process. I am a user of liquid scintillation counters and my view is from that perspective.

Not being a statistician, I chose not to dwell extensively on that subject. A detailed account of computer programming *per se* seemed inappropriate. I therefore chose to present a compilation of user considerations that could be applicable to old or new spectrometers, to old and new users, and could serve the reading audience as well as those present here. As the last plenary lecturer, it also appeared to be incumbent upon me to include a variety of loose ends which have been of concern to many as evidenced by both formal and informal discussion at this symposium.

Hicks has recently stated my general concern regarding the handling of liquid scintillation counting data when he discussed *The Computer: Is it the Solution or the Problem?* He said:

> 'Computer specifications generally emphasize the bits of precision in the floating point software, the speed of floating point hardware, the cycle time of the central processing unit, the speed and size of mass storage, etc. I submit that the discussions of such specifications are meaningless until it is known what type of problem is to be solved. The selection of an approach using small computers, a larger computer, or a mixture of the two can only be made once one has defined the problem and has some understanding of the proposed solution.'[1]

Since it is my contention that excellent data handling is often performed on inappropriate data, I have chosen to incorporate data acquisition problems and pitfalls in this discussion. The presentation will be of an historical nature since the problems of data handling are necessarily commensurate with available hardware and software. Since the problems that face the specific user are uniquely his own, I will not provide you and readers of the symposium text with specific answers to your individual problems, but hopefully will provide some considerations and approaches which you may want to apply to your own environment. Although such a survey will not allow for extensive discussion of many points, the bibliography will direct readers to many particulars.

After brief comments on a few review articles, the subject of performance parameters and instrumentation available will take our attention in a somewhat chronological fashion. We will then discuss types of quenching effects and approaches to quench correction. This will be followed by a discussion of statistical considerations and computer programs for data handling.

We will conclude with a discussion of factors influencing the data obtained in liquid scintillation spectrometry and will pose the considerations to be entertained when deciding whether to proceed with computational gear supplied with the spectrometer or to proceed with a variety of non-spectrometer computational approaches.

It is of interest to note that the 1969 Liquid Scintillation Conference in Boston[2] had sessions on data handling and the 1970 Conference in San Francisco[3] had sessions on computer programming. On the other hand, the 1957 Northwestern Conference in Chicago[4] had no such sessions since liquid scintillation counting was very much a manual procedure at that time. The various developments which have occurred since those early days have had great influences on the handling of counting data. The aforementioned symposium volumes, the book by Birks,[5] Rapkin's earlier reviews,[6,7] and his series in the Picker Nuclear Laboratory Scintillator in 1966–67, are all very useful references describing these developments.

Table 1. Developments influencing data handling.[a]

Single sample	→ Multiple samples
Visual readout	→ Printed readout
Single channel	→ Multiple channels (channels ratio)
Gross count rates	→ Background subtraction and low count reject
Lister	→ Calculator
Printed readout	→ Punched readout
Internal standardization	→ External standardization
Calculator	→ Computer

[a] Plus innumerable electronic and mechanical improvements.

INSTRUMENTATION AND PERFORMANCE PARAMETERS

Both liquid scintillation spectrometry and effective computers were born in the 1950's. By 1962 we had advanced to tritium efficiencies of 7% with a background of 52 c.p.m.[8] Some of the developments which brought us to our current abilities are shown in Table 1. Multiple sampling and printed readouts meant that we could continuously monitor samples without personally attending the counter at all times. This meant we could now spend that time generating more samples and more data to overwhelm one another in the literature. Multiple channel machines allowed for less dial-twisting for some and more dial-twisting by others. The former took advantage of the development, the latter produced developments i.e. discriminator ratio techniques and the later channels ratio approaches to quench correction. Background subtraction and low count reject features had varying effects on data handling procedures. When solenoid decks were put on printing calculators, data reduction was much more convenient and tempted investigators to do even more dial-twisting with resultant advances in quench correction techniques. With readouts being punched on cards or paper tape came the pivotal question of handling data off-line versus having the data handled on-line by the accompanying calculator or its successor, the built-in dedicated computer. The use of both external standardization and channels ratio techniques now vastly overshadow the old original internal standardization technique for quench correction.

Along with these developments came a greater degree of dependence on an extremely complex piece of apparatus that supplied us with very precise data – whether it was right or wrong. In this regard, it is appropriate to quote from Wyld's presentation[9] at the Boston symposium:

> 'There was a time in the history of liquid scintillation counting when we knew very well we could have no confidence in the results from our equipment unless we counted everything ten times or more, and unless an investigator was something of a physicist and electronics expert, he was probably not even working in this field. Today, however, the situation is changed. We pay our money and in due course receive an instrument that is really a wonder to behold. Just looking at the awe-inspiring sight generates a feeling of confidence. One or two hundred samples can be loaded at once, and the results are neatly tabulated by a typewriter and punched on tape for handling by a computer which may even be included in the package. We are tempted to feel that the automatically printed numbers must indeed represent the absolute truth, but actually our confidence must be carefully established and continually re-established. Neither men nor machines are completely or consistently reliable.'

When considering the counter itself, one initially relied on efficiency (E) and background (B) values as indices of performance and later the so-called figure of merit or E^2/B. If one is working at high count rates, the standard deviation of counting is minimized by using the highest efficiency attainable. If one is working with count rates much closer to background, then E^2/B is a better figure of merit on which to rely.[9] But if one is interested in dual-label counting, Klein and Eisler have stated that perhaps other figures of merit should be considered.[10] Using varying isotope ratios and channels ratio techniques, they devised separation efficiency numbers (S) which were then used to compute an overall performance number (P) which was the product of the S value and the efficiencies for both isotopes at their point of optimum separation. When published five years ago, they had tested ten instruments representing four manufacturers and the best performance value was one-third of that theoretically attainable. (This P value figure of merit should not be confused with the P value figure of merit discussed by Gibson in Chapter 2 and elsewhere.[11] Likewise, the merit value discussed by Fox in Chapter 15 is something completely different from any of the foregoing).

The ultimate in providing us with the best instruments for making our mistakes are the fully-automated systems with built-in computers. As one conjures potential alliance with one of these sleek sirens, consider the comment by Forrey:[12]

> 'The introduction of tape punch and teletype outputs, automatic external standardization units and internal computers in the counting instruments themselves has raised the question of how best to carry out these calculations most economically, flexibly and conveniently. The equipment manufacturers have not been explicit in the various alternatives available to the user, the cost attendant in each and the details of implementation.'

QUENCHING EFFECTS

Since our biggest common data handling problem is quench correction, let us briefly consider its types and then review the data handling techniques for its determination. If you prefer to avoid the problem altogether, avoid materials that quench. In this regard, we should not forget that the early literature provides very useful information on common quenching materials. The 1957 paper by Kerr, Hayes, Newton and Ott[13] is one which lists many of these agents and Dr. Birks has discussed their influence when using counters with the currently available better photomultiplier tubes (see Chapter 1).

In classifying quenching phenomena, the combined approach of Neary and Budd[14] and of Peng[15] is not unreasonable until we have a better classification. We therefore distinguish between chemical, colour and 'photon' quenching. Chemical quenching means the non-fluorescing molecules in the system absorb the β-energy and convert it eventually to heat instead of fluorescence. Such events occur before the emission of photons and the effect is exponential with the concentration of quencher. Colour quench is the attenuation of emitted photons. Since the path length to phototube alters pulse height distribution, this attenuation phenomenon is what distinguishes colour from chemical quench and allows this distinction to be demonstrated experimentally. Peng further distinguishes what is termed 'photon' quenching. Such quenching results from adverse geometry and insolubility phenomena occurring especially in heterogeneous systems. In such systems, the physical mass of dispersed material can prevent β-radiation interaction with the scintillation fluor.

For chemical quenching, a number of subclasses have been proposed. Neary and Budd distinguish between acid, excess fluor concentration, dilution effects, dipole–dipole interaction, and capture of secondary electrons. Those interested are referred to their discussion at the Boston symposium.[14]

Quench correction

Krichevsky *et al.*[16] have divided quench correction techniques into direct and indirect types. Direct methods include internal standardization and the extrapolation method of Peng.[15] Channels ratio and external standardization techniques are the indirect types. Although it never became popular for routine use, one could also include the isolated internal standard technique of Ross[17] as a direct method.

Before quench correction became so instrumentally simple and concurrently so mathematically complex, one used the simple formulae of Table 2(a) to calculate d.p.m. After counting the spiked sample, the formulae of Table 2(b) were used to obtain the quench corrected d.p.m. At this point it should be noted that in those early days we all knew what efficiency meant – it was the machine efficiency when a sealed, non-quenched sample of known activity was counted. Anything less was quenching. Today one must be semantically aware since many current quench correction techniques combine efficiency and quenching effects into the single term 'efficiency'. Some might prefer to call this an apparent or relative efficiency.

The discriminator ratio technique[18,19] was an early method for dual-label counting and quench correction, but still utilized internal standardization. As better multiple channel instruments became available, the pulse height shift with quenching could be determined (see Fig. 1). This led to quench determination by pulse height shift or channels ratio.[20,21] Baillie spelled out its pros and cons over ten years ago,[20] indicating its limitations as:

1. poor overall accuracy with tritium;
2. not usually being useful with strongly quenched coloured samples, and
3. having slightly inferior accuracy for very low level, quenched samples.

He listed advantages as:

Table 2. Primary computations used in programs.

(a) Basic computations

$$\text{Net c.p.m.} = \frac{\text{Total counts of sample}}{\text{Total time of sample}} - \frac{\text{Total counts of background}}{\text{Total time of background}}$$

$$\text{Efficiency} = \frac{\text{Net c.p.m. of standard}}{\text{Known d.p.m. of standard}}$$

$$\text{Net d.p.m.} = \frac{\text{Net c.p.m.}}{\text{Efficiency}}$$

(b) Internal standard quench correction

$$\text{Fraction 'seen'} = \frac{\text{Net d.p.m. with spike} - \text{net d.p.m. without spike}}{\text{d.p.m. of spike}}$$

$$\text{Corrected d.p.m.} = \frac{\text{Net d.p.m. without spike}}{\text{fraction 'seen'}}$$

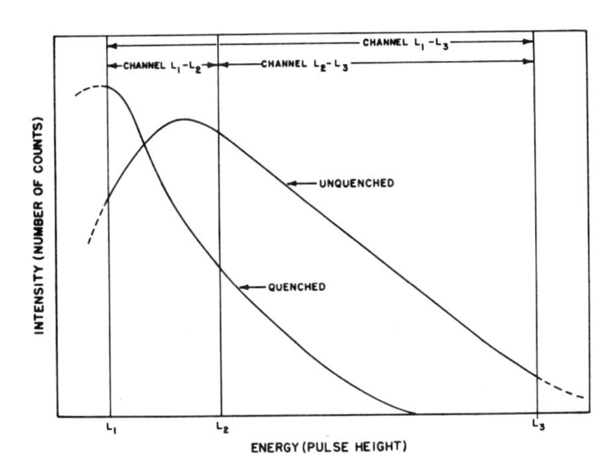

Fig. 1. Spectrometer channels in relation to β-spectra (from Ref. 21).

1. considerable time savings in routine counting;
2. improved accuracy for hot samples;
3. elimination of error due to measurement of standards, and
4. ability to recount samples at a later time since no internal standard had been added.

Five years ago, Rapkin generally concurred with this analysis,[22] but countered Baillie's time-saving statement for some by stating that the method:

> '...... does have two inter-related limitations which have generally discouraged its use for low-activity samples. When collecting adequate statistics for the channel of principal interest, the count for the weaker channel is apt to be inconclusive. And this situation worsens as quenching increases. In order to collect adequate data for the weaker channel, it is often necessary to count each sample for an excessively long time. This can make the channels ratio technique more time-consuming than the internal standard method.'

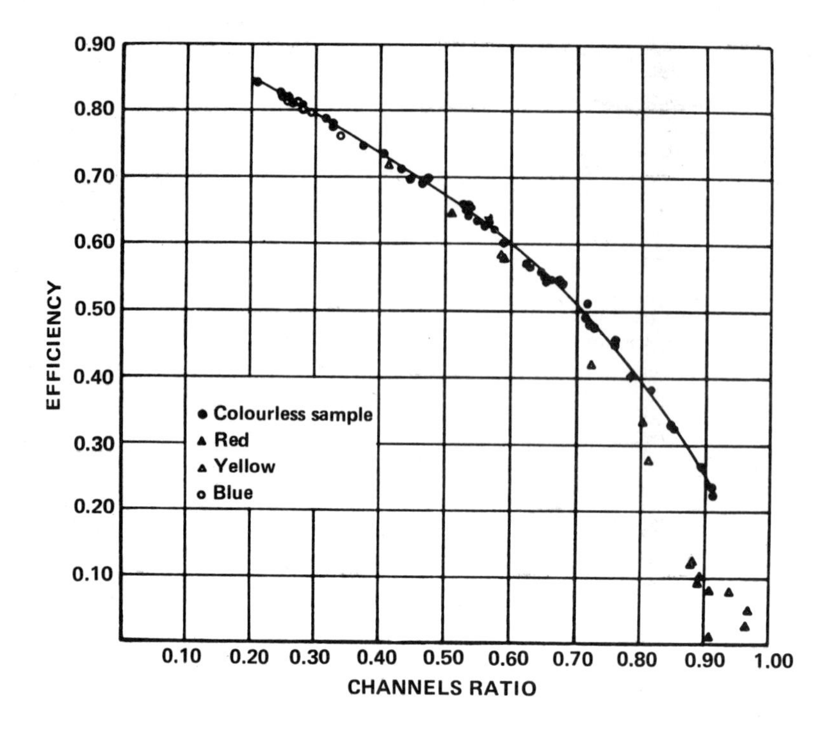

Fig. 2. Quenching curve for carbon-14 using wide channel (L_1-L_3) efficiency versus ratio of counts in channels $(L_1-L_2)/(L_1-L_3)$ (from Ref. 21).

The concern for this method with greater quench and/or low count rates is exemplified in Bush's report[21] (see Fig. 2). With multiple channel machines where one could independently set upper and lower levels for each channel came a variety of improvements on the technique and it is presently still very useful.

With the advent of automatic external standardization[23,24] we completed the major spectrometer features grossly evident to the user which have influenced data handling. Additional welcome electronic and mechanical improvements such as markedly improved photomultiplier tubes and circuitry have certainly occurred. However, many recent developments have been more concerned with a plethora of gadgetry to manipulate the count rate obtained rather than the production of the basic count rate. As such, these are developments in the data processing components of the instrumentation rather than in the spectrometer component.

Problems of quench correction for dual-label samples have received attention with many methods promulgated over the years. The early methods included the extrapolation method of Peng[25] and the channels ratio methods of Hendler.[26] These various techniques have been applied to colour[27] and other quench correction problems. The sample channels ratio method of quench correction has afforded Neary and Budd[14] the opportunity to clearly demonstrate that chemical quenching is different from colour quenching (see Figs. 3, 4 and 5).

The phenomenon of different degrees of 'spillover' between channels with varying quench (Fig. 6) is the basis for channels ratio quench correction. When two isotopes are

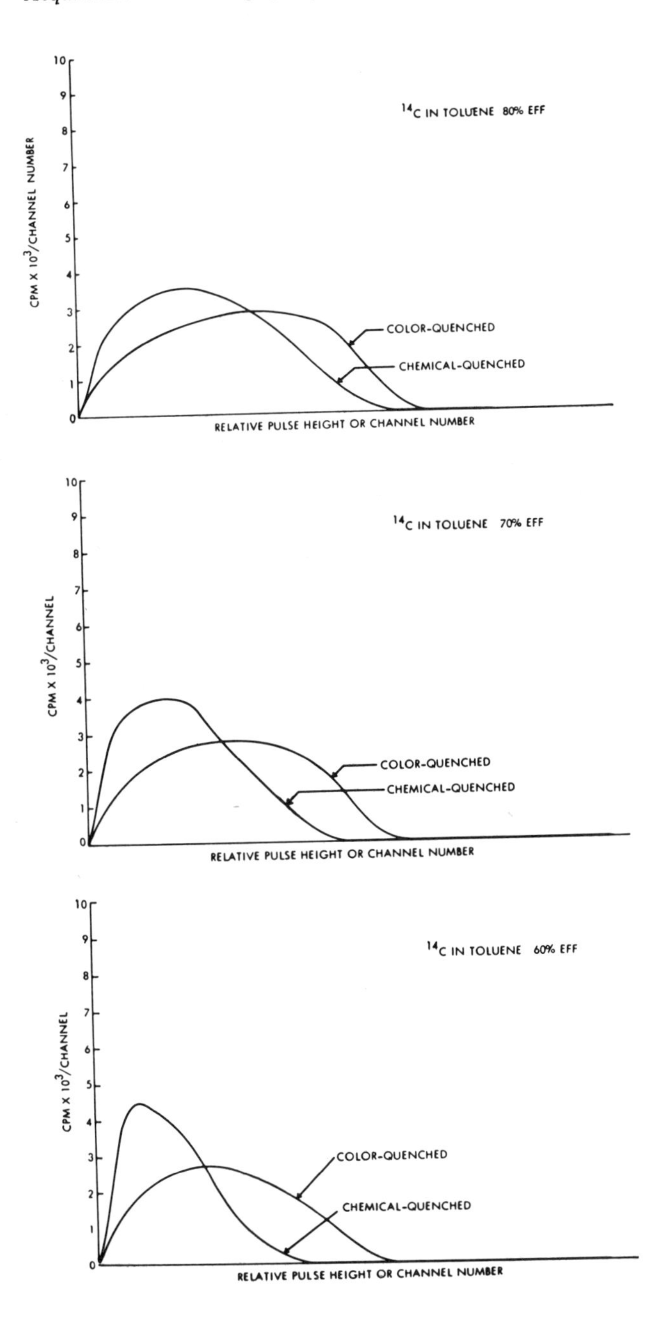

Fig. 3. Colour versus chemical quenching of carbon-14 (from Ref. 14).

used, the formulae for the d.p.m. of both the low energy isotope (D_l) and the high energy isotope (D_h) necessitate the use of five parameters (Table 3). When spillover occurs both ways, one can use the simultaneous equations and Engberg plots discussed by Kobayashi and Maudsley at the Boston symposium.[28]

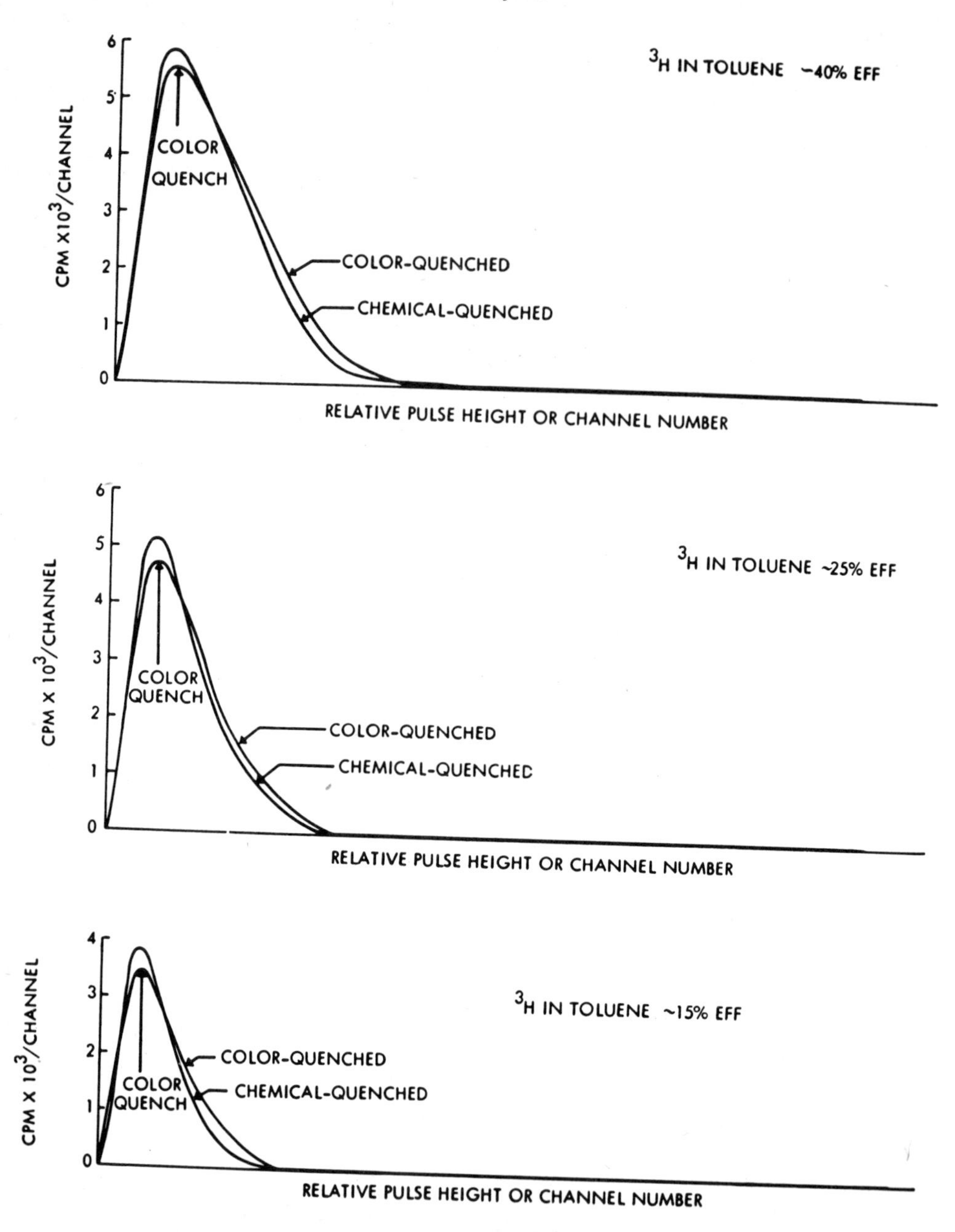

Fig. 4. Colour versus chemical quenching of tritium (from Ref. 14).

Before proceeding further, perhaps another semantic clarification is in order. When channels ratios first came into being, one was referring to the comparison of a discrete lower pulse height channel to that of a higher pulse height channel. This was the original 'sample channels ratio' or 'isotope channels ratio' concept. However, with the introduction of overlapping channels and external standardization, the literature has become more

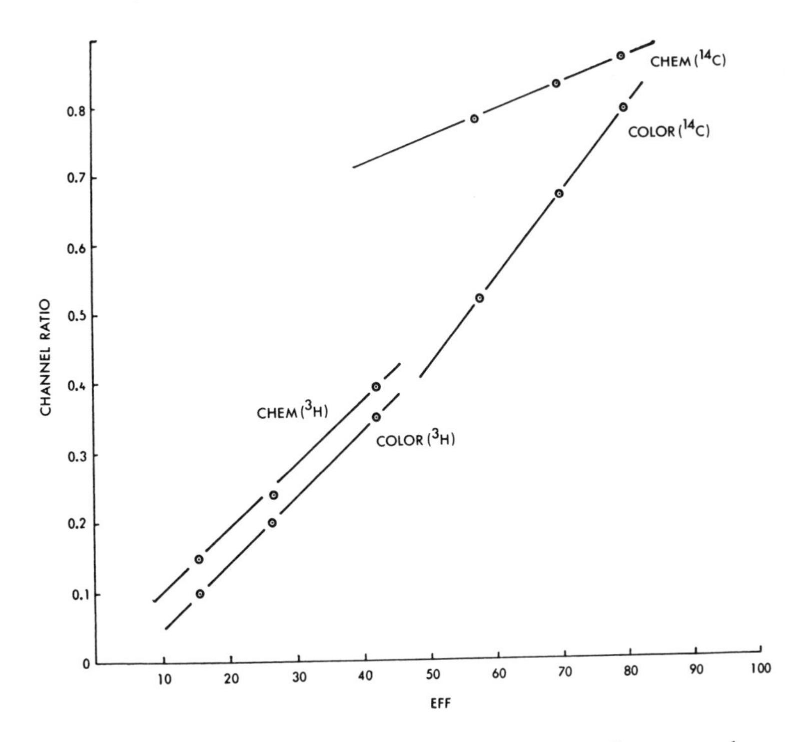

Fig. 5. Effect of differences between chemical and colour quench on quench correction of ^{14}C- and ^{3}H-toluene samples by channels ratio method (from Ref. 14).

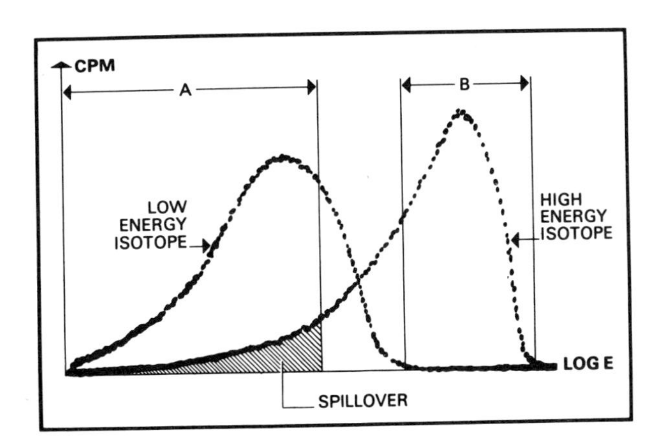

Fig. 6. Spillover of higher energy spectrum in the low energy channel (from Intertechnique brochure No. 12-Multimat-1-10/70).

confusing. Various references to channels ratios may mean sample channels ratios or external standard channels ratios – with or without overlapping pulse heights. This is an unfortunate circumstance and we should all better describe our methods for the sake of clarity.

Not too many detailed comparisons have been made between the more recent methods

Table 3. Dual isotope quench correction with 'spillover' in one direction.[a]

C_a	=	count rate low energy channel A including spillover
C_b	=	count rate high energy channel B due to high energy isotope only
E_{hb}	=	efficiency of high energy isotope in high energy channel B
E_{ha}	=	efficiency of high energy spillover in low energy channel A
E_{la}	=	efficiency of low energy isotope in low energy channel A

$$D_l = \frac{C_a - C_b(E_{ha}/E_{hb})}{E_{la}} \equiv \text{d.p.m. of low energy isotope}$$

$$D_h = \frac{C_b}{E_{hb}} = \text{d.p.m. of high energy isotope}$$

[a] After Intertechnique Brochure No. 12-MultiMat-1-10/70

of quench correction. However, when such comparisons have been made, the results have been illuminating. Noujaim *et al.* have compared automatic external standardization to channels ratio techniques.[29] They concluded that with instrumentation available a year ago the sample (or isotope) channels ratio method is quite satisfactory for reasonable counting rates with colour quenched samples. However, they found automatic external standardization (AES) correlation or external standard ratios much less satisfactory. This was true for both tritium (Fig. 7) and carbon-14 (Fig. 8).

The empirical approach on the differentiation between colour and chemical quench by Lang[30] may be of particular interest to many. He has devised a system whereby one may discriminate between colour and chemical quench on a *post hoc* basis. He has found that chemical quench and colour quench are different when apparent efficiency are plotted against the square root of the AES c.p.m. or against the AES channels ratio (Fig. 9). In his presentation at the San Francisco symposium he goes on to demonstrate how a sample may be adjudged to be chemically quenched or colour quenched by comparing its counting performance for apparent efficiency using these two correlations.[30]

It is also interesting to note that while we pay homage to the recent methods, Rogers and Moran compared internal standardization, external standardization, and sample channels ratio methods of quench correction[31] and stated that:

> 'The careful addition of internal standard is shown to give the highest accuracy in determination of the efficiency of liquid scintillation counting. The technique of channels ratio is equally accurate, when used with samples that are moderately quenched and have a high counting rate, and less accurate for highly quenched samples. Automatic external standardization is considerably less accurate than either of the other techniques.'

The necessary conclusion to be drawn from these limited comparative studies is that there is no recommended universal quench correction technique. They are all useful as long as one operates within the boundary conditions of the respective methods.

STATISTICS

Counting statistics is an obviously important consideration in data handling. Graphs, nomographs and mathematical treatments appear throughout the literature and in commercial material. Still useful early sources include those of Jarrett,[32] Loevinger and Berman,[33] and Browning.[34] The questions associated with single isotope counting versus dual isotope counting have respectively been considered by Herberg[35] and by Bush.[36] The

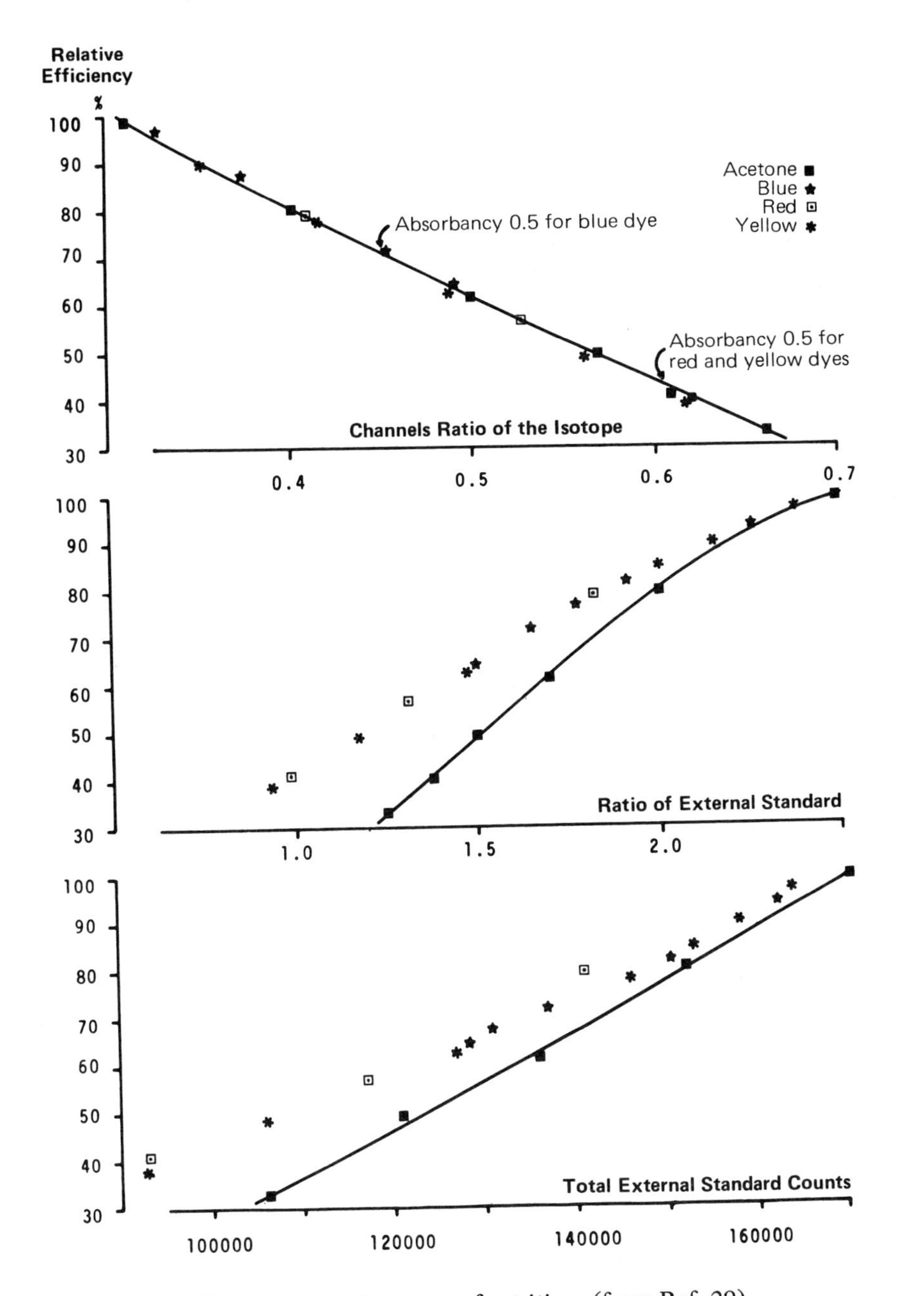

Fig. 7. Counting efficiency correction curves for tritium (from Ref. 29).

question of whether one should accept the Poisson model for estimation of counting error has been discussed by many, including Matthijssen and Goldzieher,[37] Cavanaugh[38] and Carroll and Houser.[39] In testing the Poisson model, where the square root of the count is the error of that count, Matthijssen and Goldzieher concluded that liquid scintillation spectrometers were not stable enough to support the assumption of conformance with

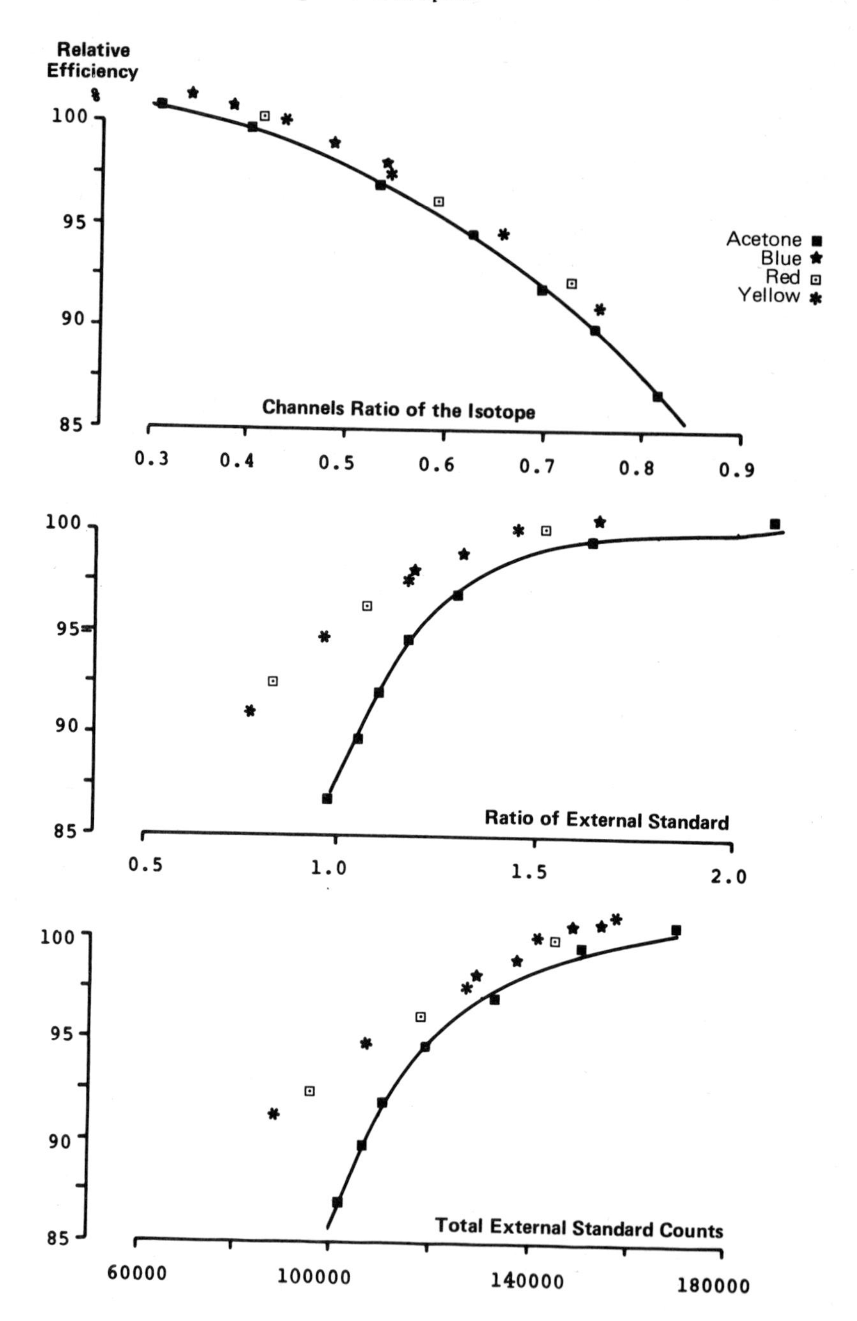

Fig. 8. Counting efficiency correction curves for carbon-14 (from Ref. 29).

the Poisson binomial distribution model.[37] This led them to state that chi-square is inappropriate for measuring instrument stability and they recommended the relative standard error instead. In discussing the statistics of external standardization methods, Cavanaugh recommends use of external standard ratios rather than the external standard counts *per se*

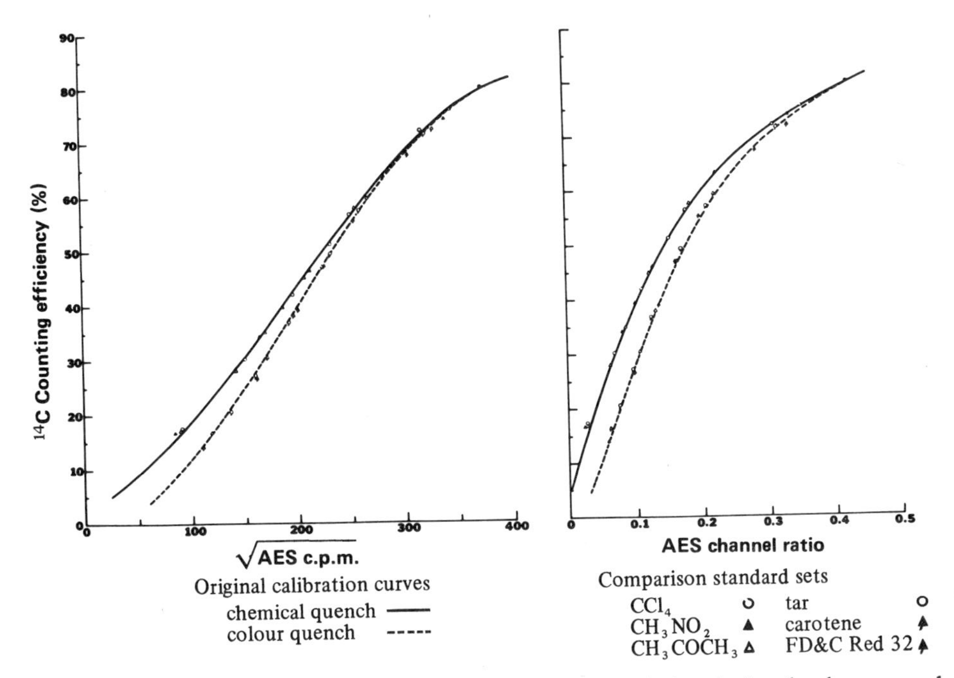

Fig. 9. Comparison of various quenching agents on carbon-14 chemical and colour quench calibration curves (from Ref. 30).

since the ratio technique quench correction curves are less dependent on volume.[38] Cavanaugh also states that averaging of sample counts from different counting cycles helps reduce non-Poisson errors.

COMPUTER PROGRAMS

In turning to the subject of computer handling of data, it was obvious quite early that liquid scintillation counting was becoming so automated that the necessary mathematical manipulation of the counts obtained was becoming the bottleneck. At the present time there is no way of knowing how many computer programs are available for such computations. The programs by Blanchard,[40] Spratt,[41] and Axelrod *et al.*[42] were among the early ones reported in the literature. Many more have subsequently appeared including an external standardization update of our program.[43] Spectrometer manufacturers have also collected an additional number of programs not necessarily described in scientific journals.

Quench techniques used in the many programs now described include internal standardization,[40,41] discriminator ratios,[44] sample channels ratios,[12,16,39,42,45,46] external standardization,[12,30,42,43,47,48] and external standard ratios.[12,30,39,45,46,48–50] Quench correction for dual isotopes in the same sample are available in many of these programs.[12,16,39,44,46,48–50] Most are used on relatively large computers, but some are designed for smaller programmable calculators or computers.[45,46,48,50] Throughout these various programs one often finds that the correlation of the quench parameter is not linear with the apparent efficiency. For this reason, a polynomial expansion to an appropriate power is often used.

The program described by Glass[45] is of particular note since all samples are quench corrected by both the sample channels ratio technique and the external standard ratio technique. This double ratio method allows for the *post hoc* selection of the most appropriate of the two quench correction techniques for each sample. In addition, it also permits a number of discrepancy checks on both the samples and the counter.

The literature describing these programs contains various comments on the question of on-line versus off-line computation.[46,48,51,52] However, whenever such comparisons are made, it must be recognized that the comparison is only valid when examined in respect to the individual user's environment. For example, Moore's comparison[51] of batch processing versus time-sharing obviously considers a number of local features. It also mixes programming concerns with production run concerns.

The instrument manufacturers are still continuing with their own gadgetry modifications which are sold as instrumentation specifically wedded to their own spectrometers. Among these modifications are the Packard Absolute Activity Analyzer and the Multi-Mat system recently announced by Intertechnique. Although the commercial literature on both of these systems refers to their ability to determine 'absolute activity', the activity of any counter should be just as 'absolute' when properly determined.

The Packard Absolute Activity Analyzer operates by defocusing the photomultiplier tubes for each unknown sample so that the external standard ratio for each sample will be identical to that of one of the previously determined ratios for a set of sealed standards.[7] The refocusing step may be repeated up to fifteen times if necessary. When the sample counting conditions match one of the standard points, the sample is counted and the d.p.m. computed. Two sources of comments on the Absolute Activity Analyzer are the presentation by Cavanaugh of Packard[38] and the independent evaluation by Herberg.[53] It should be noted that the Beckman 'Automatic Quench Correction' takes a related but somewhat different approach by the readjustment of system gain via varying amplification.[7]

From the commercial literature available, it appears that the Intertechnique Multi-Mat system involves the concurrent purchase of a small on-line computer which performs the usual quench corrections from curves which can be stored in its computer. This is the same approach that has been used both on-line and off-line by many investigators. The Multi-Mat system is somewhat different in that the on-line computer processing unit can simultaneously accept counting data from up to four spectrometers within 300 ft of the processor. This system also incorporates the feature of computing the net count error as opposed to the gross count error (Table 4). A useful discussion of the 2σ error has been provided by Wyld.[9]

Throughout all of the foregoing discussion it has been apparent that we now have all kinds of ways of handling our liquid scintillation counting data. However, we should never blindly rely on a count present 'in living colour' on nixie tubes or crisply printed in beautiful alignment by a computer. What went into the laboratory manipulations and the preparation of the sample must always be considered. Some potential pitfalls might arbitrarily be considered as sample factors, counter factors, and calculator/computer factors.

Sample factors

Sample factors (Table 5) include the vial cap and the vial itself in regard to physical composition, isotope composition, and regularity of its walls. The peroxides in dioxane and impurities in other solvents are of concern. The miscibility of the solvents at the counting temperature should, of course, be noted. When suspension systems are used, one must

Table 4. Gross versus net count rate error.[a]

(a) Measurements

S	=	gross sample count rate
B	=	background rate
t_s	=	sample counting time
t_b	=	background counting time

(b) 2σ gross count rate error

$$2\sigma_g \text{ in } \% = \frac{200}{\sqrt{S \times t_s}}$$

(c) 2σ net count rate error

$$2\sigma_n \text{ in } \% = \frac{200\sqrt{\frac{S}{t_s} + \frac{B}{t_b}}}{S - B}$$

[a] After Intertechnique brochure No. 12-MultiMat-1-10/70.

Table 5. Sample factors

Counting vial
Solvents
Miscible system
Volume
Phosphor and wavelength shifter
Chemical reactions
Volatilization
Precipitation
Fluorescence (dark adaptation)
Aberrant environmental background
Temperature (equilibration and condensation)

also consider uniformity of dispersion and the problems of potential self absorption or 'photon' quenching. Newer instrumentation includes light pipes, compound external standards and other features which minimize volume problems. However, these features are not on many counters still in use.

As mentioned earlier, excess fluor components can surprisingly cause quenching[14] whereas too little fluor or improper balance between primary fluors and wavelength shifters can result in less than optimal efficiency.

Chemical reactions between components in the sample are possible, as are volatilization or precipitation of constituents. These types of phenomena, as well as fluorescence from non-radioactive components[54] can sometimes be detected by multiple cycle counting of the same sample.

One should also not have tunnel vision in regard to environmental factors. Background is not a feature of sample and counter performance alone. Although stabilized by shielding and spectrometer electronics, it can still be influenced by radiation in the area. The temperature of the sample preparation area and the temperature during counting is another important environmental consideration. The sample should always be equilibrated to the counting temperature to minimize condensation, unanticipated precipitation, and other problems.

Table 6. Counter factors.

Photomultiplier tubes
Mechanical failure
Circuitry failure
Power failure
Positioning of AES pellet
Aberrant event (i.e. dead fly)

Counter factors

Temperature is important in counter factors as well (Table 6), especially in relation to the type of photomultiplier tubes. Mechanical and circuitry failure is to be expected with anything as complex as today's liquid scintillation spectrometers. If the event is a complete and obvious malfunction, it is often less frustrating than the intermittent, less readily detected malfunction. The same can be said for line voltage problems in the building's electrical supply. Various protective devices have been incorporated to minimize detrimental counter effects in the event of total power failure. However, power problems for both the spectrometer and the necessary air conditioning can be a constant source of concern.

A specific electro-mechanical malfunction can be the improper positioning of the external standard pellet, especially on a non-predictable basis.

My favourite malfunction concern for any instrumentation is the completely non-predictable, intermittent aberrant event which might be called the 'dead fly' or the 'grey hair' factor. This usually happens when one has a critical experiment going or needs another replicate experiment to complete the statistical comparisons in a paper to be presented the next week for which the slides have already been made. On the other hand, it never demonstrates itself when the service man is in attendance or when the counter is not being heavily used. The eventual explanation – if any – is often a dead fly or moth, some dust or lint, an old spider web, or some similar item to remind us of our complete subservience to the whims of nature.

Calculator/computer factors

If the sample has perchance miraculously survived all of the aforementioned pitfalls, you now have a count. However, this count is still subject to calculator/computer factors (Table 7) – whether on-line or off-line – before it becomes a useful datum. Electro-mechanical events and power problems can play a role, just as they do with the spectrometers. One of the favourite foibles of these devices is to print or punch an extra digit or

Table 7. Calculator/computer factors.

Circuitry failure
Mechanical failure
Power failure
Lack of parameter checks (i.e. precise nonsense)

one less digit just to be different on occasion. If enough appropriate checks on the system are not utilized, this kind of event can produce well-formated, beautifully typed sheets essentially ready for publication – except that the figures are nonsense!

Calculator/computer selection

In deciding what hardware route to pursue for liquid scintillation data handling, the first step appears to be self-evident. Except in those circumstances where only an occasional sample is counted or samples of known quench are put through a simple screening procedure for one fixed level of radioactivity, everyone is or will be using programmable calculators or computers. Quite frankly, the experiment generating the samples often has an error which deserves only slide-rule precision and we should therefore always be acutely aware of the number of significant figures we are dealing with when using various automatic computational devices. On the other hand, as I stated earlier, with the currently available multi-sample automatic spectrometers, the main bottleneck at the moment is data handling. As such, the remaining question is what direction one takes in regard to utilizing a programmable calculator, a mini-computer or a large computer. Recent developments in programmable calculators and mini-computers are making them more competitive with large computers and therefore such decisions must always be made in the investigator's local environment with the most recent information available.

Table 8. Considerations in calculator/computer selection.

Dedicated	versus	non-dedicated
Unique run	versus	production run
Single count	versus	count averaging
Off-line	versus	on-line
Batch process	versus	time sharing
Turn around time	versus	cost
Machine readability	versus	hard copy only
Laboratory	versus	institutional considerations

The individual investigator (and possibly others) must consider whether the convenience of data reduction from the liquid scintillation spectrometer may be worth the price of a computation device devoted to this exercise (dedicated) or whether this is a more centralized computer problem (non-dedicated) (see Table 8). One must also consider the uniqueness of the experiments. If one is doing only one experiment it may not be worth getting involved in computer programming. But if one is anticipating routine production runs of data after the program has been established, the computer approach may be completely satisfactory.

The question of count averaging has already been alluded to earlier where it was mentioned that Cavanaugh has shown that averaging can improve counting statistics.[38] In terms of instrumentation, such averaging is obviously handled off-line.

Off-line versus on-line is very much a local decision. The price of the gear associated with the spectrometer and the off-line local capabilities need to be considered. No one has a specific answer for these considerations except those immediately involved. Batch processing versus time sharing is also very much a local consideration, depending on the available facilities.

Turn around time is always a sticky consideration. Everyone wants immediate turn around time but very seldom do we really need this immediate turn around. The one exception is the necessity for real-time interaction such as in physiological experiments where computation will immediately influence the subsequent experimental parameters. Such needs are seldom, if ever, present in liquid scintillation spectrometry. As such, except in a possible rare circumstance of flow-through analysis, this has to be considered as a moot point. However, one must also remember that other moot points can have very real and very meaningful importance to very real individual investigators who have anticipation deadlines.

The potential cost of opting for fast turn around time has of course had an effect on many investigators. Whether one is considering operating on-line or off-line, the potential is there if one is ready and willing to pay for fast turn around.

Table 9. Advantages of machine readability.

Editing capabilities
Data storage
Data transformation
Data display
Ready statistical analysis

When one considers machine readability, a number of considerations come to mind – some of which do not always present themselves when the raw data is computed (see Table 9). Once the data are machine readable, appropriate program steps can interrogate the data to assure that input and output errors have not been made and various manipulative procedures on the data can also be performed. These options are in distinct contrast to a fixed program facility where only a hard copy with a fixed format is available.

Before concluding this presentation, it is only appropriate to recognize that the needs of a given individual investigator may not necessarily completely agree with other investigators in a given institution or the institution at large. This last stated consideration is not necessarily the least of our concerns. Everyone wants his or her own institution to flourish and yet everyone wants to do his or her own thing. As such, this consideration must be adjudged by us all, both individually and collectively. The environments and the solutions may be different, but the considerations are reasonably common to all.

In concluding, I realize that I have been variously repetitious of prior speakers and have perhaps covered too much. However, I hope this overview has set the stage for the more specific presentations to follow and, in assembling various user concerns, that it will prove beneficial for both current and future users of liquid scintillation spectrometry.

ACKNOWLEDGEMENTS

As an invited participant, the many efforts of members of the Radiochemical Methods Group has been deeply appreciated. The many beneficial conversations with other symposium participants led to a presentation continuously revised throughout the course of the meeting in an effort to fill some of the interstices. Use of the following figures was made possible by the kind permission of both authors and publishers. Figures 1 and 2 by E. T. Bush appeared in *Anal. Chem.* **35**, 1024, copyright 1963 by the American Chemical Society. Figures 3, 4 and 5 by M. P. Neary and A. L. Budd appeared in *The Current Status of Liquid Scintillation Counting,* copyright 1970 by Grune and Stratton. Figure 6 appeared in the brochure No. 12-Multi-Mat-1-10/71, copyright by Intertechnique Instruments. Figures 7 and 8 by A. Noujaim, C. Ediss and L. Wiebe, and also Fig. 9 by J. F. Lang all appeared in *Organic Scintillators and Liquid Scintillation Counting,* copyright 1971 by Academic Press.

REFERENCES

1. *A Re-examination: Centralized versus Decentralized Computers in Biomedical Environments,* Association for Computing Machinery, Chicago, Illinois, 1971.
2. E. D. Bransome (Ed.), *The Current Status of Liquid Scintillation Counting,* Grune and Stratton, New York and London, 1970.
3. D. L. Horrocks and C. T. Peng (Eds.), *Organic Scintillators and Liquid Scintillation Counting,* Academic Press, New York and London, 1971.
4. C. G. Bell and F. N. Hayes (Eds.), *Liquid Scintillation Counting,* Pergamon Press, New York and London, 1958.
5. J. B. Birks, *The Theory and Practice of Scintillation Counting,* Pergamon Press, Oxford, 1964.
6. E. Rapkin, *Intern. J. Appl. Radiation Isotopes* **15**, 69 (1964).
7. E. Rapkin, in *The Current Status of Liquid Scintillation Counting* (Ed. E. D. Bransome), Grune and Stratton, New York and London, 1970.
8. W. J. Kaufman, A. Nir, G. Parks and R. M. Hours, *Tritium Phys. Biol. Sci.* **1**, 249 (1962).
9. G. E. A. Wyld, in *The Current Status of Liquid Scintillation Counting* (Ed. E. D. Bransome), Grune and Stratton, New York and London, 1970.
10. P. D. Klein and W. J. Eisler, *Anal. Chem.* **38**, 1453 (1966).
11. J. A. B. Gibson and H. J. Gale, *J. Sci. Instr. (J. Phys. E.)* **1**, 99 (1968).
12. A. W. Forrey, in *Organic Scintillators and Liquid Scintillation Counting* (Eds. D. L. Horrocks and C. T. Peng), Academic Press, New York and London, 1971.
13. V. N. Kerr, F. N. Hayes and D. G. Ott, *Intern. J. Appl. Radiation Isotopes* **1**, 284 (1957).
14. M. P. Neary and A. L. Budd, in *The Current Status of Liquid Scintillation Counting* (Ed. E. D. Bransome), Grune and Stratton, New York and London, 1970.
15. C. T. Peng, in *The Current Status of Liquid Scintillation Counting* (Ed. E. D. Bransome), Grune and Stratton, New York and London, 1970.
16. M. I. Krichevsky, S. A. Zaveler and J. Bulkeley, *Anal. Biochem.* **22**, 442 (1968).
17. H. H. Ross, *Anal. Chem.* **37**, 621 (1965).
18. G. T. Okita, J. J. Kabara, F. Richardson and G. V. LeRoy, *Nucleonics* **15**, 111 (1957).
19. R. J. Herberg, *Anal. Chem.* **36**, 1079 (1964).
20. L. A. Baillie, *Intern. J. Appl. Radiation Isotopes* **8**, 1 (1960).
21. E. T. Bush, *Anal. Chem.* **35**, 1024 (1963).

22 E. Rapkin, *Lab. Scintillator* (Picker Nuclear Publication), **11**, 1L (1966).
23 T. Higashimura, *Intern. J. Appl. Radiation Isotopes* **13**, 308 (1962).
24 D. G. Fleishman and V. V. Glazunov, *Pribory i Tekhn. Eksperim.* **3**, 55 (1962).
25 T. C. Peng, *Anal. Chem.* **36**, 2456 (1964).
26 R. W. Hendler, *Anal. Biochem.* **7**, 110 (1964).
27 H. H. Ross and R. E. Yerick, *Anal. Chem.* **35**, 794 (1963).
28 Y. Kobayashi and D. V. Maudsley, in *The Current Status of Liquid Scintillation Counting* (Ed. E. D. Bransome), Grune and Stratton, New York and London, 1970.
29 A. Noujaim, C. Ediss and L. Wiebe, in *Organic Scintillators and Liquid Scintillation Counting* (Eds. D. L. Horrocks and C. T. Peng), Academic Press, New York and London, 1971.
30 J. F. Lang, in *Organic Scintillators and Liquid Scintillation Counting* (Eds. D. L. Horrocks and C. T. Peng), Academic Press, New York and London, 1971.
31 A. W. Rogers and J. F. Moran, *Anal. Biochem.* **16**, 206 (1966).
32 A. A. Jarrett, *U.S. At. Energy Comm., Div. Tech. Information* AECU-262 (1946).
33 R. Loevinger and M. Berman, *Nucleonics* **9**, 26 (1951).
34 W. E. Browning, *Nucleonics* **9**, 63 (1951).
35 R. J. Herberg, *Anal. Chem.* **35**, 786 (1963).
36 E. T. Bush, *Anal. Chem.* **36**, 1082 (1964).
37 C. Matthijssen and J. W. Goldzieher, *Anal. Biochem.* **10**, 401 (1965).
38 R. Cavanaugh, in *The Current Status of Liquid Scintillation Counting* (Ed. E. D. Bransome), Grune and Stratton, New York and London, 1970.
39 C. O. Carroll and T. J. Houser, *Intern. J. Appl. Radiation Isotopes* **21**, 261 (1970).
40 F. A. Blanchard, *Intern. J. Appl. Radiation Isotopes* **14**, 213 (1963).
41 J. L. Spratt, *Intern. J. Appl. Radiation Isotopes* **16**, 439 (1965).
42 L. R. Axelrod, C. Matthijssen, J. W. Goldzieher and J. E. Pulliam, *Acta Endocrinol. Suppl.* **49**, 99S (1965).
43 J. L. Spratt and G. L. Lage, *Intern. J. Appl. Radiation Isotopes* **18**, 247 (1967).
44 E. D. Plotka, E. G. Stant Jr., F. A. Waltz, V. A. Garwood and R. E. Erb, *Intern. J. Appl. Radiation Isotopes* **17**, 637 (1966).
45 D. S. Glass, in *Organic Scintillators and Liquid Scintillation Counting* (Eds. D. L. Horrocks and C. T. Peng), Academic Press, New York and London, 1971.
46 R. L. Little, in *The Current Status of Liquid Scintillation Counting* (Ed. E. D. Bransome), Grune and Stratton, New York and London, 1970.
47 M. Strolin-Benedetti, P. Strolin and B. Glasson, *Computers Biomed. Res.* **2**, 461 (1969).
48 M. F. Grower and E. D. Bransome, in *The Current Status of Liquid Scintillation Counting* (Ed. E. D. Bransome), Grune and Stratton, New York and London, 1970.
49 S. K. Figdor, *Computers Biomed. Res.* **3**, 201 (1970).
50 E. L. Forker and D. Wycoff, *Anal. Biochem.* **45**, 107 (1972).
51 R. Moore, *Curr. Mod. Biol.* **1**, 204 (1967).
52 J. L. Spratt, in *The Current Status of Liquid Scintillation Counting* (Ed. E. D. Bransome), Grune and Stratton, New York and London, 1970.
53 R. J. Herberg, in *Organic Scintillators and Liquid Scintillation Counting* (Eds. D. L. Horrocks and C. T. Peng), Academic Press, New York and London, 1971.
54 R. J. Herberg, *Science* **128**, 199 (1958).

DISCUSSION

E. Rapkin: I have two comments to make; firstly the Intertechnique counters are capable of storing repetitive counts and averaging results. Also, those Intertechnique counters which are capable of computing d.p.m. can average results of repetitive computations, thereby effectively averaging results of repetitive external standardization.
Secondly, too many users trust d.p.m. values based on calibration curves constructed from a single external standard count (30 s in/30 s out) and then a single external standard count of their test samples.

J. L. Spratt: In regard to your first point, once the first cycle count has been determined and stored, one is operating off-line in regard to further data processing of that number. This is true whether the computer facility sits on the counter or is a detached computer some distance away. Either way, I always appreciate having the option of multiple cycle counting and data handling which you mention.
Your second point is a quite valid concern for all of us.

J. H. Deterding: Would you like to comment on what is to me, the surprisingly uncritical acceptance of the performance number of P. D. Klein and W. J. Eisler (*Anal. Chem.* **38**, 1453 (1966))? I believe that there may be alternative, less arbitrary, criteria for the comparison of the performance of instruments used in experiments involving the counting of two isotopes.

J. L. Spratt: I share your concern on how much reliability we should place on Klein and Eisler's performance numbers. However, their approach was mentioned for the sake of completeness and to indicate that instrument performance can be viewed in many ways, depending on the instrument's use. They may be criticized for being somewhat arbitrary, but they cannot be faulted for at least attempting to devise criteria other than efficiency (E), background (B) and E^2/B.

J. L. Spratt: Comment: From various concerns expressed elsewhere at these meetings, perhaps it is appropriate for us to return to a dissociation between the counter and the sample in reference to efficiency. Both the manufacturers and users would then be in a better position to debate the relative merits of the basic instrumentation versus the uses to which it is put.

Chapter 22

Liquid Scintillation Counting of Biological Samples using External Standardization and Automatic Data Processing

P. Johnson, P. A. Rising and T. J. Rising

Wellcome Research Laboratories, Langley Court, Beckenham, Kent, England

INTRODUCTION

The purpose of this communication is to describe the method by which we routinely process biological samples by liquid scintillation counting. We have used three different counters to compare standardization methods for a variety of samples with wide degrees of quenching. The samples are processed through one of the following counters: the Packard Tri-Carb model 3324, the Beckman model LS-200B or the latest Intertechnique model ABAC SL40-4K, all of which are fitted with a teletype for computer processing. We found that the automatic external standard (AES) channels ratio method, which is a built-in feature of the Beckman, was unsatisfactory for carbon-14 samples with counting efficiencies lower than 70%. We therefore modified the counter using the auxiliary power supply and fitting an external switch to drive the caesium-137 γ-source into position to obtain gross counts as required. Both the Tri-Carb and Intertechnique counters are fitted with an AES, but in addition the Intertechnique has its own built-in 4K computer. This gives the option with this counter of using either its own or our ICL 1903A computer which is used routinely for off-line applications.

The AES process in the modified Beckman is different from that in the Tri-Carb and Intertechnique counters. In the modified Beckman, the samples are all counted and then the AES source is placed in position and each sample is counted again. In the Tri-Carb and Intertechnique, after the β-emission of each individual sample has been recorded, the γ-source is automatically brought out of its lead shielding and the total external standard counts are measured for that particular sample in a separate channel before counting the next sample. To overcome earlier troubles caused by variable positioning of the AES source, a magnetic latch system is now available for the Tri-Carb and has been fitted to our machine. Thus the geometry of counting varies in the three systems.

In our laboratory the punched tape output from the counters is relayed to the ICL 1903A computer by an off-line system, the program language is Fortran IV and the average turn round time is 1 h.

METHOD

The basic method depends on the production of a quench curve (Fig. 1). Eight

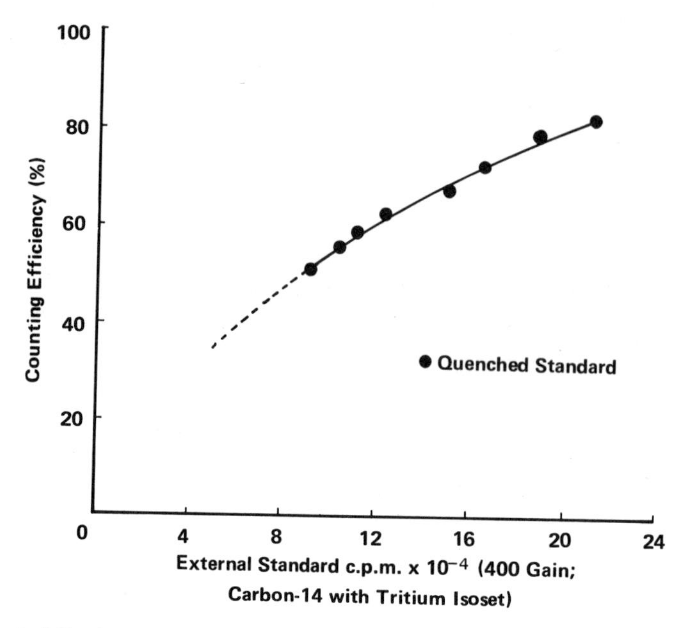

Fig. 1. A typical Beckman external standard quench curve – Hyamine scintillator (10 ml) plus bile/water mixtures (1 ml) plus ^{14}C-hexadecane.

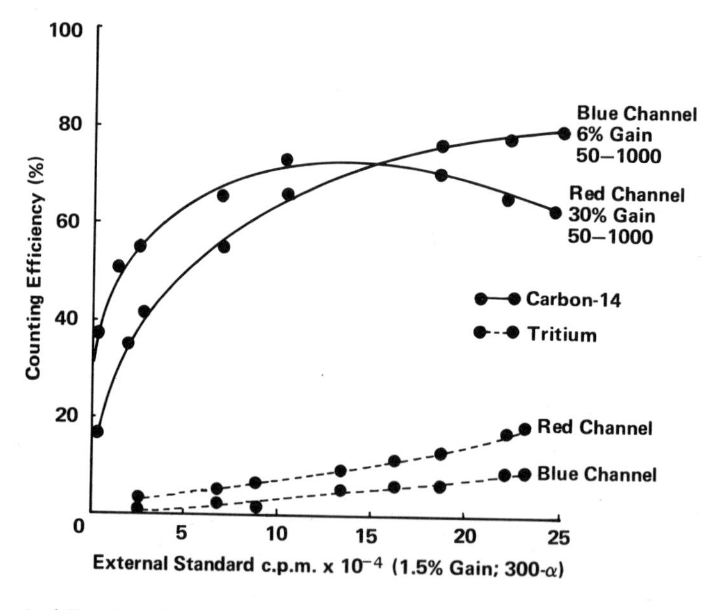

Fig. 2. A typical Tri-Carb dual-labelled quench curve.

standards (or sixteen for dual-labelled determinations) are prepared containing the same volume of scintillator and 10 μl of ^{14}C- or ^{3}H-labelled hexadecane. These are counted to ensure that the accuracy of pipetting of the hexadecane is within ±1%. These standards are quenched to varying degrees either by using non-radioactive aliquots of the biological sample to be counted or with carbon tetrachloride to produce a curve in which the counting efficiency is plotted against the external standard counts observed.

The counting efficiencies of the eight standards are calculated and equated with their corresponding external standard counts to give a best fit curve by the method of least squares. For single-labelled samples this is achieved by using the exponential of a third order of the logarithms. The curves obtained from the standards for a typical dual-labelled isotope experiment are shown in Fig. 2. The polynomial function used is the square of a fourth order of the square roots and the d.p.m. values are determined by solving two simultaneous equations. The values of the coefficients obtained from these expressions are entered into the data store for the subsequent calculations.

Table 1. Routine biological samples, scintillators and quenching agents.

Sample	Scintillator	Quenching Agent
Urine Plasma Bile	Triton X-100/xylene phosphor[a] (1 : 2) or 14.5% (w/v) hyamine chloride in butyl-PBD–toluene[b]	Dilutions of the biological sample
$^{14}CO_2$	Ethanolamine/methyl Cellosolve (3 : 7) in butyl-PBD–toluene[c]	CO_2 and CCl_4
^3H_2O	Koch-Light scintillator 354[d]	CCl_4

[a]DPO (0.6%) + POPOP (0.12%) in AR xylene.
[b]Butyl-PBD (0.75%) + Oxitol (7.5%) in AR toluene.
[c]Butyl-PBD (1.0%) in toluene/methyl alcohol (2: 1).
[d]1,4-dioxan based.

Choice of quencher

Quench curves of the type described above are generated for various biological samples and scintillator systems as illustrated in Table 1. The correct choice of quenching agent is of prime importance. We have compared colour and chemical quenching and have shown a distinct difference between the two at low counting efficiencies (Fig. 3). Walter and Purcell[1] came to a similar conclusion using carbon tetrachloride and lycopene, whereas Higashimura *et al.*[2] found no difference using acetone and Sudan III. This is probably due to the choice of quencher. However, there can also be a difference for the counting of biological samples (e.g. urine and plasma) with or without a chemical quencher (Fig. 4). It can be seen that there is a variation between the curves especially at low counting efficiencies. If, however, the activities of urine samples are calculated from a standard curve prepared from the urine of different species, the results are acceptable (Table 2). We obtain high counting efficiencies with human urine which is probably due solely to the cleaner conditions of collection of such samples, with one or two exceptions! A particularly important example of the effect of the biological sample itself has been experienced in the counting of expired $^{14}CO_2$. Using ethanolamine as trapping agent from small animal metabolic studies it has been found that the ethanolamine quenches differently from the carbamate formed during the carbon dioxide absorption. To overcome this, the

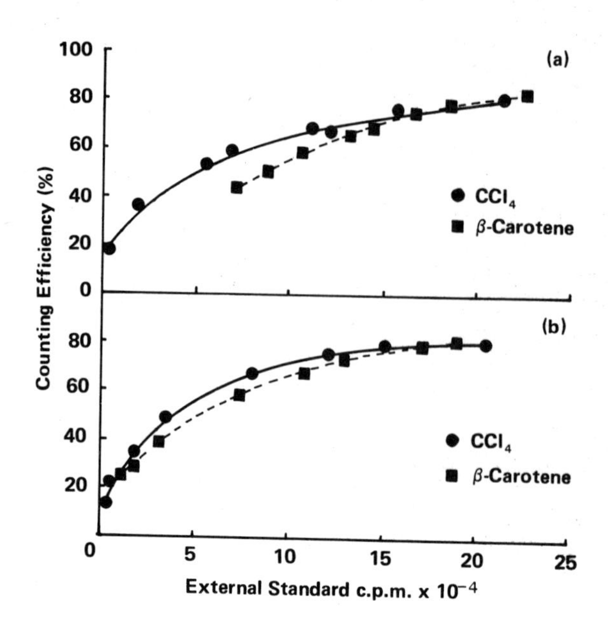

Fig. 3. Colour versus chemical quenching. (a) Beckman – Hyamine scintillator (10 ml) plus 0.5 ml urine plus ^{14}C-hexadecane (10 μl); (b) Tri-Carb – Koch-Light scintillator 354 (10 ml) plus 0.5 ml water plus ^{14}C-hexadecane (10 μl).

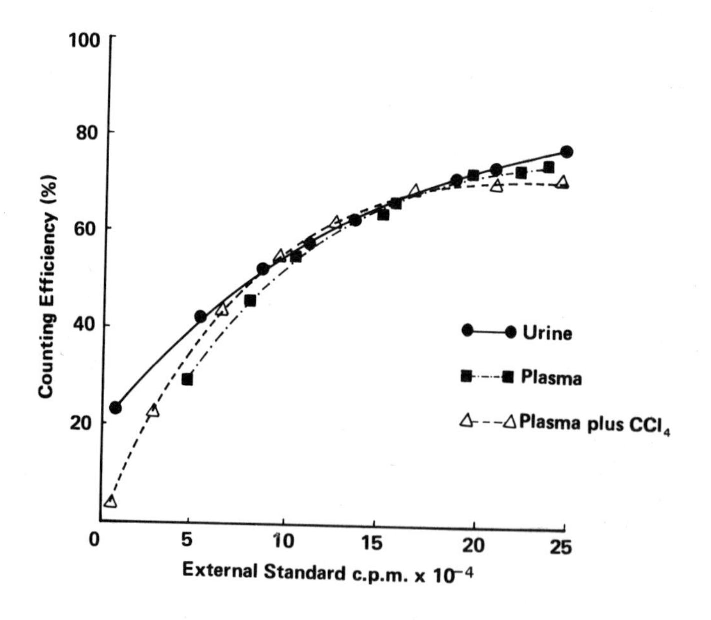

Fig. 4. A comparison of urine, plasma and carbon tetrachloride quench curves. Tri-Carb – Triton X-100 scintillator (10 ml) plus urine or plasma (1 ml) plus ^{14}C-hexadecane (10 μl).

Table 2. Calculation of d.p.m. values from a combined urine quench curve.

Species	Counting efficiency	d.p.m.	% Error
Sheep	76.8	18061	−0.1
	67.0	17769	−1.7
	60.4	18054	−0.2
	52.8	18168	+0.5
	42.0	18059	−0.1
Rat	74.3	18176	+0.5
	67.0	18140	+0.3
	60.8	17515	−3.1
	54.3	18107	+0.1
	46.3	18198	+0.6
	35.1	17644	−2.4
Human	81.3	17906	−0.4
	78.3	18177	+0.5
	75.1	18574	+2.7
	74.9	18499	+1.7
	76.7	17727	−1.9
	75.7	17792	−1.9

The standard curve was prepared from hyamine scintillator (10 ml) plus dilutions (1 ml) of rat, sheep and human urine. Each sample contained urine (1 ml) plus ^{14}C-hexadecane (18,083 d.p.m.) and was counted in the Beckman to an accuracy of 1%.

Table 3. The effect of expired carbon dioxide on the ethanolamine/EGME quench curve.

Sample number	CCl_4 standards			$(CCl_4 + CO_2)$ standards		
	Efficiency	d.p.m.	% Error	Efficiency	d.p.m.	% Error
1	57.6	14256	−21.8	45.4	18110	−0.7
2	26.6	19375	+ 6.2	28.4	18122	−0.6
3	25.9	19578	+ 7.3	27.9	18168	−0.4
4	17.9	17692	− 3.0	17.8	17698	−3.0
5	10.5	17990	− 1.4	10.1	18330	+0.5
6	8.7	18429	+ 1.0	8.9	18016	−1.2

The standard curves were prepared from ethanolamine/EGME absorber plus butyl-PBD scintillator and contained either CCl_4 or $CCl_4 + CO_2$ as quenching agent. All samples were authentic pre-dose absorber/scintillator mixtures containing ^{14}C-hexadecane (18,240 d.p.m.) and were counted in the Tri-Carb.

carbon-14 standards must be saturated with carbon dioxide as well as containing carbon tetrachloride as quencher (Table 3).

The shelf-lives of various sets of quenched scintillator standards have also been investigated (Table 4). It appears that the stability of these standards is dependent on the scintillator, the biological sample and the quenching agent and in general is independent of storage temperature. Unstable samples such as bile are normally used only on the day

Table 4. The shelf-lives of scintillator standards.

Sample	Scintillator	Isotope	Shelf-life 0°	R.T.
Plasma	Triton X-100	Carbon-14	8 weeks	8 weeks
		Tritium	2 weeks	1 week
Urine	Hyamine	Carbon-14	> 12 weeks	> 12 weeks
	Triton X-100	Tritium	–	12 weeks
Bile	Hyamine	Carbon-14	–	1 day
Carbon dioxide	Ethanolamine/butyl-PBD	Carbon-14	4 days	4 days
Water	Koch-Light No. 354	Tritium	–	> 8 weeks

All samples were stored in the dark, either at room temperature (R.T.) or at 0 to 4°C (0°). The shelf-lives were determined by counting freshly labelled hexadecane and using the original standard scintillator sets to calculate the d.p.m. values. The shelf-life normally refers to the maximum period for which the standards gave a counting accuracy of ±2.5%.

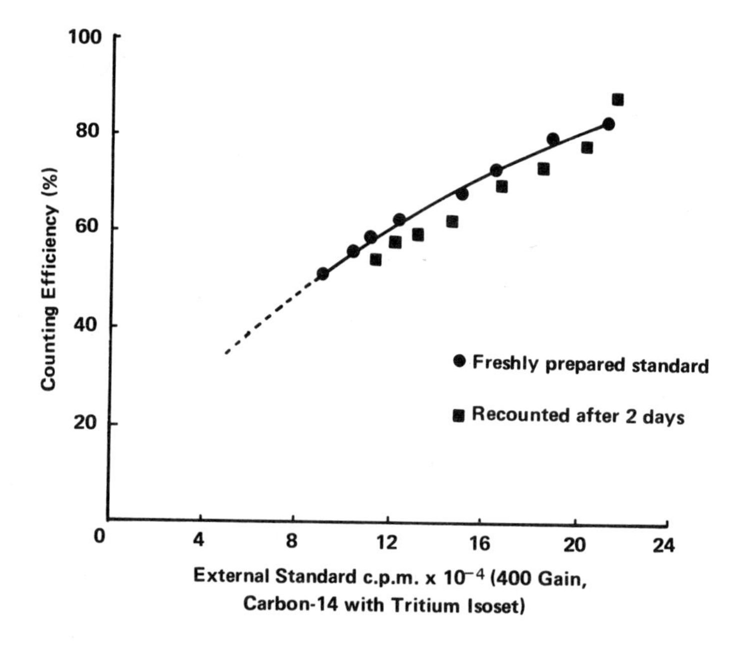

Fig. 5. The shelf-life of bile quenched standards.

of preparation (Fig. 5). For the stable scintillator sets both the counting efficiency and external standard values decrease with increasing storage time. An example of this is shown in Fig. 6. It might be argued that the coefficients derived from these stable sets could be stored in the program and this would certainly be possible given stable counting conditions. However, this would involve the inclusion of various sub-routines in the program and time and expense to remove and replace programs in the computer library file when the standards needed to be changed. One method reported to have overcome these difficulties associated with a stored standard curve is to make slight adjustments of the attenuation

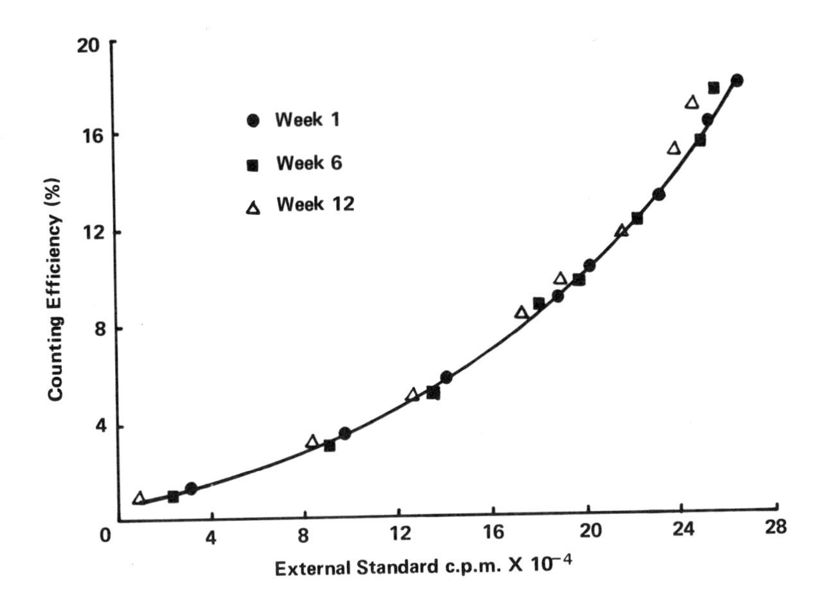

Fig. 6. Shelf-life of ^{3}H-hexadecane/urine standards. Beckman – Triton X-100 scintillator (10 ml) plus urine dilutions (1 ml) plus ^{14}C-hexadecane (10 μl).

settings (Figdor,[3] Jordan, Kaczmar and Köberle[4]). However, in our opinion the best way to overcome any day-to-day variations and drift of the liquid scintillation counter is to count the relevant standards on each occasion.

Print outs

A typical print out for a single isotope experiment is shown in Table 5. It can be seen that there is a minimal deviation between measured and curve efficiency which shows a correct choice of polynomial function – the same is true for dual-labelled determinations. The d.p.m. values of the hexadecane standards are written into the program with a half-life correction for tritium. All our samples are counted in duplicate and the average d.p.m. for each sample is calculated. For the print out of d.p.m. from dual-labelled experiments (Table 6) a cautionary message was incorporated into the program. This was required very occasionally when the tritium counting efficiency was below 1% and the corresponding curve was extrapolated by the computer to infinity rather than to zero.

Comparison of methods

As yet there is no all-purpose quench correction method and like other methods the use of the external standard gross count method has been reported to have certain disadvantages. These have been adequately resolved for our purposes:

1. The problem of variable external standard geometry does not really arise in the Beckman. The standard deviation of the external standard count rate under several quenched conditions in the Tri-Carb has been observed. When calculated in terms of efficiency this deviation is minimal at low quenching and at high quenching (20% efficiency for carbon-14 and 2% efficiency for tritium) is approximately ±1% for carbon-14 and ±0.1% for tritium.

Table 5. A typical printout for a single isotope experiment.

PROGRAM RRC6 LINK WRL8 PAR
00 0 DATE 23/06/71 TIME 12/32/44

14C URINE TEST SAMPLES IN TRITON SCINTILLATOR
C 18083.0

CALIBRATION CURVE

COUNTS STD. IN	COUNTS STD. OUT	MEASURED EFFICIENCY	CURVE EFFICIENCY
190838.	14024.	77.55	77.59
176439.	13663.	75.56	75.80
107952.	11510.	63.65	63.73
171393.	13705.	75.79	75.12
128805.	12376.	68.44	68.05
70366.	9897.	54.73	54.86
143657.	12655.	69.98	70.79
86015.	10650.	58.90	58.65

BACKGROUND = 50.5

SAMPLE NO.	COUNTS	EFFICIENCY	D.P.M.
11	1425.1	76.55	1795.7
12	1482.7	78.97	1813.7
MEAN			1804.7
13	1478.8	78.61	1817.0
14	1388.5	73.68	1815.9
MEAN			1816.4

Table 6. The d.p.m. of the ^{14}C- and ^{3}H-labelled samples.

BACKGROUND, RED CH. = 48.0 , BLUE CH. = 41.0

SAMPLE NO.	RED CH. C.P.M.	BLUE CH. C.P.M.	RED C EFF	BLUE C EFF	RED T EFF	BLUE T EFF	C D.P.M.	T D.P.M.
1	17293.0	15770.0	64.53	75.55	17.45	7.73	18270.3	33227.9
2	17257.0	15707.0	64.86	75.31	17.19	7.56	18347.4	34943.1
STD. FOR NEXT LINE IS OUTSIDE RANGE OF ONE OF CALIBRATION CURVES								
3	17756.0	17025.0	62.04	76.94	18.87	8.60	18356.9	33743.4
4	14708.0	11745.0	67.67	60.74	7.29	2.24	18092.8	33794.0
5	13680.0	10871.0	64.79	54.93	6.16	1.93	18017.9	32056.1
6	12359.0	9234.0	60.75	49.09	4.56	1.37	18003.5	31205.8
STD. FOR NEXT LINE IS OUTSIDE RANGE OF ONE OF CALIBRATION CURVES								
7	12810.0	10046.0	61.51	50.09	4.92	1.14	19624.3	15031.2

2. Since the AES count rate depends also on the sample vial material and wall thickness being the same, we always use the same type of counting vial and it has been shown that a 10% variation in wall thickness causes a count rate error of less than 0.2% (Higashimura *et al.*[2]).
3. Variation in the scintillator volume causes large errors. In our system the scintillator volume is therefore kept constant.
4. New calibration standards are prepared for each new scintillator batch, thus overcoming small batch variation.
5. Since the external standard method is not normally suitable for emulsions or suspension counting it is not used under these conditions.

Table 7(a). A comparison of standardization methods – 20 urine samples in Triton X-100 scintillator, each containing 1858 d.p.m., were standardized by five different methods.

Method	External standard gross count Tri-Carb	Beckmann	Intertechnique	External standard channels ratio	Internal
Average	1877	1845	1873	1863	1847
Range	1811-1989	1788-1928	1798-1943	1761-1996	1755-1919
SEM	9.2	8.7	9.6	12.0	10.5
Average % deviation from theoretical value	1.90	1.93	1.95	2.00	2.12

Table 7(b). A comparison of standardization methods – the d.p.m. values reported represent a range of 65 ^{14}C-labelled rat urine samples in Hyamine scintillator from a typical metabolism experiment.

External standard gross count Tri-Carb	Beckman	Intertechnique	External standard channels ratio	Internal
57198	57336	56637	57183	55952
37233	36825	36195	36770	35951
20855	20531	20069	20315	20860
6269	6178	6305	6211	6391
399	451	424	435	468

We have compared our method of external standardization with internal standardization for most of the biological systems used and have generally found agreement to within 3% or better. Examples of these comparisons are shown in Tables 7(a) and 7(b). In Table 7(a) the average percentage deviation from the theoretical value, which might appear relatively high, was calculated on the basis that the activity of all samples was identical. The calculated d.p.m. values of a number of samples were for all methods consistently either above or below the theoretical activity. Also all samples were only counted to an accuracy of ±1%. Therefore this percentage error would probably be lower under more suitable conditions. The values in Table 7(b) represent five typical samples from a metabolic experiment covering the whole range of activities. Again there were only small variations in the d.p.m. values between the different methods. Some counters (e.g. the Intertechnique used in the present study) use a combination of the channels ratio and external standard methods. This removes the disadvantages of the channels ratio method and also reduces some of the shortcomings of the external standard method since the fluctuations in wall thickness, the distance between the specimen and the γ-source and the scintillator volume change the total count rate more than the channels ratio (Frampton[5]). Figdor[3] using this method for counting various carbon-14 and sulphur-35 biological samples on a Nuclear Chicago model 6860 reported a percentage difference from internal standardization of −6.3% to +4.0%. From our own studies (unreported) on the external standard channels ratio method using the Tri-Carb, this error seems large, and we are more in agreement with Parmentier and ten Haaf[6] who in a recent review concluded that for modern counters this combination method was the best for standardization. However, one of the disadvantages of using this method with the older models of counter is the alteration of window and gain settings which would be involved when both carbon-14 and tritium samples are continuously being counted.

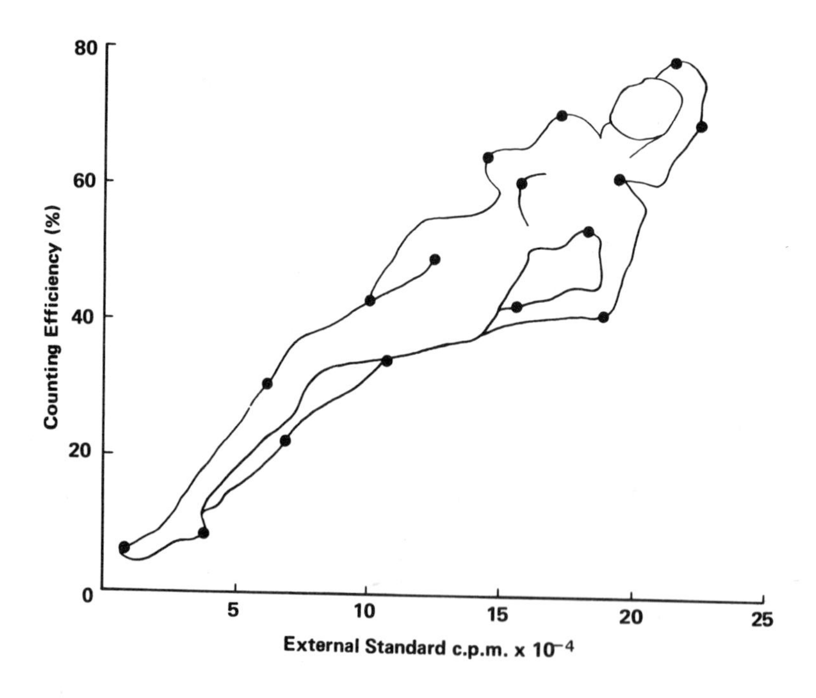

Fig. 7. Miss Plot 1971.

CONCLUSIONS

There is a wide range of quench correction methods available. For routine counting of biological samples in older models of counter we recommend the use of the method described; that is, the external standard gross count method. Other than the obvious need for computer facilities, the success of the method depends on the careful preparation of the quenched standards. If, however, care is not taken in the preparation of the quenched standards, our computer puts its own interpretation on the data input and the quench curves shown in Fig. 7 are obtained.

ACKNOWLEDGEMENT

The authors wish to thank Mr. G. C. Sheppey for writing the computer programs.

REFERENCES

1 W. H. Walter and A. E. Purcell, *Anal. Biochem.* **16**, 466 (1966).
2 T. Higashimura, O. Yamada, N. Nohara and T. Shidei, *Intern. J. Appl. Radiation Isotopes* **13**, 308 (1962).
3 S. K. Figdor, *Comput. Biomed. Res.* **3**, 201 (1970).
4 P. Jordan, U. Kaczmar and P. Köberle, *Nucl. Instr. Methods* **60**, 77 (1968).
5 J. A. Frampton, Fachtagung Nuclex, Basel, 8-14.9, Vortrag 7/25 (1966).
6 J. H. Parmentier and F. E. L. ten Haaf, *Intern. J. Appl. Radiation Isotopes* **20**, 305 (1969).

DISCUSSION

F. E. L. ten Haaf: Comment: This paper has shown that external standardization can be an excellent method provided the necessary precautions are observed.

Chapter 23

Automatic Processing of Data from Liquid Scintillation Counters Illustrated by Drug Distribution Studies

H. E. Barber and G. R. Bourne

Department of Pharmacology, University of Liverpool, Liverpool L69 3BX, England.

Liquid scintillation spectrometry is now widely used in biological laboratories for the assay of radioactive isotopes. To facilitate automatic processing of the output from scintillation spectrometers, instruments are available which include small computers in their specifications. Most systems process the data to give only disintegrations/min (d.p.m.) whereas our laboratory takes advantage of a large central computer to process the data in final results form. Computer programs are written in ALGOL and stored in the computer.

In experiments on the distribution of drugs ^{14}C- and/or ^{3}H-labelled drugs are used. The drugs are given to the animal as a rapid intravenous injection at zero time and thereafter blood and urine samples are collected at known time intervals. Samples are analysed by liquid scintillation counting for either drugs or metabolites after appropriate separations. Our data processing technique (see Fig. 1) will be illustrated with reference to the analysis of plasma data in terms of an open two-compartment model of distribution.[1]

The scintillation counter is a small three-channel instrument equipped with a barium-133 external standard, lister and 8-hole paper tape punch. The output consists of sample number, counting time and counts in three channels and is punched simultaneously with the listing. Different experiments are automatically separated on the paper tape so more than one user of the system is possible at any time. The paper tape output is spliced to an identifying lead tape and is then processed to final form by the central computer. The efficiency of counting and d.p.m. of the samples are determined by the 'external standard channels ratio' method.[2] For counting, quenched tritium and carbon-14 standards follow a background bottle. The test samples are placed in the counter after these standards. These latter are made up in a similar manner to the test vials and the number of tritium standards is equal to the number of carbon-14 standards. The degree of quenching of the most quenched tritium standard is equal to that of the similar carbon-14 standard.

Generally samples exhibit a minimum variable degree of quenching and a quadratic function is used to express the relationship between counting efficiency and barium channels ratio of the standards. The coefficients of the quadratic are derived by the method of least squares. Two sets of barium ratios are calculated – for tritium and carbon-14. The highest (maxa) and lowest (mina) ratios of the sets are found and their standard deviation

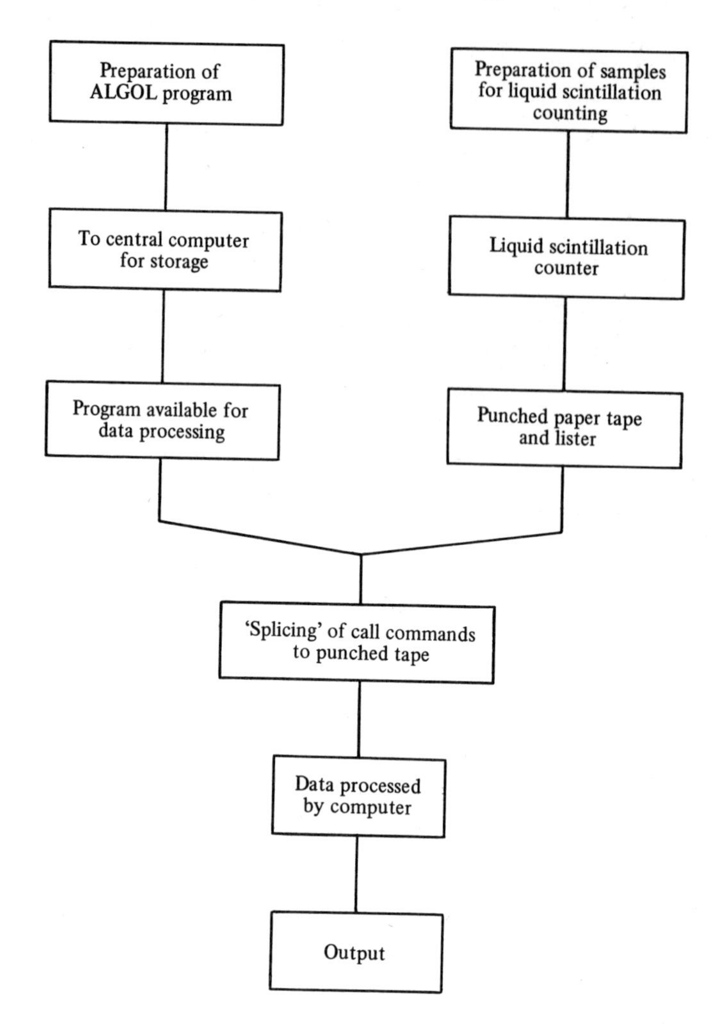

Fig. 1. General procedure.

(SD) calculated. A maximum (max) and minimum (min) allowed ratio is computed as:

$$\text{min} = \text{mina} - (2 \times \text{SD of mina})$$
$$\text{max} = \text{maxa} + (2 \times \text{SD of maxa})$$

If any test ratio is found to be outside the limits of min or max the computer prints a warning message. However, the d.p.m.'s of these vials are still calculated but the extrapolation of the quadratic function is noted. Single or doubly labelled samples are processed by this program and final output includes:

1. the input from the paper tape, which can be compared with the listed output from the scintillation counter as a check for errors in punching or reading of the tape,
2. d.p.m. of the isotopes,

3. SD of tritium samples, and
4. the efficiency of counting of the isotopes in the three channels.

These calculations are performed by the operation of the program SCINTILL, subsequent calculations are worked by other programs in the sequence – DPMTWOCPT which includes SCINTILL. The d.p.m.'s from SCINTILL are used to evaluate concentration terms which are then used as further input for the calculation of the parameters of the two-compartment model of distribution. For the plasma samples, least squares fits are calculated using a biexponential function. The computation is done by the method of residuals[3] and all combinations of biexponential fits to the data are calculated. From these fits the biexponential which has the minimum logarithmic sums of squares of deviations of experimental points to the calculated line is chosen as the biexponential function which best describes the data. The parameters of this function are evaluated in terms of the model – calculations of first order distribution rate constants, fractions of the drug which are located in each compartment with time, etc.

```
begin       declarations


            programme body



end
```

Fig. 2. Schematic representation of an ALGOL program.

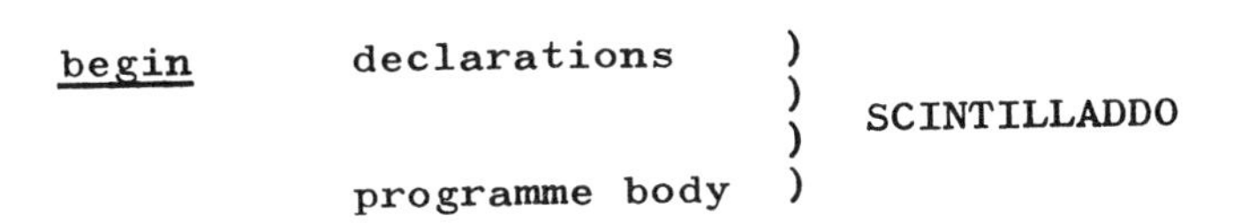

Fig. 3. Schematic representation of SCINTILLADDO program.

The principle of the data processing technique and the correct 'ordering' of programs is now explained. In Fig. 2 is shown a schematic representation of an ALGOL program. This program has complete block structure[4] and would perform a production sequence. Figure 3 represents a program SCINTILLADDO; this is designed to perform the calculation of d.p.m. from scintillation counting data. However the absence of the final 'end' statement destroys the block structure and this program is not capable of producing output.

In Fig. 4 the program SCINTILL is depicted. The final 'end' is appended to SCINTILLADDO to complete the block arrangement and the running of the program SCINTILL with appropriate data causes the relevant calculations to be executed as pre-

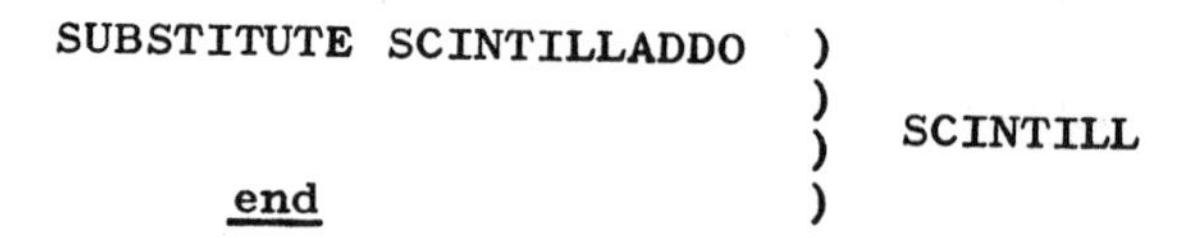

Fig. 4. Schematic representation of SCINTILL program.

```
DPMTWOCPT:

          SUBSTITUTE   SCINTILLADDO

          SUBSTITUTE   ARRAY

          SUBSTITUTE   BITWO

     [ Any minor finishing statements ]
     [      that may be required.     ]
```

Fig. 5. Schematic representation of DPMTWOCPT program.

viously described. It is, of course, possible to insert between SCINTILLADDO and 'end' any other program the user may desire providing the rules of block structure in ALGOL are adhered to.[4] This is depicted in Fig. 5 with the program DPMTWOCPT, the working of which was explained. SCINTILLADDO determines d.p.m. as before, ARRAY causes d.p.m.'s to be converted to concentration terms and BITWO the calculations of the biexponential function and parameters of the compartmental model. Minor finishing statements would include those for the completion of block structure, e.g. 'end' to link SCINTILLADDO. The system is flexible and the same programs can be included in different sequences and used for different purposes decided by the user. The simple rules of block structure in ALGOL facilitate this procedure and SCINTILL has been linked to programs which process clearance data from plasma and urine samples.

With this computerized approach to data processing, laboratory results are evaluated in their final form with no manual calculations. The system is simple and can be operated and used with very little training in the techniques of computer programming. Without the aid of the computer, evaluation of results would be time-consuming and laborious with attendant human error.

REFERENCES

1 S. Riegelman, J. C. K. Loo and M. Rowland, *J. Pharm. Sci.* **57**, 117 (1968).
2 T. C. Hall and C. J. Weiser, *Anal. Biochem.* **17**, 294 (1966).
3 D. S. Riggs, *The Mathematical Approach to Physiological Problems,* Williams and Wilkins Co., Baltimore, 1963, Chapter 6.

4 R. Wooldridge and J. F. Ratcliffe, *An Introduction to Algol Programming*, The English Universities Press Ltd., London, 1963, Chapter 6.

DISCUSSION

B. Scales: It is very interesting to see a program in use for the routine processing of labelled urine and plasma samples with the final output in the form of the required pharmacokinetic constants. However, it must be appreciated that valid results can only be obtained when the radioactivity in the samples is present as a single chemical entity. When more than one labelled chemical species is present, as is often the case due to rapid metabolism even after *intravenous* dosing, then the calculated pharmacokinetic data can be quite meaningless.

H. E. Barber: We always take great care and I am sure that these comments do not apply to our results.

B. Scales: I am not suggesting that they do but as a general comment, so often people use highly sophisticated techniques and programs, completely ignoring the fact that their initial data are inadequate or inappropriate.

H. E. Barber: Yes, we are aware of these problems and check our early and late samples to ensure that only one labelled component is present; otherwise the different labelled components are separated prior to the initial counting procedures.

P. Johnson: Comment: Dr. Scales' point has been made at other times during this meeting and cannot be emphasized too strongly.

Chapter 24

Determination of Absolute Radioactivity in Multi-Labelled Samples using External Standardization or Channels Ratio: A Fortran IV Program

P. E. Stanley

Department of Agricultural Biochemistry, The Waite Agricultural Research Institute, University of Adelaide, Glen Osmond, South Australia, 5064.

INTRODUCTION

Data from liquid scintillation spectrometers are nowadays readily obtained since the processing of samples is completely automated. The analysis of such data and their reduction to more tractable levels can be conveniently accomplished by use of the computer operated either ON-LINE or OFF-LINE. In the former case a small dedicated computer is frequently built into the spectrometer or the instrument is used ON-LINE to a main computer installation. In the OFF-LINE configuration data may be processed by a small electronic desk calculator or at a large computer via the use of punched cards or tape. Each of these four modes has its own merits and demerits but it is not proposed to compare and contrast them here. The present paper is concerned with OFF-LINE operation to a large computer, a CDC 6400, on the basis of its availability, reliability, storage capacity (core) as well as its processing speed and fast INPUT/OUTPUT system.

SCOPE OF THE PROGRAM

The program I wish to discuss concerns the measurement of amounts of individual radioisotopes in multilabelled samples containing up to three species. It is essential that the isotopes have energies which give rise to pulse height spectra the ratios of which exceed 1 : 4 for any pair of isotopes with adjacent energies. This allows for adequate pulse height analysis to separate the isotopes. The commonly encountered trio, tritium, carbon-14 and phosphorus-32, have ratios of energies of 1 : 10 : 100 approximately and are thus adequately separated for this purpose.

The degree of quenching in samples may be assessed by using the external standardization procedure (up to three radioisotopes) or by sample channels ratio of the isotope having the higher energy (two radioisotopes in the case of a three-channel spectrometer).

The program can be used for samples which are either all chemically quenched *or* all colour quenched. It cannot be used to deal with samples quenched by both processes, basically because each type of quenching gives rise to entirely different shaped pulse height spectra.[1]

In using a large computer it has been possible to accomplish not only the straightforward arithmetical computation of disintegrations/min of the isotopes but also to make a statistical analysis of the results obtained from what are, after all, estimates of counting rates which exhibit a Poisson distribution (neglecting instrument drift and other factors to be mentioned later). An attempt has also been made to assess the errors associated with the efficiency–quench correction curves and to incorporate these into a final error term. To date such statistical analysis has been largely neglected in the field of data processing in liquid scintillation spectrometry. Notable exceptions are the work of Horton and Tait[2] and that of Carroll and Houser.[3]

In using the program it is assumed that the liquid scintillation process to be measured is free from spurious events.

Because of the restriction on time no discussion is given here for setting up the spectrometer to provide the optimum separation of isotopes. The reader is referred instead to the paper of Kobayashi and Maudsley.[4]

EXTERNAL STANDARDIZATION

Before presenting the basic approach adopted in designing the program it is necessary to discuss some of the problems encountered in obtaining a useful estimate of the external standard ratio/count when this technique is used to assess the degree of quenching in a sample.

Sources of error in external standardization

Most commercial spectrometers permit replicate counting of samples. Thus it is possible to detect variation in external standard values which are associated with one or more of the following parameters:

1. Lack of reproducibility in positioning the external standard source.
2. Irregular geometry of sample vials.
3. Vial to vial variation.
4. Instrument drift.

Evaluation of errors

The effects of (1) and (2) above have been assessed for a Packard Tricarb Model 3375 which has a compound external standard source of americium-241 and radium-226. Using a single sealed background vial a series of external standard ratios were obtained for the following three situations:

A. Both the vial and the external standard source were kept in the same unaltered position (constant source and vial geometry).

B. The vial was kept in a constant position but the external standard source was allowed to cycle IN/OUT as is usual in automatic operation (constant vial geometry).

C. An external standard measurement was made as in (B) and the vial unloaded, cycled around the sample changer before another such measurement. This process was repeated a number of times so that the vial position in the detector was randomized.

The results presented in Fig. 1 indicate that although there is a small error associated with source geometry (non-reproducible positioning) there is a larger one resulting from the irregular vial geometry. Consequently this must be taken into account when estimating the external standard ratio/count.

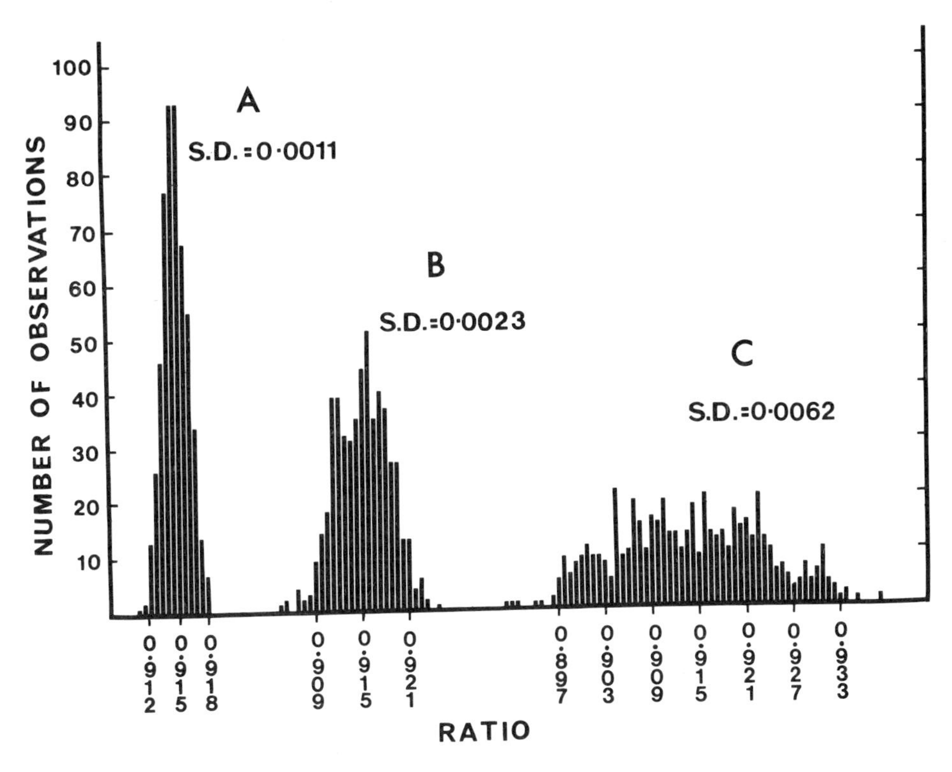

Fig. 1. Distribution of external standard ratios; A = constant source and vial geometry, B = constant vial geometry, C = normal automatic operation (see text for further details). S.D. = standard deviation.

The long term drift of this particular instrument over a period of three weeks was negligible, for after comparing the distribution and means of external standard ratios of the same sealed background vial (50 observations) on six occasions during this time there was no significant difference between the means and distributions. Thus the error associated with the external standard value is, in large part, due to irregular vial geometry and vial positioning within the detector.

Influence of vial geometry on external standard value

To confirm this variation in external standard value as a function of vial geometry the same sealed background vial was placed in twelve different positions within the detector such that each position was at an angle 30° to the radial position of the previous one. Ten assessments of external standard ratio were made for each position. A statistical analysis showed that the means of the external standard ratios in each of the twelve positions were significantly different as is shown in Fig. 2. It is of interest that external standard counts as well as sample channels ratio also show this dependency on vial orientation. Variation in vial geometry has been confirmed in vials from three independent suppliers and consequently in estimating the external standard ratio it is necessary to take this into account by making evaluations in random vial positions. This is conveniently done by the cycling technique.

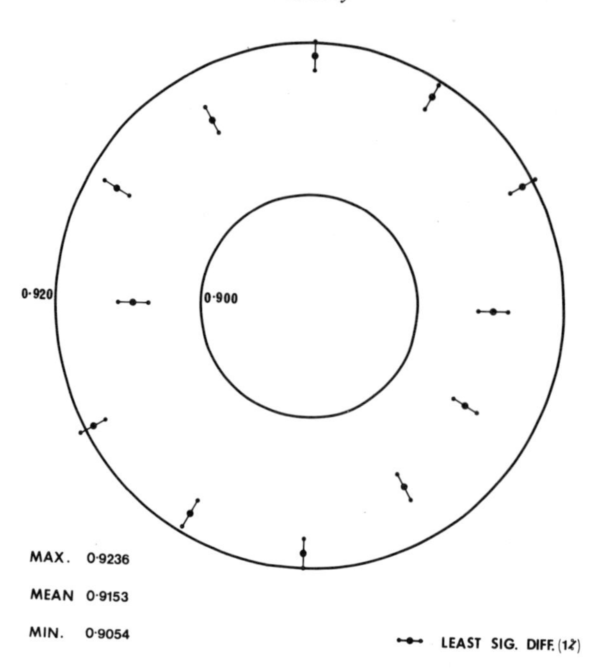

Fig. 2. The effect of vial orientation on external standard ratio.

Selection of vials

Inter-vial variation is reduced by selecting vials containing standard fluor under unquenched, moderately and heavily quenched conditions. At the same time it is also convenient to select vials for uniform background values and this is especially important if low activity samples are to be measured.

SOURCES OF ERROR IN ASSESSING DISINTEGRATIONS/MIN (d.p.m.)

The following sources of error must be taken into account when assessing the d.p.m. of a quenched sample:

1. The external standard value of both the standards and the samples.
2. Poisson statistics associated with the accumulated counts.
3. Subtraction of the background.
4. Curve fitting for the efficiency versus quenching (external standard).
5. Vial to vial variation for both external standard and background.

DESIGN OF THE COMPUTER PROGRAM

The program is written in FORTRAN IV and conforms to USASI standards. The program runs on a CDC 6400 computer which has 400_8K core of which 140_8K core is available to the user. Data are supplied via punched cards.

Constraints within the program

The capacity of the program is limited as follows:

1. Up to three isotopes may be measured.

2. Up to 25 standards (quenched) for each isotope.
3. Up to 200 samples (standards and unknown samples).
4. Up to ten background standards (quenched).
5. All backgrounds counted for the same time.
6. All samples must remain in the same sample belt position from cycle to cycle.

The background standards are run before the samples since the former will generally require more counting time. A check on the stability of the background of the equipment can be made by inserting background samples amongst the unknown samples.

Design of the program and its operating requirements

The program is divided into two parts:

1. Program XSTD
2. Program GETDPM.

The first program, XSTD, which reads into core all data and rearranges them, requires a large amount of core (100_8K) but a short amount of central processor time. It makes two files to be processed by GETDPM. The second program deals with these two files and processes the standards, one isotope at a time, sets up efficiency versus quench curves and then processes the unknown samples one at a time. Program GETDPM requires a modest amount of core (40_8K) but more central processor time.

Basic details concerning program XSTD

In this program the information first read in concerns the parameters associated with this experiment. It includes the number of isotopes, channels, standards, unknown samples, cycles, background samples, background cycles, together with the number of standard deviations to be employed in the statistical tests and the names of the isotopes as well as the number of the channel to be used for numerator and for denominator if the sample channels ratio technique is to be employed.

The type of quench assessment is then read which may be one of the following:

1. External standard ratio or counts in separate external standard channels.
2. External standard ratio or counts in the isotope counting channels.
3. Channels ratio of the sample.

The INPUT format for the background data (if any) is then read which allows the user the flexibility of choosing the INPUT format. The background standards (quenched or unquenched) are then read and all values stored before obtaining means of c.p.m. in each channel. A check is then made to ascertain if data lie within the user's preset statistical limits. Those that do not can be rejected or retained as required with an informative diagnostic printed out.

The INPUT format to be employed in reading standards and unknown samples is read in, followed by data concerning standards and samples. Data are read in cycle by cycle. All values are stored before subtraction of background values from each according to the degree of quenching associated with it. A simple linear interpolation procedure is employed. The mean c.p.m. for each standard and each unknown sample in each channel is then calculated.

The next step is to calculate the times elapsed between the start of counting of the first sample in the first cycle and counting of every sample in every cycle. This is valuable to determine the d.p.m. of fast decaying isotopes. The elapsed times can be conveniently calculated by taking account of the times for sample loading, counting, printing, unloading and the sample belt movement respectively. Vacant positions are accounted for during their bypass. The time for the various types of external standard cycle is included as necessary.

The final part of the program creates two files, one with all the data for the standards and the other with all the data required for processing the unknown samples. The data are arranged sample by sample.

Basic details concerning program GETDPM

This program processes the standards data and then the unknown samples. In addition it reads from cards the dates and time of day when the standards for each isotope were prepared as well as the time at which counting commenced. The d.p.m.'s for the standards are also read in. This program has a library of decay constants for commonly used isotopes. The d.p.m.'s are then corrected for decay during this period except in the case of long-lived radioisotopes such as carbon-14.

Mean external standard ratios/counts or sample channels ratios are then calculated and the appropriate values are tested to ascertain if they lie within the preset statistical limits. An informative diagnostic is printed if values exceed the preset limits and these can be either rejected or retained.

Prior to curve fitting for efficiency versus quenching, the external standard or channels ratio values are scaled by dividing all values by the largest in the array. The largest value then takes a value of one. This is necessary so as to avoid dealing with a large range of numbers in the curve fitting procedure when their squared values cannot be readily accommodated even in the CDC 6400 which allows fifteen significant digits to be handled in single precision. When employing a smaller computer it would be advisable to work in double precision.

The d.p.m. for each standard in each channel at a preselected clock time is then calculated, a plot of the log c.p.m. of each standard in each channel versus elapsed time (during each cycle) is made and a regression line of slope λ (decay constant for that isotope) is obtained to test for c.p.m.'s outside the preset statistical limits. The efficiencies at the preselected clock time are then calculated. This procedure is very valuable when dealing with a large number of samples containing short-lived isotopes when the counting time may involve several days.

The next event in GETDPM is the fitting of orthogonal polynomials to generate curves of the reciprocal of the counting efficiency (of an isotope in a channel) against the external standard value (or channels ratio value). By employing the reciprocal it is possible to avoid an efficiency matrix inversion with its attendant mathematical complexity of deriving an estimation of the error.

GETDPM then deals with the unknown samples one at a time (all samples). The means of the external standard or channels ratio values are then calculated and values exceeding the preset errors are dealt with as before. The d.p.m.'s for each isotope in the sample are then computed and two error terms are determined. The first error is associated with curve fitting and the external standard (or channels ratio) value of the unknown samples. The second one gives an error for the Poisson statistics associated with radioactive decay.

The results may be recorded in units other than d.p.m. e.g. mg/ml, provided a conversion factor is included.

CONCLUSION

The program has been designed to give maximum flexibility compatible with accommodating quenched samples containing up to three radioisotopes together with an assessment of the error associated with d.p.m. values. It is likely that this program will be of value to biochemists and radiochemists using more than one radioisotope at a time in their work.

ACKNOWLEDGEMENTS

I am grateful to Mr. P. J. Malcolm of the Biometry Section of this Institute for his assistance with the programming and statistical analysis.

I would like also to thank the Australian Wheat Industry Research Council for generous financial support and Professor D. J. D. Nicholas for his encouragement throughout this work.

REFERENCES

1 M. P. Neary and A. L. Budd, in *The Current Status of Liquid Scintillation Counting* (Ed. E. D. Bransome), Grune and Stratton, New York and London, 1970, p. 273.

2 R. Horton and J. F. Tait, *J. Clin. Endocrinol. Metab.* **27**, 79 (1967).

3 C. O. Carroll and T. J. Houser, *Intern. J. Appl. Radiation Isotopes* **21**, 261 (1970).

4 Y. Kobayashi and D. V. Maudsley, in *The Current Status of Liquid Scintillation Counting* (Ed. E. D. Bransome), Grune and Stratton, New York and London, 1970, p. 76.

DISCUSSION

J. H. Deterding: I have a question, Mr. Chairman, which may seem rather detailed but does show the sort of problem which comes up occasionally. I would like to ask Dr. Stanley if he has ever found it necessary in his Fortran programs to use 'double-precision'? Recently I had to find the standard deviation of the mean of the three counts 190, 189 and 191 with a Fortran program which is used routinely; one way of finding the answer (about 0.6) involves the subtraction of two six-figure numbers, and 'double-precision' was needed. Has Dr. Stanley had similar experiences?

P. E. Stanley: No I have not yet found it necessary to use 'double-precision'.
In integer subtraction and addition on the computer used (CDC 6400) eighteen decimal digits are significant in single precision. Integer division and multiplication are conducted in floating point notation. In all floating point manipulation fifteen digits are significant in single precision and twenty-nine in double precision. In work to date single precision with fifteen digits has been adequate but I agree with Mr. Deterding that six digit manipulation could lead to problems as he has found. This would be noticeable in computers employing 36 or 48 bit words. The CDC 6400 is a 60 bit word instrument.

Chapter 25

Determination of Statistical Precision of Tritium d.p.m. in Dual Labelled Samples with Variable Isotope Ratios and Quenching

J. Assailly, C. Bader,* J.-L. Funck-Brentano and D. Pavel

Unité de Recherches sur l'Application des Radio-Eléments à l'Etude des Maladies Métaboliques, INSERM U.90, Hôpital Necker, Paris, France

INTRODUCTION

The statistics of liquid scintillation counting are a function of numerous parameters.[1] In dual isotope counting the principal ones are the isotope ratio and the degree of quenching. These two parameters affect statistics simultaneously. The quantitation of statistics pertaining to the activity of the weaker isotope takes on a special significance when the isotope ratio and quenching vary simultaneously for a given sample series. Many authors[2,3] have investigated experimental series of dual isotope samples (tritium, carbon-14), allowing the isotope ratio to vary while keeping quenching at a constant level and *vice versa.* This type of study permits the definition of a lower isotope ratio limit below which the weaker isotope may not be counted under statistically significant conditions.

In this communication we have set as our goal the automatic estimation, using a computer, of the relative standard deviation (RSD) of tritium activity in d.p.m. of a series of dual isotope samples (tritium, carbon-14), where isotope ratio and quenching vary simultaneously. The method of estimation is based on the implicit relationship between the RSD of the tritium activity, the count rate (c.p.m.) ratio in the two counting channels A and B, quenching and the counting statistics of channels A and B. This relationship may be estimated graphically while the computer program permits us to estimate in particular the RSD of computed tritium activity of each sample by processing data furnished by a liquid scintillation counter.

The method finds application in the fields of cellular incorporation where two labelled precursors are used simultaneously (thymidine, uridine, etc.). In fact, in these cases the isotope ratio is not known *a priori* and one may observe important variations in quenching from one sample to the next.[4]

MATERIALS AND METHODS

Materials

The following materials were used in our investigations:

1. SL 40 liquid scintillation counter†

---

* Attaché de Recherches.
† Intertechnique, 78 Plaisir, France.

2. PDP–10 computer*
3. Labelled toluene –tritium 2.58×10^6 d.p.m./mg
   carbon-14 5.95×10^5 d.p.m./mg
4. Carbon tetrachloride

Methods

Principle of estimating the RSD of tritium activity. The relative standard deviation (RSD) of tritium d.p.m. in the case of dual isotope counting of tritium/carbon-14 by means of screening tritium from the carbon-14 channel is given by the following equation:

$$\text{d.p.m.}^3\text{H} = \frac{(\text{c.p.m.A} - \text{BA}) - (\text{c.p.m.B} - \text{BB})\,(E[^{14}\text{C}(^3\text{H})]/E[^{14}\text{C}])}{E[^3\text{H}]}$$

$$\sigma_R[\text{d.p.m.}^3\text{H}] = \frac{\sigma[\text{d.p.m.}^3\text{H}]}{\text{d.p.m.}^3\text{H}} = \frac{\sqrt{\sigma_{RA^2}\cdot\text{c.p.m.A}^2 + \sigma_{BA^2}\,\text{BA}^2 + (E[^{14}\text{C}(^3\text{H})]/E[^{14}\text{C}])^2\cdot(\sigma_{RB^2}\cdot\text{c.p.m.B}^2 + \sigma_{BB^2}\cdot\text{BB}^2)}}{\text{c.p.m.}^3\text{H}} \quad (1)$$

where:

$E[^3\text{H}]$	=	tritium efficiency in channel A
$E[^{14}\text{C}]$	=	carbon-14 efficiency in channel B
$E[^{14}\text{C}(^3\text{H})]$	=	carbon-14 efficiency in channel A
σ_{RA}	=	relative standard deviation of counts per minute (c.p.m.A) in channel A
σ_{RB}	=	relative standard deviation of counts per minute (c.p.m.B) in channel B
σ_{BA}	=	relative standard deviation of background in channel A
σ_{BB}	=	relative standard deviation of background in channel B
BA	=	background in channel A
BB	=	background in channel B

In Eqn. (1) each factor is known for a given sample, but it is not possible to calculate the value $\sigma_R[^3\text{H}]$ d.p.m. from a single value of tritium c.p.m. The count rates in channels A and B express both the activity of each isotope and quenching; we have taken the count ratio A/B as a computation parameter.

To estimate the $\sigma_R[^3\text{H}]$ d.p.m. of a dual isotope sample counted once, using Eqn. (1), different values of activity for each isotope are generated to obtain the ratio A/B identical to that of the sample. From then on using Eqn. (1) the $\sigma_R[^3\text{H}]$ d.p.m. may be estimated taking into account the sample characteristics ($E[^3\text{H}]$, $E[^{14}\text{C}]$, $E[^{14}\text{C}(^3\text{H})]$, σ_{RA}, σ_{RB}, σ_{BA}, σ_{BB}) and the values of the activity of the two isotopes corresponding to an A/B ratio identical to the sample under study. Thus, the $\sigma_R[^3\text{H}]$ d.p.m. estimation is not connected to the computed absolute values of the activity of the two isotopes.

Figure 1 traces the identification process and the calculations. For example, if a tritium d.p.m. value is fixed at 10000, taking account of the efficiencies for each sample measured, the value Y_i (carbon-14 d.p.m.) is found which gives a ratio A/B identical to the sample measured. With the statistical conditions of the sample measured, $\sigma_R[^3\text{H}]$ d.p.m. is computed using Eqn. (1).

---

* Digital Equipment Corporation, Maynard, Massachusetts 01754, U.S.A.

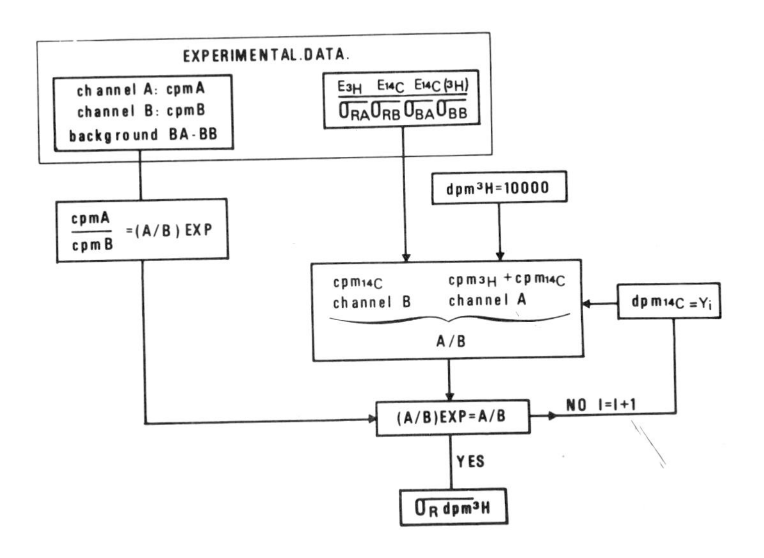

Fig. 1. Iterative calculations for RSD of tritium d.p.m.

Fortran IV program. We have written a Fortran program for computing the RSD of tritium d.p.m. for a series of dual isotope samples. Counting data is provided by the liquid scintillation counter on punched tape while the same tape provides the external standard ratio, E_1/E_2, from which counting efficiencies are computed. Quench curves (efficiencies versus E_1/E_2) are entered in the computer in the form of a third order polynomial of which the coefficients have been previously established by the method of least squares.

The computing process is shown by Fig. 1 while Fig. 2 illustrates the general flow chart of the program.

The equivalence of the experimental and computed ratios c.p.m.A/c.p.m.B is fixed within an interval of ±0.001 × (c.p.m.A/c.p.m.B) (experimental). This value of 0.001 is that which experimentally corresponds to a minimum identification interval of two (c.p.m.A/c.p.m.B) values and computed, taking into account the slow rate of increase of the value of DHC – (DHC = ^{3}H d.p.m./^{14}C d.p.m.). In order to reduce the number of iterations, an approximate value of the DHC ratio is calculated from which the iterative process starts.

The computer calculates the RSD of tritium d.p.m. and carbon-14 d.p.m. as well as the ratio ^{3}H d.p.m./^{14}C d.p.m.

Verification of results. We have experimentally checked the graphs and the results obtained with the Fortran program by repetitive counting of a series of samples with different isotope ratios and quench levels. The calculation of the true RSD based on several measurements has permitted us to evaluate the estimate which we make of $\sigma_R[^3H]$ d.p.m. from a single measurement. Samples were measured thirty times for each quench level and were prepared according to Table 1. They were quenched progressively with carbon tetrachloride so as to give the efficiencies shown in Table 2.

RESULTS

We established for different quench levels the curves relating the tritium d.p.m. RSD

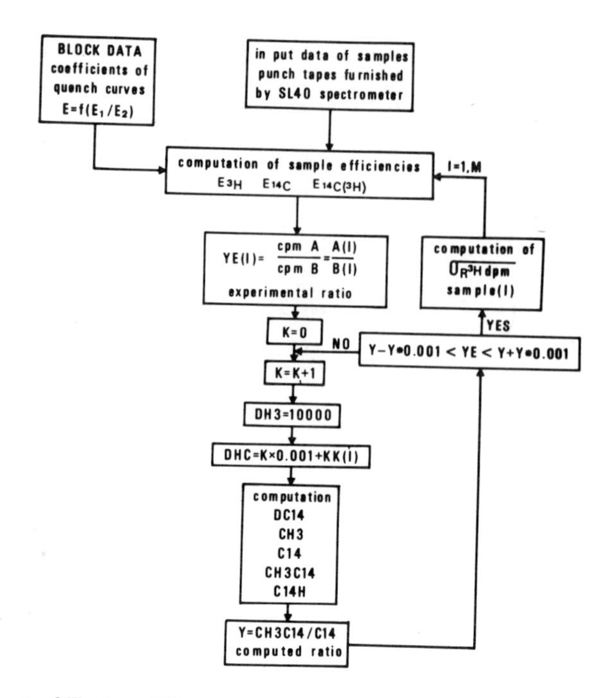

Fig. 2. Flow chart of Fortran IV program.
M number of samples . DHC = DH3/DC14 = ^{3}H d.p.m./^{14}C d.p.m.
KK: initial value of DHC for the sample (I) compute as a function of YE, E^{3}H, E^{14}C, E^{14}C(^{3}H).
c.p.m.A = CH3C14 = C14H + CH3 . c.p.m.B = C14.
Y = c.p.m.A/c.p.m.B (computed).

Table 1. Preparation of experimental samples (tritium, carbon-14).

Number	Toluene standard ^{3}H (μl)	Toluene standard ^{14}C (μl)	d.p.m.^{3}H/d.p.m.^{14}C
1	0	100	0
2	10	800	0.046
3	10	400	0.101
4	100	800	0.536
5	100	400	1.011
6	100	200	2.066
7	100	100	4.122
8	100	50	7.939
9	100	20	21.899
10	100	0	∞

to the (^{3}H d.p.m./^{14}C d.p.m.) ratio (see Fig. 3) and to the channels count rate ratio A/B (see Fig. 4). The curves take two parameters into account, quenching and activity ratio. When quench ranges are very wide and the statistical counting conditions of the samples are different, the curves are not useful.

Table 2. Quench and efficiency values of each isotope for the sample series prepared according to Table 1.

E_1/E_2	E^3H	$E^{14}C$	$E^{14}C(^3H)$
5.20	0.46	0.68	0.17
3.85	0.34	0.575	0.26
2.91	0.24	0.42	0.34
1.94	0.135	0.175	0.52

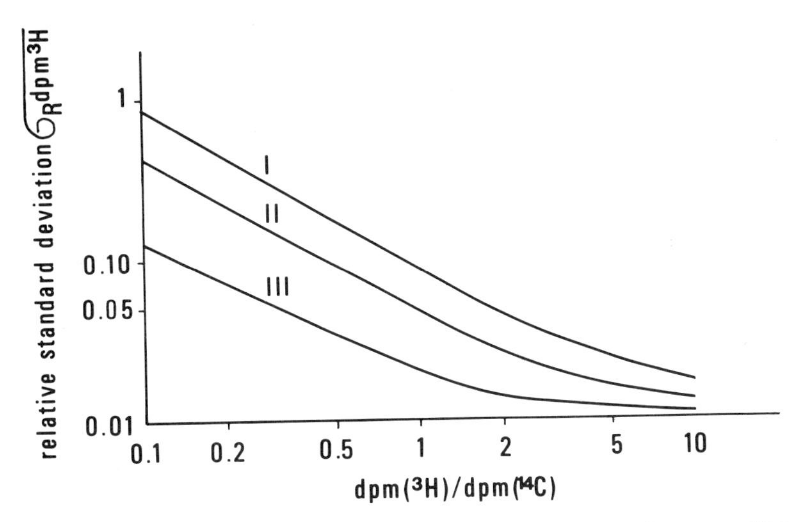

Fig. 3. Relative standard deviation σ_R [d.p.m. ^{3}H] as a function of ^{3}H d.p.m./^{14}C d.p.m. The relative standard deviation of c.p.m. in the channels A and B are equal at 1% ($\sigma_{RA} = \sigma_{RB} = 1\%$).Curve I, quenching $E_1/E_2 = 1.6$; curve II, quenching $E_1/E_2 = 2.23$; curve III, quenching $E_1/E_2 = 3.5$.

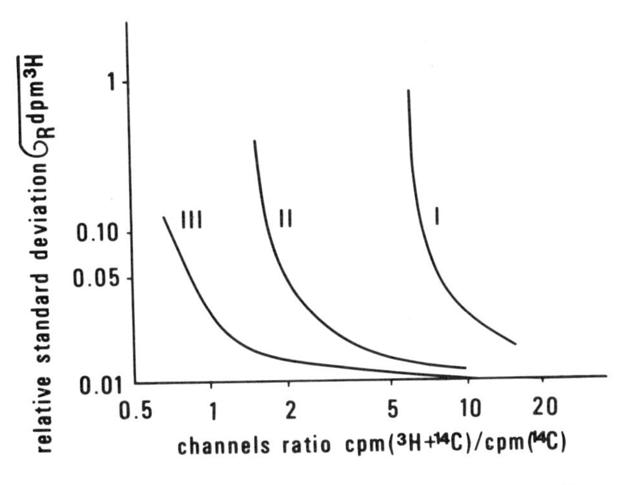

Fig. 4. Relative standard deviation σ_R [d.p.m.^{3}H] as a function of ^{3}H c.p.m./^{14}C c.p.m. The computation parameters are as those of Fig. 3.

```
**************************************************************************
* N  *  T  *   CPM A    * RA   *   CPM B    * RB   *    E1 /E2
*    *     *   DPM A    * RSD  *   DPM B    * RSD  *  DPM A /DPM B
**************************************************************************
*  1 * 0.12*   81975.4  *  1.0*   14245.9  *  2.4*   1.79000
*    *     *  212526.6  *  6.3*  104225.2  *  2.4*   2.039
**************************************************************************
*  2 * 0.16*   62317.7  *  1.0*  116971.9  *  0.7*   3.83000
*    *     *   26484.9  *  8.2*  205908.3  *  0.7*   0.129
**************************************************************************
*  3 * 0.08*  116880.8  *  1.0*  251647.1  *  0.6*   4.02300
*    *     *   33399.6  * 11.4*  426472.3  *  0.6*   0.078
**************************************************************************
*  4 * 0.10*   99830.3  *  1.0*   61207.5  *  1.2*   3.88600
*    *     *  212178.4  *  1.4*  106506.9  *  1.2*   1.992
**************************************************************************
*  5 * 0.09*  117948.1  *  1.0*   21816.0  *  2.3*   1.55900
*    *     *   29370.1  * 87.1*  197787.8  *  2.3*   0.148
**************************************************************************
```

Fig. 5. Example of the computer print out for different dual-labelled samples (tritium, carbon-14).

Table 3. Comparison of relative standard deviation σ_R d.p.m. tritium between experimental and calculated values for different isotope ratios and quenching. E = experimental values; C = computed values. $\sigma_{RA} > 0.6\%$; $\sigma_{RB} > 0.6\%$.

No.	d.p.m. ^{3}H / d.p.m. ^{14}C	σ_R d.p.m. tritium (%) $E_1/E_2 = 5.20$		$E_1/E_2 = 3.85$		$E_1/E_2 = 2.91$		$E_1/E_2 = 1.94$	
		E	C	E	C	E	C	E	C
1	0	78.0	83.0	86.0	873.0	73.2	431.0	–	–
2	0.046	16.4	15.1	18.2	17.0	23.5	31.0	–	–
3	0.101	7.1	6.5	10.7	9.2	13.8	17.2	30.0	87.0
4	0.536	2.4	2.0	2.4	2.5	7.5	5.9	14.5	12.5
5	1.011	1.9	1.6	2.21	1.9	2.4	2.3	9.3	8.0
6	2.066	1.5	1.3	1.5	1.4	1.6	1.3	6.4	6.4
7	4.122	1.2	1.1	1.4	1.2	1.4	1.2	2.3	2.5
8	7.939	1.1	1.0	1.2	1.1	1.2	1.1	1.1	1.1
9	21.899	1.1	1.0	1.1	1.0	1.17	1.0	1.1	1.0
10	∞	1.1	1.0	1.1	1.0	1.14	1.0	1.1	1.0

We have written a computer program to obtain an estimate for each sample of the relative standard deviation of the net count rate of the weaker isotope. Figure 5 shows an example of the print out for different dual-labelled samples (tritium, carbon-14).

We have checked the results of the program for estimating statistical error by a series of samples defined previously and counted repetitively to obtain the true relative standard deviation of computed tritium d.p.m. Table 3 gives the results obtained for various (^{3}H d.p.m./^{14}C d.p.m.) activity ratios and quench levels where tritium efficiencies vary from 0.135 to 0.46.

Measurements made where σ_{RA} and σ_{RB} were less than 0.006 (gross count rate error in channels A and B) resulted in a computed RSD lower than the experimental values obtained in repeat measurements. The experimental RSD includes the total of statistical fluctuations while the computed value includes only the σ_{RA} and σ_{RB} as counting con-

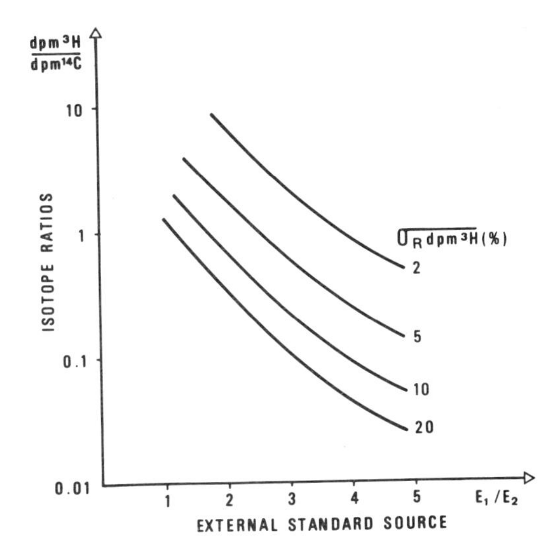

Fig. 6. Isotope ratios ^{3}H d.p.m./^{14}C d.p.m. as a function of quenching for different values of σ_R [d.p.m.^{3}H] . $\sigma_{RA} = 1\%$; $\sigma_{RB} = 0.6\%$.

ditions. We have observed that for values of σ_{RA} and σ_{RB} greater than 0.6%, the experimental and computed results for $\sigma_R[^3H]$ d.p.m. are identical.

Our results permit the construction of a curve network relating the isotope ratio (^{3}H d.p.m./^{14}C d.p.m.) to quenching (E_1/E_2 – count rate ratio of external standard) for different values of $\sigma_R[^3H]$ d.p.m. (see Fig. 6). While it is not imperative to accompany each measurement by the tritium d.p.m. RSD value, one may use the curves to distinguish between samples having an $\sigma_R[^3H]$ d.p.m. greater or less than a given value.

DISCUSSION

The estimate of tritium d.p.m. RSD made by the computer does not take into account the quality of quench curve fitting. In practice the fits are performed with a precision of 1% and may certainly be neglected within the framework of our $\sigma_R[^3H]$ d.p.m. estimates. However, we have verified the accuracy of the estimate as regards variations in values produced by the computer for repetitively counted samples (Fig. 7). The estimate of $\sigma_R[^3H]$ d.p.m. based on a single measurement remains constant up to a value of about 20%. Above this value, excessive fluctuation occurs. This phenomenon is not a problem when results of tritium d.p.m. obtained with a statistical error much above 20% must be reconsidered.

The experimental results which we have obtained agree with those found in the literature.[5] Bush[2] has determined optimum counting conditions to obtain minimal statistical error in dual isotope counting. With large series of samples one is obliged to adopt fixed isotope window settings. Thereafter it becomes necessary to evaluate the statistical precision of the measurement of each sample. This procedure is particularly useful for experiments involving cellular incorporation but also when the order of magnitude of the isotope activities is unknown or when the quench range is very wide. Such is the case when counting plasma dissolved directly in liquid scintillator with the aid of emulsifiers.[6]

In conclusion, the estimate of the relative standard deviation of the activity of the

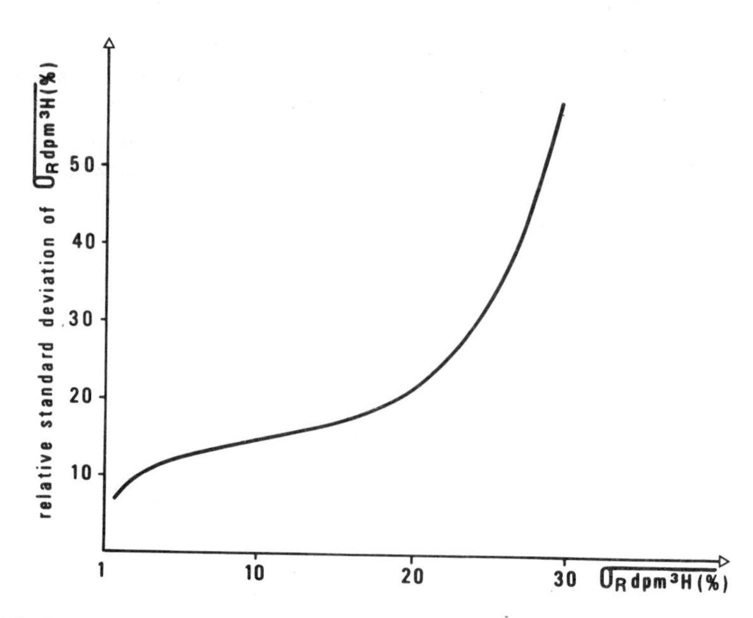

Fig. 7. Relative standard deviation of σ_R [d.p.m.^{3}H] as a function of σ_R [d.p.m.^{3}H].

weaker isotope in each dual isotope sample facilitates and increases the reliability of the dual isotope technique using liquid scintillation counting when activity ratios and quenching vary widely:

(Copy of the program written in Fortran IV is available through the research laboratory INSERM U.90, Hôpital Necker, Paris).

ACKNOWLEDGEMENTS

The authors wish to express their appreciation for helpful discussions with Mr. C. Palais (Intertechnique, France), and thank Mr. D. Gomez for technical assistance on this study.

REFERENCES

1 C. Matthijssen and J. W. Goldzietter, *Anal. Biochem.* **10**, 401 (1965).
2 E. T. Bush, *Anal. Chem.* **36**, 1083 (1964).
3 M. I. Krichevsky, S. A. Zaveler and J. Bulkeley, *Anal. Biochem.* **22**, 442 (1968).
4 C. Bader, *La Transplantation Rénale,* Flammarion Medecine-Sciences, Paris, 1971, p. 371.
5 Y. Kobayashi and D. V. Maudsley, in *The Current Status of Liquid Scintillation Counting* (Ed. E. D. Bransome), Grune and Stratton, New York, 1970.
6 J. Assailly, C. Bader, J.-L. Funck-Brentano and D. Pavel, *Liquid Scintillation Counting Volume 2*, Heyden and Son, London, 1972, p. 223.

DISCUSSION

H. E. Barber: Your formula shows all radioactivity counts as c.p.m. Did the calculation of standard deviation of tritium operate on c.p.m. or on total counts, conversion to c.p.m. being made at the end of the statistical calculation thus taking advantage of any prolonged counting time on the precision of the estimation?

J. Assailly: Of course, the relative standard deviation of c.p.m. or total counts is the same. In the formula we have expressed the relative standard deviation of d.p.m. tritium activity, and we can simplify the denominator by replacing $E[^3H]$ d.p.m. tritium with c.p.m. tritium. In this formula, all the factors are experimental data. But, of course we cannot calculate the $\sigma_R[^3H]$ d.p.m. by means of a single measurement, it is the estimation program which performs this.

F. E. L. ten Haaf: I could not follow your explanation of your last slide (Fig. 7). Can you explain this again?

J. Assailly: This slide shows the variations of the relative standard deviation of $\sigma_R[^3H]$ d.p.m. as a function of $\sigma_R[^3H]$ d.p.m. In fact this curve is a graphical picture of the stability of our estimation program of the d.p.m. tritium precision.

Chapter 26

Computer Data Handling for the Radiochemical Immunoassay of Insulin

G. Ayrey and K. L. Evans

Isotope Unit, Queen Elizabeth College,
University of London, London W8 7AH, England

INTRODUCTION

The concentration of insulin in serum is commonly measured by a substoichiometric isotope dilution technique. The most widely used method is based on the original work of Hales and Randle[1] and is routinely implemented using kits supplied by the Radiochemical Centre.[2] A known quantity of labelled iodinated insulin is mixed with unknown insulin solution and the two insulins are then allowed to compete for an insufficiency of insulin antibody which forms an insoluble complex. The radioactivity of the isolated precipitate then gives a measure of the extent of dilution of the labelled by the unlabelled insulin and hence the concentration of the latter may be determined. Originally iodine-131 was used to label insulin but more recently the longer-lived and less hazardous iodine-125 has been introduced. This isotope decays by electron capture but also yields low energy internal conversion electrons which may be measured by liquid scintillation counting under conditions similar to those used for tritium. Insulin–antibody complexes are collected on membrane filters which are dried and then suspended in a suitable fluor solution.

CALCULATION OF INSULIN CONCENTRATION

A calibration procedure, preferred by many biochemists, and recommended in the manual supplied with each immunoassay kit,[2] is to assay a series of standard insulin solutions and plot the experimental data on a curve relating count rate to insulin concentration as shown in Fig. 1 (solid line). Concentrations of unknown insulin solutions are then determined by direct reference to the calibration curve. Most measurements are performed in triplicate and the mean count rate used for subsequent calculations. Thus, for analysis of large numbers of samples, manual computation is both laborious and time consuming.

MATHEMATICAL MODELS FOR COMPUTATION

A first approach to this problem was to attempt a mathematical model fit to the curve shown in Fig. 1. The most straightforward method is a linear fit between adjacent points (Fig. 1, broken line). For good experimental data, deviations from the operator

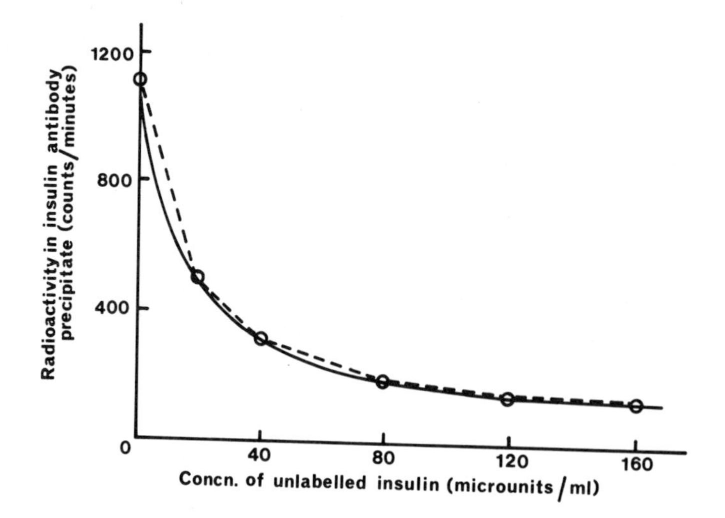

Fig. 1. Calibration curve relating count rate of precipitated insulin–antibody complex to concentration of insulin. (——) operator drawn curve; (--------) linear between pairs fit.

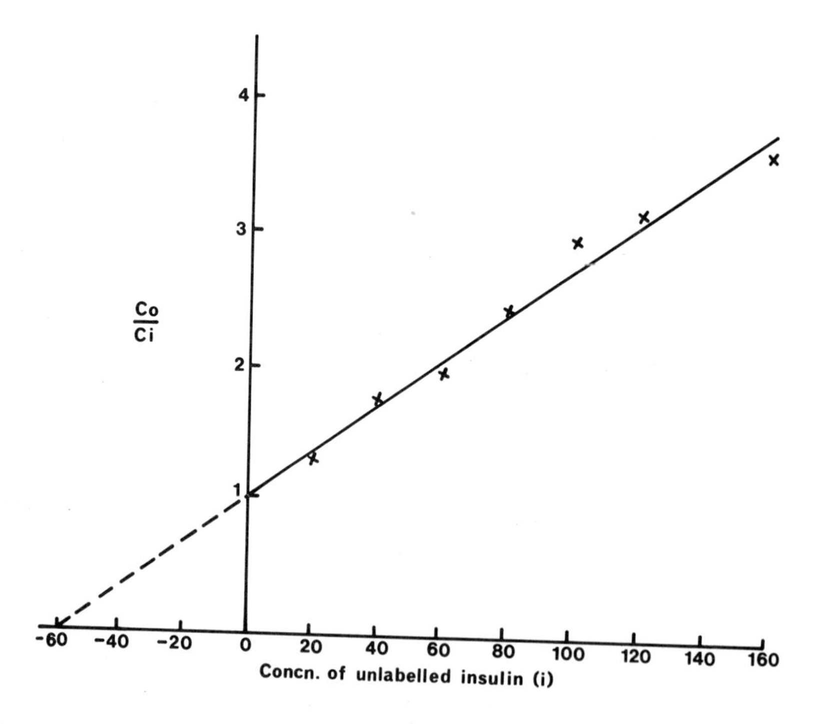

Fig. 2. Calibration curve relating C_0/C_i with insulin concentration.

drawn line are not large. The magnitude of the errors involved in this model could be considerably reduced by using an increased number of standards. The time involved in preparing additional standards was negligible when compared with the time saved by using a computer to process the results. The model was quite satisfactory if a skilled technician performed the experimental work. However, it was subsequently shown that unacceptable deviations could occur if the experimental calibration results were subject to rather wider variations (see, for example, Fig. 10).

Application of the principles of isotope dilution by Hales and Randle[1] yielded the equation:

$$C_0/C_i = i/i_0 + 1$$

where i and i_0 are the concentrations of unlabelled and labelled insulin respectively and C_0 and C_i are the count rates of the precipitated insulin–antibody complexes when unlabelled insulin concentrations were equal to 0 and i respectively. Theoretically, therefore, there should be a linear relationship between C_0/C_i and i with a slope of $1/i_0$. Unfortunately the isotope dilution principle was not strictly obeyed because the amount of insulin bound by antibody varied with the insulin concentration. Nevertheless, Hales and Randle[1] showed that C_0/C_i plots were linear, at least over the range of insulin concentrations normally encountered, though the slopes of the lines varied with the type of antisera used.

We have confirmed that this linearity holds and have demonstrated that a linear least squares fit to the experimental standard determinations gives a reliable calibration curve which may be used for computational purposes. The line representing the rather poor data displayed in Fig. 10 is shown in Fig. 2. More reliable data, which are most usually available, give a closer fit.

COMPUTER PROGRAMS

For either of the above models, computational processes are similar and may be represented on a simplified flow diagram (Fig. 3). The programs were written in ALGOL 60 and computation was carried out on an Elliott model 903C digital computer having an 8K store. The programs were devised so that either routine or non-routine runs could be accommodated, the technician merely having to supply answers to a few simple questions which are posed by the on-line teleprinter. These are best illustrated by consideration of a specific example where the analyst had prepared eight standards in triplicate and had five samples of unknown insulin concentration of which three had been prepared in triplicate, one in duplicate and one as a single sample. In practice, batches of fifty to a hundred unknowns are dealt with and duplicates or single samples only result from occasional accidents or possibly extremely small samples of serum. Prepared samples are counted for a fixed time and the data collected on punched tape. For convenience, all calculations are performed on total counts rather than c.p.m.

The program is first loaded into the computer and questions are asked which the operator answers on the on-line teleprinter as shown in Fig. 4. (In Figs. 4 to 6, the answers supplied by the operator to questions posed by the teleprinter, have been underlined for clarity). In the routine run illustrated on Fig. 4 all necessary data relating to replication and concentration of the standards, background count, sample number of the first standard etc. are contained on the initial data tape requested in the final demand. For a non-routine run the operator is asked for additional data as shown in Fig. 5. In either case further data are requested for the unknown samples (Fig. 6). Most of these questions are self explanatory. In our laboratory the liquid scintillation counters are multi-user instru-

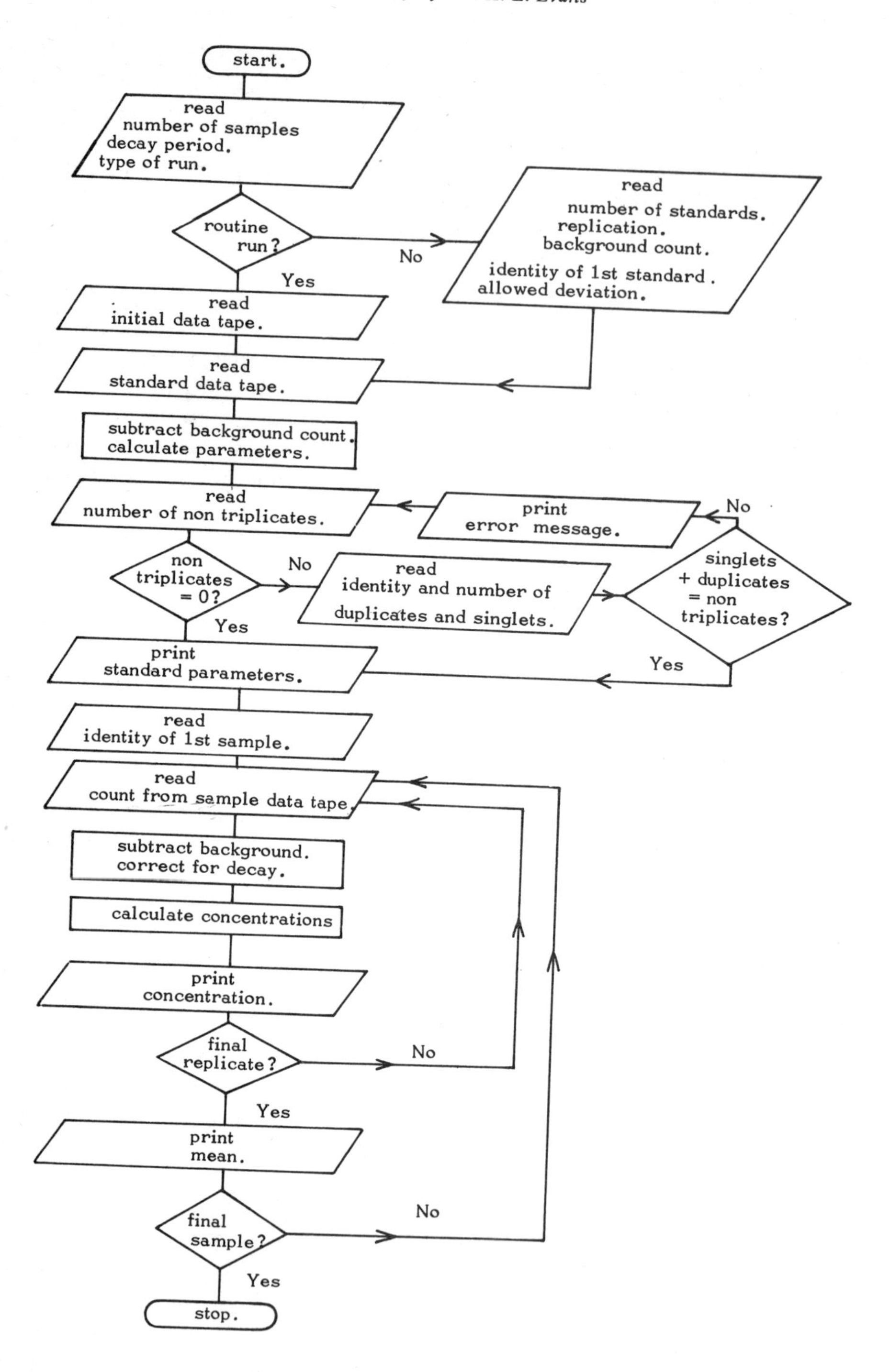

Fig. 3. Flow diagram for computer programs.

```
NUMBER OF SAMPLES=5

NUMBER OF DAYS BETWEEN STANDARD AND SAMPLE COUNT=2

FOR ROUTINE RUN PRINT 1. FOR NON ROUTINE RUN PRINT 2. 1

LOAD INITIAL DATA TAPE AND CONTINUE AT 9.

LOAD STANDARD DATA TAPE AND CONTINUE AT 9.
```

Fig. 4. Format of input for a routine computer run.

```
NUMBER OF SAMPLES=5

NUMBER OF DAYS BETWEEN STANDARD AND SAMPLE COUNT=2

NUMBER OF STANDARDS=8

NUMBER OF REPLICATES=3

BACKGROUND COUNT=347

IDENTIFY STANDARD 1,1 1

STANDARD CONCENTRATIONS=0 20 40 60 80 100 120 160

ALLOWED PERCENTAGE VARIATION BETWEEN INDIVIDUAL COUNTS AND MEAN=10

LOAD STANDARD DATA TAPE AND CONTINUE AT 9
```

Fig. 5. Format of input for a non-routine computer run.

```
NUMBER OF SAMPLES NOT IN TRIPLICATE=2

NUMBER AND IDENTITY OF SAMPLES IN DUPLICATE 1 1

NUMBER AND IDENTITY OF SAMPLES AS SINGLETS 1 4

IDENTITY OF SAMPLE 1,1 25
```

Fig. 6. Format of requests for sample data.

ments and the punched tape output contains much data irrelevant to immunoassay – hence the demand for identification of the first standard. The tape is then automatically searched for this sample before data are taken into the computer store.

Output of results is slightly different for the two programs. For the linear between pairs program (Fig. 7) the triplicate standards are averaged and the mean value used in subsequent computation on unknowns. Where large single deviations within triplicates occur, the program provides for rejection of these results according to a pre-selected maximum deviation from the mean (see, for example, Fig. 5). Such rejections are indicated by the suffix R in Fig. 7, and the mean value recorded is that calculated from the two remaining results.

In the least squares program, standard triplicates are not averaged. All the individual results are fed into the least squares sub-routine for calculation of the best straight line. Net counts and per cent deviations of each point are then printed together with the slope and intercept of the best straight line (Fig. 8). These parameters are then used to calculate

```
                 STANDARD COUNTS.

REPL.   1           2            3          MEAN        CONC.
NO
1     15038       15877        15988        15634          0
2     12190       12652        10589 R      12421         20
3      9074        8498         9142         8904         40
4      8031        7724         7977         7910         60
5      6276        6020         6762         6352         80
6      4982        5252         5385         5206        100
7      4731        5151         4673         4851        120
8      4107        5045 R       3596         3851        160

LOAD SAMPLE DATA TAPE AND CONTINUE AT 9.

SAMPLE NO    1      7679      62
                    7702      61
                                                 MEAN=   62
SAMPLE NO    2      4357     135
                    4402     134
                    4372     135
                                                 MEAN=  135
SAMPLE NO    3     11123      27
                   10544      30
                   11109      27
                                                 MEAN=   28
SAMPLE NO    4     16563       0 OUTSIDE RANGE
                                                 MEAN=    0
SAMPLE NO    5      3859     155
                    3968     151
                    3800     154
                                                 MEAN=  153

FINISH
```

Fig. 7. Output from a computer run using the linear between pairs program.

insulin concentrations for the unknown samples the final results being presented as shown in Fig. 9.

COMPARISON OF MODELS

Experimental data from a run of 25 unknown samples were used to calculate insulin concentration (i) by the traditional operator drawn line and manual calculation, (ii) by computer using the linear between pairs approximation, and (iii) by computer using the least squares fit to a linear C_0/C_i plot. The calibration standards for this comparison were a specially selected *poor* set of data (i.e. the data represented in Fig. 10) chosen so as to emphasize differences between the models. Considering this selection, the results presented in Table 1 show little variation with the method of calculation though in general, method (i) and method (iii) give the closest agreement.

Calibration curves drawn manually through scattered points as shown in Fig. 10 tend to be subjective and no two operators will draw precisely the same curve. A 'best' line through the points of Fig. 10 was obtained via back calculation of C_i values from the C_0/C_i ratios obtained from the least squares best straight line (Fig. 2) and it is clearly seen that this deviates appreciably from the operator's concept of a calibration curve. Thus, in addition to greatly speeding calculation of results, the computer has the added advantage of eliminating operator prejudice from interpretation of calibration data.

STANDARD NO.	INSULIN CONC.	NET COUNT	C0/CI OBS.	RELATIVE ERROR (PERCENT.)
1 1	0	15038	1.040	2.6
1 2	0	15877	0.985	-2.8
1 3	0	15988	0.978	-3.5
2 1	20	12190	1.283	-6.9
2 2	20	12652	1.236	-11.0
2 3	20	10589	1.476	7.1
3 1	40	9074	1.723	-0.5
3 2	40	8498	1.840	5.9
3 3	40	9142	1.710	-1.2
4 1	60	8031	1.974	-7.4
4 2	60	7724	2.024	-3.3
4 3	60	7977	1.960	-6.7
5 1	80	6276	2.491	1.7
5 2	80	6020	2.597	5.7
5 3	80	6762	2.312	-6.0
6 1	100	4982	3.138	10.5
6 2	100	5252	2.977	5.6
6 3	100	5385	2.903	3.2
7 1	120	4731	3.305	4.1
7 2	120	5151	3.035	-4.4
7 3	120	4673	3.346	5.3
8 1	160	4107	3.807	-2.1
8 2	160	5045	3.099	-25.4
8 3	160	3596	4.348	10.6

GRADIENT= 0.0180 INTERCEPT= 1.0123

Fig. 8. Output of standard data and calibration curve parameters for least squares program.

SAMPLE NO.	NET COUNT	C0/CI OBS.	CALCULATED CONC.	MEAN CONC.	COMMENT
1 1	7679	2.036	57		
1 2	7702	2.030	57		
				57	
2 1	4357	3.588	143		
2 2	4402	3.552	141		
2 3	4372	3.576	143		
				142	
3 1	11123	1.406	22		
3 2	10544	1.483	26		
3 3	11109	1.407	22		
				23	
4 1	16563	0.944	0		OUT OF RANGE
				0	
5 1	3859	4.051	169		OUT OF RANGE
5 2	3968	3.940	163		OUT OF RANGE
5 3	3900	4.009	167		OUT OF RANGE
				166	

Fig. 9. Output of sample data from least squares program.

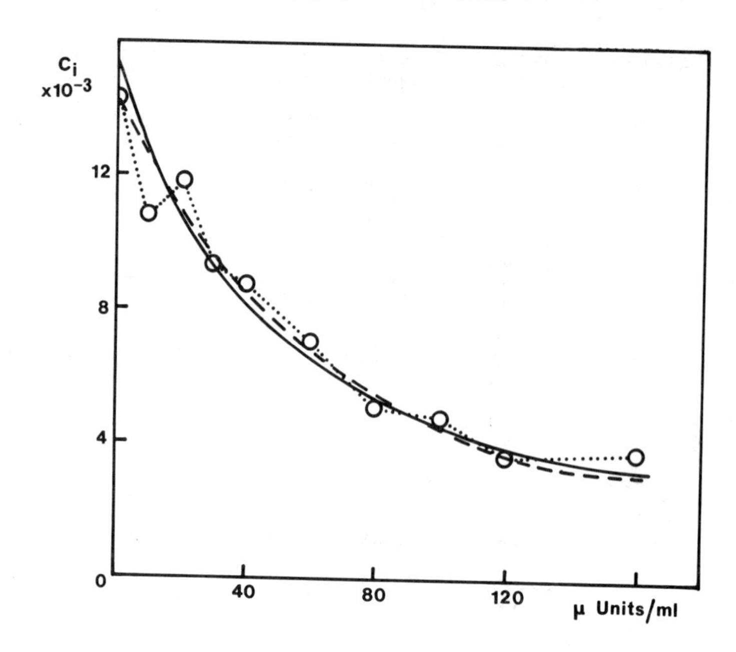

Fig. 10. Comparison of operator drawn line (---------) with least squares fit (———) and linear between pairs fit (.........).

Table 1. Comparison of results calculated via three different methods.

Sample number	Calculated Insulin Concentrations (microunits/ml)		
	Operator drawn (Method (i))	Linear between points (Method (ii))	Least squares (Method (iii))
1	3	3	7
2	7	4	10
3	8	4	10
4	8	5	10
5	8	5	11
6	9	5	11
7	10	5	12
8	11	6	13
9	11	7	13
10	13	7	15
11	19	14	19
12	24	26	23
13	28	29	27
14	39	42	36
15	45	51	43
16	46	52	42
17	58	61	52
18	64	65	57
19	64	65	58
20	81	75	74
21	87	78	80
22	87	78	81
23	99	96	95
24	111	111	111
25	120	120	128

CONCLUSIONS

In this preliminary communication it has been demonstrated that use of a computer can greatly speed calculation of results for insulin immunoassay. Consideration of the calibration function has shown that although it is possible to obtain a good approximation to the conventional calibration curve by a linear between points approximation, a more satisfactory procedure involves a least squares approximation for the best straight line in a plot of C_0/C_i against insulin concentration. Data handling by digital computer also improves reliability by reducing possible errors due to operator prejudice or fallibility.

ACKNOWLEDGEMENT

We gratefully adknowledge the help of Mr. T. A. Adejumo who performed the experimental part of the insulin immunoassay.

REFERENCES

1 C. N. Hales and P. J. Randle, *Biochem. J.* **88**, 137 (1963).
2 Insulin Immunoassay Kit, The Radiochemical Centre, Amersham, Bucks., England.

DISCUSSION

D. Bowyer: For a given number of points, if one increases the order of the polynomial, one may improve the fit of the curve at the points, but of course between the points one may get a poor fit.

K. L. Evans: Yes, I accept this point but I do not feel it is relevant in the context of our work.

J. W. MacMillan: I agree with Mr. Evans that the discussion of polynomial fitting is irrelevant in the context of this paper since the isotope dilution expression is an exact mathematical description which correlates the data and is, therefore, the preferred one to use for the least squares fit.

F. E. L. ten Haaf: Did you use a least square fit to the absolute or to the relative deviations?

K. L. Evans: We used a least square fit to the absolute deviations.

B. Seaton: I would like to draw attention to the fact that a straight line is merely a first order polynomial and is, therefore, subject to the same general limitations as nth order polynomials. Secondly, it is useful to remember the rule of thumb that when fitting any function having *n parameters* one needs $2n$ experimental data points to avoid producing the kind of useless curve referred to by Dr. Bowyer. I would also emphasize that an nth order polynomial has $n + 1$ parameters:

e.g. $A + Bx + Cx^2$ = *2nd* order with *three* parameters

and that it is the number of parameters which must be taken into account.

K. L. Evans: We prefer to use more than four points for the immunoassay calibration.

Chapter 27

Experience in Off-Line Computer Processing of Liquid Scintillation Counting Data

J. H. Deterding

Shell Research Ltd., Thornton Research Centre, Chester CH1 3SH, England

INTRODUCTION

Liquid scintillation counting has developed rapidly since the work of Kallmann and Furst,[1] so that now, with spectrometers fitted with automatic sample changers, it is a technique that can give rise to a large mass of numerical results; it is therefore an obvious field for the application of electronic computers. Many workers have found it worthwhile to apply a computer to liquid scintillation counting (either 'on-line',[2] with a 'desk-top' computer[3] or with a large 'off-line' computer). Often the computer is used in a data processing role to make very simple calculations quickly and reliably on the results of counting large numbers of samples, but sometimes also to carry out more complex calculations.

In this chapter I shall concentrate on the work done at Thornton Research Centre, having chosen this approach deliberately, not with any intention of belittling the work of others active in this field, but because it may be of interest if I describe how this application of computers has developed in one laboratory. The equipment and type of computer program will be considered, and examples will be given of the use of a computer in determining the stability of instruments and in fitting curves to experimental data. Finally, I shall mention briefly some relevant published work and try to define some of the future needs of radiochemists.

The earliest applications of liquid scintillation counting at Thornton were carried out with a home made apparatus containing a single photomultiplier tube. The increasing numbers of samples involved in a wide variety of radiotracer experiments led to the installation in 1961 of a Packard TriCarb automatic liquid scintillation spectrometer (Model 314 EX), to which was added a punched paper tape output. This instrument has been in use for nearly ten years, but was supplemented in 1969 by a more modern counter, a Packard TriCarb Model 3320, which is a three-channel, 200-sample instrument with external standard and teletype output. Paper tape has been found to be an entirely satisfactory medium for the collection of data from these instruments for off-line processing, as we are dealing with low rates of collection of data (of the order of fifty characters after each counting period of, say, 10 min).

It was decided to use off-line computing of results because a Ferranti Pegasus computer had been installed at Thornton for other purposes. Since 1965, links to larger central computers have been used, first to a Univac 1107 run by a computer service bureau in Birmingham and more recently to a Univac 1108 at Shell Centre in London. The programs for radioactivity counting experiments have been written in FORTRAN (FORmula TRANslation) computer language. The use of such a widely recognized programming language has the advantage that the programs will be readily applicable to other types of computer.

THE COMPUTER PROGRAM

The general form of the program is dictated by the fact that we have at first to carry out many standard operations (of taking in counting data in a standard form) but that, after this stage, we must make different calculations depending on the nature of the experiment (for instance, whether it involves internal standard, external standard, or channels ratio methods of determining counting efficiency). The appropriate form of program for this is a general routine to be used in each application, together with a collection of optional subroutines appropriate to particular experiments. The various stages of the program will now be considered, beginning with the form of the counting data, which of course depends on the type of scintillation counter/interface/output device used. For simplicity the following sections have been confined to the system Packard TriCarb 3320/Teletype ASR 33.

Data are produced in a standard form as sample number (three digits), time (six characters, including a decimal point) and three six-digit numbers for the counts in each of three channels, which may be followed by three further six-digit numbers if the external standard is used. These data are punched automatically on paper tape in the teletype, groups of digits being separated by spaces and the printout from each sample being terminated by separating symbols ('carriage return' and 'line feeds'). Eight-hole paper tape is used in which each digit is represented by two, four or six punched holes, so that a punching error causing the addition or omission of one hole produces a major error, rather than the substitution of one digit for another (parity check). This is the so-called ASCII (American Standard Code for Information Interchange).

Parameter data

Additional information is fed into the computer to identify the counting data, to specify the calculations required, to define the settings used on the counter, and to set limits of precision. This information can be punched manually on two data cards, which precede the count data, or as a heading on the paper tape on which the counts are to be recorded. It consists of the following:

1. Date of counting.
2. Counter used.
   (These two parameters have been found to be sufficient to identify the counting data, which may have been produced from instruments other than a liquid scintillation counter, e.g. a γ-ray scintillation counter or Geiger counter).
3. Subroutines to be used for processing the data.
   (Up to six different subroutines can be used with one set of data).
4. Number of samples counted.
5. Number of cycles.
   (Each cycle consists of one set of counts on each sample).
6. Time for sample counts.
   (This may be in seconds, in which case a typical printout from the 3320 would be

10.0, or in minutes, which would be printed as 10.00).

7. Preset counts.
8. Use (digit 1) or absence (digit 0) of external standard.
   (Parameters (4) to (8) provide all the information required on the settings of the counter).
9. Limits of precision for counts from samples.
10. Limits of precision for counts from the external standard.
   (These limits are required for tests, which will be described later, on the differences between statistical and experimental standard deviations of the counts).

11-13. Digits 1 or 0 to signify whether the external standard counts in each of the three channels exceeds 10^6.

Form of computer programs

FORTRAN computer programs have been written for use with any counter that can produce data in the form described above (or in a similar form such as that produced on five-hole paper tape by counters similar to the Packard 314). These programs have been found useful with a γ-ray spectrometer and with a Geiger counter, as well as with a liquid scintillation counter, since the same basic calculations are required for data from any of these radioactivity detectors. The program contains a short *main program*, which reads and stores data and organizes the 'calling' of required subroutines. This is followed by *compulsory subroutines,* which are used to interpret every set of data and to carry out basic calculations. A number of *optional subroutines* (fifteen have been written to date) are used to carry out further calculations on any particular set of data and to print out results in the required form.

In FORTRAN, data and results can easily be transferred from one program or subroutine to another. By using this form of program, it has been found possible for individual users to write their own subroutines, which can be added to the existing program and can use the results of calculations carried out in other subroutines. For a particular project a subroutine can be written in such a way that results are printed out in the most suitable form.

The program and subroutines are stored on magnetic tape or on a 'drum', ready for reading into a computer. About 32000 36-bit words of core storage are used for program and subroutines together with the data (for which storage is 16000 six-digit numbers).

Main program

The main FORTRAN program, RADIOC, is short and is concerned with the storage of data and organization of subroutines; no calculations are included in this program. The following functions are performed:

1. Computing starts when a call is made to 'execute' the main program.
2. Data are read, stored and made available for use in subroutines.
3. Compulsory subroutines are 'called'.
4. Optional subroutines are 'called', as listed on parameter cards.
5. Computing ends.

Compulsory subroutines

The compulsory subroutines are used with all sets of counting data, which must be in a standard form as described above. Information from the parameter cards and from the liquid scintillation counter are read into the computer; the counts and times are stored in a three-dimensional array, in which the first index specifies the counting channel (counting time being included as a channel), the second is the sample number, and the third is the number of the cycle. The information in this array is made available for use in any subroutine. The mean count rates in each of the three count channels are calculated for each sample and stored in a two-dimensional array, in which the first index specifies the channel and the second is the sample number.

The standard deviations of the mean count rates are calculated, expressed as a percentage. One value (SDS) of each standard deviation is found from the statistics of radioactive disintegrations. This standard deviation is found from the square root of the total counts, and is stored in a two-dimensional array. A second value (SDE) of each standard deviation is calculated when a sample has been counted more than once. This standard deviation is found from the differences from the mean of each individual count rate, and is again stored in a two-dimensional array. Finally, a test is made to determine whether or not there is a significant difference between the values of SDS and SDE. For this purpose the F-test[4] is used at 99% significance.

The ratio of the standard deviations is recorded as significant if $(SDE/SDS)^2$ exceeds the critical F value, and if SDE exceeds the fixed limit as given in the data on parameter cards. The results of these tests are stored in a fourth two-dimensional matrix.

Optional subroutines

The subroutines available for use with the main program RADIOC include the following:

1. Printout of edited data.
2. Printout of mean count rates and standard deviations.
3. (Used for γ-counting).

4.&5. Net count rates per unit weight.

6. Water transfer rates (permeability experiments).[5]
7. Count rates corrected for radioactive decay.

8.&9. Channels ratio (Baillie's method).[6]

10. Channels ratio (parabolic curve fitting).[7]

11.&12. Net count rates × factor/count rate from standard sample.

13. Disintegration rates (internal standard method).
14. Oil consumption rates (tritium tracer method).[8]
15. Disintegration rates (external standard ratio (ESR) method).

Up to six of these subroutines can be used with any one set of data. It has been found that subroutines (1) and (2) are used for almost all sets of data, for checking counting results, and these are usually followed by one or two of the other subroutines. In most cases calculations are made on net count rates of samples above a background count rate, so that the order of samples and background in the counter is fixed for each subroutine, (except (1) and (2)). The maximum numbers of samples and cycles of counts are defined (200 samples for the Packard TriCarb 3320). An increase in the numbers of samples would require only minor changes in the program, but would involve more storage space in the computer. For all these subroutines (except (1), (2) and (11)) additional data (e.g. weights) are required; such data can be punched manually on parameter cards, or on paper tape, which accompany the count data to the computer.

In most subroutines the mathematics involved are straightforward, and most of the instructions are concerned with the manipulation of data and with the production of tables of results. Examples of the output from subroutines, used for processing liquid scintillation counting data, are shown in Figs. 1 and 2 from which the calculations involved are obvious. As an added example, further details are given below of one subroutine.

TRICARB 29-30/6/71

GROSS COUNT RATES AND PER CENT STANDARD DEVIATIONS

RACK	RED CHANNEL			GREEN CHANNEL		
NO.	CPM	SDS	SDE	CPM	SDS	SDE
1	26.5	3.55	4.20	23.1	3.80	5.24
2	654.8	.71	.43	570.8	.76	.65
3	12460.9	.16	.30	10854.9	.18	.28
4	620.1	.73	.19	542.0	.78	.17
5	12139.0	.17	.15	10577.2	.18	.10
6	714.2	.68	.91	624.6	.73	1.12
7	12323.1	.16	.14	10742.3	.18	.12
8	626.6	.73	.82	543.8	.78	.55
9	11568.8	.17	.45	10087.4	.18	.45
10	529.1	.79	.79	460.4	.85	.61
11	11599.5	.17	.44	10119.7	.18	.39
12	663.8	.71	.35	581.0	.76	.52
13	11617.1	.17	.13	10131.4	.18	.15
14	681.3	.70	.53	594.3	.75	.64
15	11925.8	.17	.16	10390.0	.18	.18
16	503.9	.81	.64	439.4	.87	.38
17	10276.8	.18	.19	8940.4	.19	.11
18	64.1	2.28	3.59	55.2	2.46	4.25
19	349.2	.98	.60	309.4	1.04	.78
20	17713.2	.14	.29	15772.5	.15	.32
21	371.7	.95	.98	329.0	1.01	.68
22	17445.0	.14	.23	15505.4	.15	.30
23	383.1	.93	.50	338.7	.99	.45
24	17361.0	.14	.48	15432.2	.15	.53
25	379.3	.94	1.59	335.7	1.00	1.55
26	17327.0	.14	.43	15409.8	.15	.45
27	365.0	.96	.54	324.3	1.01	.42
28	17315.9	.14	.49	15397.1	.15	.51
29	348.7	.98	.58	307.9	1.04	.87
30	17295.2	.14	.05	15376.7	.15	.07
31	351.3	.97	1.31	308.5	1.04	1.27
32	17393.4	.14	.33	15460.4	.15	.30
33	345.0	.98	.98	306.4	1.04	1.41
34	17454.9	.14	.20	15537.7	.15	.18
35	11771.4	.17	.27	7102.7	.22	*46.83
36	41.6	2.83	4.41	36.7	3.02	3.47

* ... THE COUNTS SHOULD BE CHECKED FOR ERRORS

RACK NO.	CHANNEL	CYCLE 1	2	3
35	GREEN	103977	104599	4505

Fig. 1. Output from subroutine (2).

Subroutine (14): Oil consumption rates. This subroutine is used to determine oil consumption rates from samples of tritiated water which have been condensed from the exhaust of an engine run with tritiated oil. The technique was originated by Guinn and Coit[8] in 1959 and since then has been developed considerably.

The disintegration rates of tritium in the samples have been measured in a liquid

TRICARB 29-30/6/71

DISINTEGRATION RATES (RED CHANNEL) AND STANDARD DEVIATIONS

RACK NO.	SAMPLE NO.	SAMPLE NET CPM	SDM	STANDARD CPM	EFFICIENCY PER CENT	SAMPLE DPM	SDM
2	WG1	628	5	11806	15.1	4173	35
4	WG2	594	5	11519	14.7	4040	33
6	WG3	688	7	11609	14.8	4644	46
8	WG4	600	5	10942	14.0	4300	43
10	WG5	503	4	11070	14.1	3560	35
12	WG6	637	5	10953	14.0	4561	36
14	WG7	655	5	11244	14.3	4565	35
16	WG8	477	4	9773	12.5	3829	35
19	OG1	285	4	17364	26.5	1077	16
21	OG2	308	4	17073	26.0	1182	17
23	OG3	319	4	16978	25.9	1233	18
25	OG4	315	6	16948	25.8	1220	26
27	OG5	301	4	16951	25.8	1164	17
29	OG6	285	4	16947	25.8	1102	16
31	OG7	287	5	17042	26.0	1106	20
33	OG8	281	4	17110	26.1	1077	16

SDM ... ABSOLUTE STD. DEV. (MAX. OF STATISTICAL AND EXPERIMENTAL)

Fig. 2. Output from subroutine (13).

scintillation counter by the internal standard method – for accuracy this was preferred to external standard methods (*cf.* Rogers and Moran)[9] owing to the presence of both colour and chemical quenching, and the channels ratio method was not used owing to the low levels of radioactivity. Results showing the count rates and efficiencies are shown in Figs. 1 and 2.

The oil consumption rates have been calculated in subroutine (14) from the relation:

$$R_0 = \frac{S_w}{S_0} \times R_f (K_f + A \times H_a)$$

where R_0 = rate of oil consumption (g/h)
R_f = rate of fuel consumption (g/h)
S_w = specific activity of water (d.p.m./g)
S_0 = specific activity of oil (d.p.m./g)
K_f = water/fuel weight ratio, on combustion of the fuel to water
A = air/fuel weight ratio
H_a = water/air weight ratio, depending on the relative humidity. This is calculated from wet and dry bulb temperatures from a table of relative humidities, which is included in subroutine (14).

The results of these calculations, for the eight pairs of oil and water samples (36 vials in all, including standards and backgrounds) are shown in Fig. 3. Computer processing time for this operation (giving outputs as shown in Figs. 1, 2 and 3) is about 1.5 s – compared with many man-hours of 'hand' calculations or several minutes with a 'desk-top' computer.

Two examples will now be given of further uses of an off-line computer for processing liquid scintillation counting data.

```
RESULTS OF OIL CONSUMPTION TEST , WITH TRITIATED OIL

ENGINE                                        D.11
FUEL                                          10
TEST NO.                                      327
DATE OF TEST                                  28/6/71
AIR:FUEL RATIO                                23.0

OIL                                           B.4635.T
FRACTION TRITIATED                            HVI 160B
WT. PER CENT TRITIATED                        91.50
GMS. WATER OF COMBUSTION PER GM. OF FUEL      1.215
```

SAMPLE	SPECIFIC ACTIVITY 3H-COMPONENT OF OIL DPM/MG	SD(ABS)	WATER DPM/G	SD(ABS)	FUEL FLOW KG/HR	HUMIDITY CORRN. PER CENT	3H-LABELLED OIL CONSUMED GMS. PER HR.	SD(ABS)
J1	567	8	4173	35	2.81	12.2	28.6	.5
J2	573	8	4040	33	2.81	11.9	27.3	.4
J3	597	8	4644	46	2.81	10.5	29.7	.5
J4	591	12	4300	43	2.81	10.4	27.7	.6
J5	564	8	3560	35	2.81	9.8	23.9	.4
J6	573	8	4561	36	2.81	10.4	30.3	.5
J7	575	10	4565	35	2.81	10.8	30.4	.6
J8	560	8	3829	35	2.81	10.8	26.1	.5

FIGURES UNDER THE HEADINGS SD ARE THE ABSOLUTE STANDARD DEVIATIONS CALCULATED FROM THE COUNT RATES.
FURTHER ERRORS MAY BE INTRODUCED DURING SAMPLE PREPARATION.
THE PRECISION OF THE RESULTS IS ESTIMATED TO BE 5 PER CENT OR SD WHICHEVER IS THE GREATER.

Fig. 3. Output from subroutine (14).

Table 1. Repeatability of count rates from tritium samples.

Sample	Counter	Time of counting (min)	Temperature (approx) (°C)	Mean counts/min (c.p.m.)	SD of mean c.p.m. (%)	
					Statistical	Experimental
PH3	314 EX	250 X 10	− 3	12573	0.018	0.055
NPH3	3320	900 X 10	+10	83119	0.0037	0.0056

Precision of measurements

Experience has shown that instability of samples often limits the precision of results of liquid scintillation counting more than instrumental fluctuations or statistical considerations of radioactive disintegration; however, it is necessary to determine the stability of the instruments in use, and this can easily be done with the use of a computer.

As radioactive disintegration is a random phenomenon, it is necessary to accumulate a large number of counts in order to assess small instrument drifts, and this is possible with the use of a computer. To find the stability of Packard TriCarb counters over several days, under normal operating conditions, sealed standards of tritium were counted in both the Models 314 and 3320 at balance point. Ten minute repeat counts were recorded on paper tape from each of these instruments – more than one week of counting time was needed. Results of these experiments are shown in Table 1, and were obtained from the computer for which a simple FORTRAN program was written similar to that used in a RADIOC subroutine. This shows remarkably stable operation of both instruments. It would not be practicable to check by hand the calculation of the standard deviation of the mean of 900 six-digit numbers, so this was done (successfully) on a randomly selected dozen of the repeat counts.

It was found that the stabilities of each of the two channels of the ten year old instrument were similar, and that all three channels of the 3320 were even more stable by a factor of about ten than those of the 314. In routine work it is possible also to check the repeatability of operation of the external standard and we have found that, at the channel settings used by us, the ESR counts in one minute are usually steady for any given sample with a standard deviation of the mean of less than 1%.

It is concluded that instrumental drift can be negligible compared with other errors; however, reliability is all-important for computer processing of data. The ten year old instrument now produces so many errors on punched tape that it is being superseded by the modern instrument, which has a teletype output and has so far this year produced only one error in more than 500000 characters.

Curve fitting

When FORTRAN programs and a Univac 1108 computer have been used, it has been found that the fitting of curves to experimental determinations of ESR and efficiency is straightforward for the radiochemist, particularly when existing subroutines are available.[10] One important advantage of our off-line computer for curve fitting is that different types of curve can be fitted to experimental points, and it is possible to select the equation appropriate to the experimental conditions. This is illustrated in Fig. 4 which shows the fit of quadratic, cubic and logarithmic curves to experimentally determined values of carbon-14 efficiency and ESR. It is evident that, for the instrument settings used and the

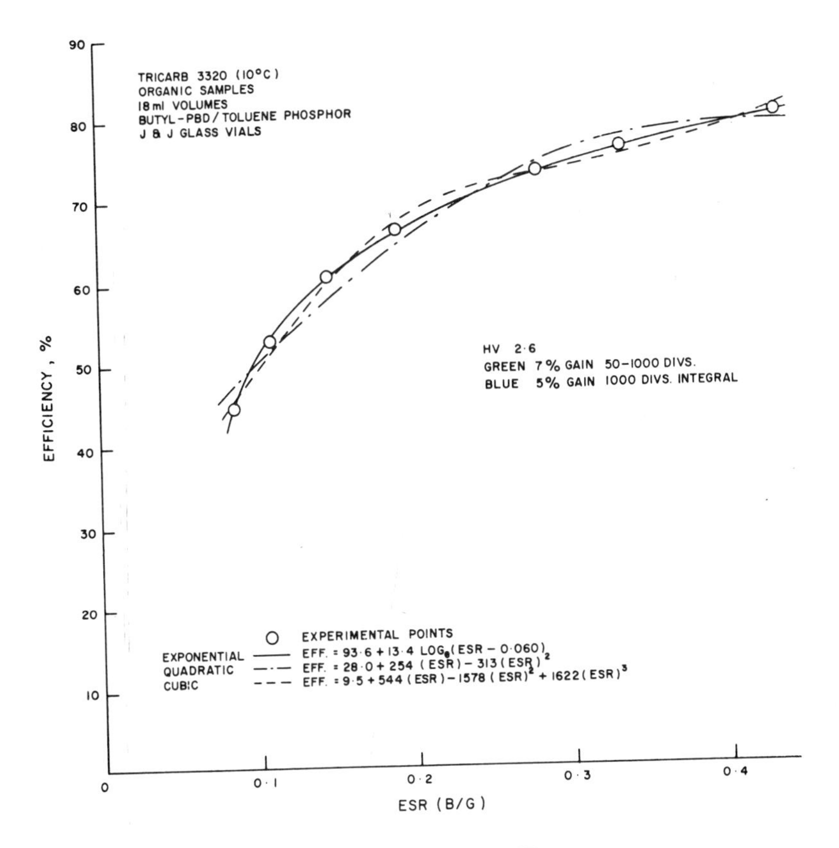

Fig. 4. Relation between carbon-14 efficiency and ESR.

range of quenching with which we are concerned, a cubic does not give as good a fit as the simple three constant expression:

$$\text{Efficiency} = A + B \log_e (\text{ESR} - C)$$

where A, B and C are constants.

DISCUSSION

Published work

Many papers have been published, particularly in the last few years, on the subject of data processing of liquid scintillation counting data and only a few can be mentioned here. Blanchard[11] described an early general program in ALGOL, which included a data checking stage and various modes of operation for the determination of background and counting efficiency. Spratt[12] described another general program in FORTRAN IV, which was designed for use with punched card output from the spectrometer. An unusual feature was that for application of the internal standard method two separate runs through the computer were required, for the results on the samples alone and for results on the samples plus the internal standard. This was extended to include the use of an automatic external standard.[13] Other recent studies have been made by Clegg,[14] Reich[15] and Glass.[16]

The extensive studies of Gordon and co-workers[17,18] covered both the channels ratio and external standard methods, and they show several interesting features. These workers recognized that when a computer is used it is unnecessary to find a setting giving a linear relation between counting efficiency and channels ratio, as has usually been preferred for graphical methods; instead they expressed the counting efficiency as a cubic in the logarithm of the channels ratio. They also examined instrumental drift with the object of avoiding the need to redetermine the whole calibration curve regularly. They found it was valid to measure just one standard each day, the instrument being then brought back on a standard calibration curve by adjustment of the gain. Their application of the external standard was to a dual-labelling experiment, and thus involved four counting efficiencies (two nuclides × two channels). Four calibration curves were used so that the four efficiencies could all be determined from the measured external standard count rate.

The value of having a standard error for every result is emphasized by Nodine.[19] In his programs the standard error is carried through the calculation, i.e. standard errors are calculated explicitly for intermediate results such as net counting rates. We have also considered it important that final results should have their standard errors. The calculations involved are generally longer than those used in obtaining the results alone, and are often omitted as being too tedious for manual calculation. In the case of the channels ratio method, the need for standard error calculations has led to different ways of treating the information from the two channels.

The use of the channels ratio method for dual-labelling experiments can lead us into non-linear simultaneous equations which may give multiple solutions: this approach has been studied by Krichevsky and co-workers.[20]

Today Spratt and others have described to us their methods of data processing of liquid scintillation counting data; we have heard about the processing of data from multi-labelled samples, and it seems evident that the computer is rapidly becoming an indispensable instrument for the radiochemist.

Advantages of computer processing

The main advantage in the computer processing of liquid scintillation counting data is the saving of time. As illustrated above, results can be obtained in a few seconds of computer processing time, compared with hours for manual processing. However, the time needed for writing a correct subroutine is of the order of a few days, and the procedure is only worthwhile for experiments which are repeated many times, or where large numbers of samples are involved.

Over a period of about eight years, we have found that a further advantage of computer processing is in the elimination of errors. Many checks have been made between results from the computer and those found by calculation with a desk calculator; differences, once a correct program has been written, have always come from errors in the manual calculation.

By the use of a computer many more detailed calculations can be made on count data, calculations which would be too tedious manually, particularly in checking count data. Samples can be counted several times without the need for tedious manual averaging, so that comparisons can be made of statistical and experimental standard deviations of mean count rates, as described earlier. These have shown up several types of error, some of these being as follows.

Occasionally, errors occur in the operation of the counter; particularly, a numeral is incorrect in both the printed and the paper tape output. (One such error is shown in Fig. 1).

Systematic drifts have occurred in repeat counts of long-lived isotopes. These have been traced to adsorption of part of a sample on the walls of the counting vessel, long-term phosphorescence of dioxan-based samples due to exposure to sunlight, and increasing quenching in samples due to slow colour-producing reactions.

Finally, it has been found possible by computer processing to produce tables of results in easily readable form. Results from some subroutines (e.g. those shown in Figs. 1 and 2) are of use to the radiochemist making the measurements, and others (as shown in Fig. 3) can be given directly to the scientist in another discipline who wishes to make use of the results; in either case a well-edited table greatly assists clear communication – nowadays some scientists seem only to believe results which are produced by 'the computer'.

Future developments

There is considerable interest today in the use of small on-line computers in the laboratory. I have found that access to a large off-line computer is very convenient for radiochemical work, particularly for liquid scintillation counting. It is easier to write a FORTRAN program for a large computer with plenty of storage space than it is to write non-standard programs for small computers. It may well be more economical to use a few seconds per day on a large computer than longer times on a small computer. However, the amount of money available and the nature of the work will determine the relative merits of the off-line and the on-line computer. One of these types of computer is becoming essential; ideally, the radiochemist would like both.

REFERENCES

1. H. Kallmann and M. Furst, *Phys. Rev.* **79**, 875 (1950).
2. J. H. Parmentier and F. E. L. ten Haaf, *Intern. J. Appl. Radiation Isotopes* **20**, 305 (1969).
3. B. F. Scott, *J. Radioanal. Chem.* **1**, 61 (1968).
4. O. L. Davies, *Statistical Methods in Research and Production,* Oliver and Boyd, Edinburgh and London, 1947.
5. J. H. Deterding, D. W. Singleton and R. W. Wilson, *Farbe Lack* **70**, 35 (1964).
6. L. A. Baillie, *Intern. J. Appl. Radiation Isotopes* **8**, 1 (1960).
7. B. D. Caddock, P. T. Davies and J. H. Deterding, *Intern. J. Appl. Radiation Isotopes* **18**, 209 (1967).
8. V. P. Guinn and R. A. Coit, *Nucleonics* **17**, 112 (1959).
9. A. W. Rogers and J. F. Moran, *Anal. Biochem.* **16**, 206 (1966).
10. A. Jones, *Computer J.* **13**, 301 (1970).
11. F. A. Blanchard, *Intern. J. Appl. Radiation Isotopes* **14**, 213 (1963).
12. J. L. Spratt, *Intern. J. Appl. Radiation Isotopes* **16**, 439 (1965).
13. J. L. Spratt and G. L. Lage, *Intern. J. Appl. Radiation Isotopes* **18**, 247 (1967).
14. B. Clegg, *IEEE Trans. Nucl. Sci.* **NS-15**, 138 (1968).
15. A. R. Reich, *J. Radioanal. Chem.* **6**, 437 (1970).
16. D. S. Glass, *Intern. J. Appl. Radiation Isotopes* **21**, 531 (1970).
17. B. E. Gordon, W. T. Shebs, D. H. Lee and R. U. Bonnar, *J. Amer. Oil Chemists' Soc.* **43**, 525 (1966).
18. B. E. Gordon, W. T. Shebs and R. U. Bonnar, *J. Amer. Oil Chemists' Soc.* **44**, 711 (1967).
19. J. H. Nodine, *The Nucleus* No. 19, 3 (1965) [Nuclear Chicago Corporation].

20 M. I. Krichevsky, S. A. Zaveler and J. Bulkeley, *Anal. Biochem.* 22, 442 (1968).

DISCUSSION

P. Stanley: Were your measurements made at 'balance-point' or not?

J. H. Deterding: Yes indeed, all samples were measured at 'balance-point'.

P. Stanley: Balance-point operation will nullify the effects of spectrometer drift. To observe spectrometer drift it would be perhaps better to set two channels to monitor the top 5% and the whole of the energy range of the pulse height spectrum and then look for any significant change in the ratio of the two rates. A fairly hot standard (say 5×10^5 d.p.m.) would be necessary if one wished to observe short-term drift since it would be necessary to accumulate a statistically acceptable number of counts in a short time for a valid test.

Subject Index

Index to Contributors

Bold type denotes page numbers of chapters.
Asterisk denotes participant in discussion only.